NanoBioTechnology

NANOBIOTECHNOLOGY

BioInspired Devices
and Materials of the Future

Edited by

ODED SHOSEYOV

*The Institute of Plant Science and Genetics in Agriculture
and The Otto Warburg Center for Agricultural Biotechnology,
The Hebrew University of Jerusalem, Rehovot, Israel*

and

ILAN LEVY

Intel Research Israel, Intel Electronics, Jerusalem, Israel

HUMANA PRESS
TOTOWA, NEW JERSEY

© 2008 Humana Press Inc.
999 Riverview Drive, Suite 208
Totowa, New Jersey 07512

www.humanapress.com

This publication is printed on acid-free paper. ∞
ANSI Z39.48-1984 (American National Standards Institute)

Production Editor: Michele Seugling.

Cover design by Nancy Fallatt.

Cover Illustration: Figure 1, Chapter 6, "Effective Model for Charge Trandport in DNA Nanowires," by Rafael Gutierrez and Gianaurelio Cuniberti, and Figure 2, Chapter 13, "Nano-Sized Carriers for Drug Delivery," by Sajeeb K. Sahoo, Tapan K. Jain, Maram K. Reddy, and Vinod Labhasetwar.

For additional copies, pricing for bulk purchases, and/or information about other Humana titles, contact Humana at the above address or at any of the following numbers: Tel.: 973-256-1699; Fax: 973-256-8341; E-mail: humana@humanapr.com; or visit our Website: www.humanapress.com

Printed in the United States of America.   10   9   8   7   6   5   4   3   2   1

e-ISBN: 978-1-59745-218-2

Library of Congress Cataloging-in-Publication Data

Nanobiotechnology: bioinspired devices and materials of the future/edited by Oded Shoseyov and Ilan Levy.
    p. ; cm.
    Includes bibliographical references and index.
    ISBN-13: 978-1-58829-894-2 (alk. paper)
    1. Nanotechnology. 2. Biotechnology. I. Oded Shoseyov. II. Ilan Levy.
    [DNLM: 1. Biotechnology. 2. Nanotechnology. 3. Nanomedicine. QT 36.5 N183 2007]
TP248.25.N35N28 2007
660.6– –dc22   2007005772

PREFACE

Research and applied science, as we see it today, has advanced to a place in which, instead of manipulating substances at the molecular level, we can control them at the atomic level. This exciting operational space, where the laws of physics shift from Newtonian to quantum, provides us with novel discoveries, which hold the promise of future developments that, until recently, belonged to the realm of science fiction.

Nanobiotechnology is a multidisciplinary field that covers a vast and diverse array of technologies from engineering, physics, chemistry, and biology. It is expected to have a dramatic infrastructural impact on both nanotechnology and biotechnology. Its applications could potentially be quite diverse, from building faster computers to finding cancerous tumors that are still invisible to the human eye. As nanotechnology moves forward, the development of a 'nano-toolbox' appears to be an inevitable outcome. This toolbox will provide new technologies and instruments that will enable molecular manipulation and fabrication via both 'top-down' and 'bottom-up' approaches.

This book is organized into five major sections; 1. Introduction, 2. Bio-templating, 3. Bionanoelectronics and Nanocomputing, 4. Nanomedicine, Nanopharmaceuticals and Nanosensing, and 5. *De Novo* Designed Structures.

Section 1 is an introductory overview on nanobiotechnology, which briefly describes the many aspects of this field, while addressing the reader to relevant sources for broader information overviews.

Biological materials can serve as nanotemplates for 'bottom-up' fabrication. In fact, this is considered one of the most promising 'bottom-up' approaches, mainly due to the nearly infinite types of templates available. This approach is demonstrated in Section 2.

The convergence of nanotechnology and biotechnology may combine biological and man-made devices for the design and fabrication of bio-nanoelectronics and for their use in nanocomputing. This area is addressed in Section 3, which covers the use of biological macromolecules for electron transfer and computation.

One of the main reasons nanobiotechnology holds so much promise is that it operates at the biological size scale. Biological molecules (such as enzymes, receptors, DNA), microorganisms and individual cells in our bodies are all nano-sized. Engineered ultrasmall particles that are made in the exact size needed to perform specific tasks, such as drug release in particular locations in the body, drug delivery into the blood stream, or to pinpoint malfunctioning tissues (cancerous tissue, for example), are examples of the new medical discipline termed 'nanomedicine'. Section 4 gives a brief look at this extensive and rapidly growing field.

The fact that nanobiotechnology embraces and attracts many different disciplines, encompassing both researchers and business leaders, has produced many examples of bio-inspired *de novo* designed structures. Each scientific group approaches the molecular level with unique skills, training, and language, and a few examples are presented in Section 5. Cross-talk and collaborative research among academic disciplines, and between the researchers and their counterparts in business, are critical to the advancement of nanobiotechnology and constitute the foundation for the new material generation.

Working at the molecular or atomic level allows researchers to develop innovations that will dramatically improve our lives. The new territory of bionanotechnology holds the promise of improving our health, our industry, and our society in ways that may even surpass what computers and biotechnology have already achieved.

Ilan Levy and Oded Shoseyov

CONTENTS

vii

CONTRIBUTORS

KENJI ARINAGA, *Walter Schottky Institut, Technische Universitaet Muenchen, Garching, Germany, and Fujitsu Laboratories Ltd., Atsugi, Japan*

LANE A. BAKER, *Departments of Chemistry and Anesthesiology, University of Florida, Gainesville, FL*

ROBERT R. BIRGE, *Department of Molecular and Cell Biology, and Department of Chemistry University of Connecticut, Storrs, CT*

STEFAN H. BOSSMANN, *Kansas State University, Department of Chemistry, Manhattan, Kansas*

JIN-HO CHOY, *Center for Intelligent NanoBio Materials (CINBM), Department of Chemistry and Division of Nanoscience, Ewha Womans University, Seoul 120-750, Korea*

GIANAURELIO CUNIBERTI, *Molecular Computing Group, Institute of Theoretical Physics, University of Regensburg, Regensburg, Germany*

ROSA DI FELICE, *National Center on nanoStructures and bioSystems at Surfaces (S3) of INFM-CNR, Modena, Italy*

EVA-M. EGELSEER, *Center for NanoBiotechnology, University of Natural Resources and Applied Life Sciences, Vienna, Austria*

JOSEPH FARFEL, *Department of Computer Science, Duke University, Durham, NC*

ERAN GABBAI, *Do-Coop Technologies Ltd, Or Yehuda, Israel*

EHUD GAZIT, *Department of Molecular Microbiology and Biotechnology, Tel Aviv University, Tel Aviv, Israel*

RAFAEL GUTIERREZ, *Molecular Computing Group, Institute of Theoretical Physics, University of Regensburg, Regensburg, Germany*

TETSUYA HARUYAMA, *Department of Biological Functions and Engineering, Kyushu Institute of Technology, Fukuoka, Japan*

JASON R. HILLEBRECHT, *Department of Molecular and Cell Biology, University of Connecticut, Storrs, CT*

KEWAL K. JAIN, *PharmaBiotech, Basel, Switzerland*

TAPAN K. JAIN, *Department of Pharmaceutical Sciences, College of Pharmacy, University of Nebraska Medical Center, Omaha, NE*

KATHARINE JANIK, *Kansas State University, Department of Chemistry, Manhattan, Kansas*

NATAŠA JONOSKA, *University of South Florida, Department of Mathematics, Tampa, FL*

ANDREAS KAGE, *Charité Universitätsmedizin Berlin, Zentralinstitut für Laboratoriumsmedizin und Pathobiochemie, Berlin, Germany*

JEREMY F. KOSCIELECKI, *Department of Chemistry, University of Connecticut, Storrs, CT*

MARK P. KREBS, *Department of Ophthalmology, College of Medicine, University of Florida, Gainesville, FL*

VINOD LABHASETWAR, *Department of Biomedical Engineering, Lemer Research Institute, Cleveland Clinic, Cleveland, OH*

KATARZYNA LAMPARSKA-KUPSIK, *City of Hope National Medical Center and Beckman Research Institute, Duarte, CA*

ILAN LEVY, *Intel Research Israel, Intel Electronics, Jerusalem, Israel*

LOREN LIMBERIS, *Department of Engineering, East Carolina University, Greenville, NC*

CHARLES R. MARTIN, *Departments of Chemistry and Anesthesiology, University of Florida, Gainesville, FL*

MICHAEL NIEDERWEIS, *University of Alabama at Birmingham, Department of Microbiology, Bevill Biomedical Research Building, Birmingham, AL*

JAE-MIN OH, *Center for Intelligent NanoBio Materials (CINBM), Department of Chemistry and Division of Nanoscience, Ewha Womans University, Seoul 120-750, Korea*

MAN PARK, *Center for Intelligent NanoBio Materials (CINBM), Department of Chemistry and Division of Nanoscience, Ewha Womans University, Seoul 120-750, Korea*

MEGH RAJ POKHREL, *Central Department of Chemistry, Tribhuvan University, Kirtipur, Kathmandu, Nepal*

DANNY PORATH, *Physical Chemistry Department and Center for Nanoscience and Nanotechnology, The Hebrew University of Jerusalem, Israel*

DIETMAR PUM, *Center for NanoBiotechnology, University of Natural Resources and Applied Life Sciences, Vienna, Austria*

ULRICH RANT, *Walter Schottky Institut, Technische Universitaet Muenchen, Garching, Germany*

MARAM K. REDDY, *Department of Biomedical Engineering, Lemer Research Institute, Cleveland Clinic, Cleveland, OH*

JENNIFER SAGER, *Department of Computer Science, University of New Mexico, Albuquerque, NM*

SANJEEB K. SAHOO, *Institute of Life Sciences, Nalco Square, Bhubaneswar, Orissa, India*

Margit Sára, *Center for NanoBiotechnology, University of Natural Resources and Applied Life Sciences, Vienna, Austria*

Bernhard Schuster, *Center for NanoBiotechnology, University of Natural Resources and Applied Life Sciences, Vienna, Austria*

Oded Shoseyov, *The Institute of Plant Science and Genetics in Agriculture and The Otto Warburg Center for Agricultural Biotechnology, Faculty of Agricultural, Food and Environmental Quality Sciences, The Hebrew University of Jerusalem, Rehovot, Israel*

Uwe B. Sleytr, *Center for NanoBiotechnology, University of Natural Resources and Applied Life Sciences, Vienna, Austria*

Steven S. Smith, *City of Hope National Medical Center and Beckman Research Institute, Duarte, CA*

Darko Stefanovic, *Department of Computer Science, University of New Mexico, Albuquerque, NM*

Russell J. Stewart, *Department of Bioengineering, University of Utah, Salt Lake City, UT*

Jeffrey A. Stuart, *W. M. Keck Center for Molecular Electronics, Syracuse University, Syracuse, NY*

Marc Tornow, *Institute of Semiconductor Technology, Technical University of Braunschweig, Braunschweig, Germany*

Hans-Achim Wagenknecht, *University of Regensburg, Institute of Organic Chemistry, Regensburg, Germany*

Kevin J. Wise, *Department of Molecular and Cell Biology, University of Connecticut, Storrs, CT*

I
INTRODUCTION

1

Nanobiotechnology Overview

Oded Shoseyov and Ilan Levy

Summary

Nanobiotechnology is a multidisciplinary field that covers a vast and diverse array of technologies coming from engineering, physics, chemistry, and biology. It is the combination of these fields that has led to the birth of a new generation of materials and methods of making them. The scope of applications is enormous and every day we discover new areas of our daily lives where they can find use. This chapter aims to provide the reader with a brief overview of nanobiotechnology by describing different aspects and approaches in research and application of this exciting field. It also provides a short list of recently published review articles and books on the different topics in nanobiotechnology.

Key Words: Nanobiotechnology; nanocomputing; nanoelectronics; nanofabrication; nanomedicine; nanotechnology.

1. INTRODUCTION TO NANOSCIENCE AND NANOTECHNOLOGY

The prefix *nano* is derived from the Greek word *nanos* meaning "dwarf," need and today it is used as a prefix describing 10^{-9} (one billionth) of a measuring unit. Therefore, nanotechnology is the field of research and fabrication that is on a scale of 1 to 100 nm. The primary concept was presented on December 29, 1959, when Richard Feynman presented a lecture entitled "There's Plenty of Room at the Bottom" at the annual meeting of the American Physical Society, the California Institute of Technology (this lecture can be found on several web sites; *see* ref. *1*). Back then, manipulating single atoms or molecules was not possible because they were far too small for available tools. Thus, his speech was completely theoretical and seemingly far-fetched. He described how the laws of physics do not limit our ability to manipulate single atoms and molecules. Instead, it was our lack of the appropriate methods for doing so. However, he correctly predicted that the

From: *NanoBioTechonology: BioInspired Devices and Materials of the Future*
Edited by: Oded Shoseyov and Ilan Levy © Humana Press Inc., Totowa, NJ

time for the atomically precise manipulation of matter would inevitably arrive. Today, that lecture is considered to be the first landmark of science at the nanolevel.

The first 30 yr or so of the nanosciences were devoted mainly to studying and fabricating materials at the nanolevel. In those studies, much effort was devoted to shrinking the dimension of fabricated materials. It was also a time when the two basic fabrication approaches were defined: "bottom-up" and "top-down." The bottom-up approach seeks the means and tools to build things by combining smaller components such as single molecules and atoms, which are held together by covalent forces. Theoretically, it can be exemplified by molecular assemblers, where nanomachines are programmed to build a structure one atom or molecule at a time or by self-assembly, where these structures are built spontaneously. The advantage of the bottom-up design is that the covalent bonds holding a single molecule together are far stronger than the weak interactions that hold more than one molecule together. The top-down approach refers to the molding, carving, and fabricating of small materials and components by using larger objects such as mechanical tools and lasers, such as is used today in current photolithographic approaches in silicon chip fabrication. Currently, techniques using both approaches are evolving, and many applications are likely to involve combination approaches. However, the bottom-up approach, at least theoretically, holds far more practical and applicative future potential.

Nanoscience is therefore a multidisciplinary field that seeks to integrate mature nanoscale technology of fields such as physics, biology, engineering, chemistry, computer science, and material science.

2. THE "NANO"–"BIO" INTERFACE

Biosystems are governed by nanoscale processes and structures that have been optimized over millions of years. Biologists have been operating for many years at the molecular level, in the range of nanometers (DNA and proteins) to micrometers (cells). A typical protein like hemoglobin has a diameter of about 5 nm, the DNA's double helix is about 2 nm wide, and a mitochondrion spans a few hundred nanometers. Therefore, the study of any subcellular entity can be considered "nanobiology." Furthermore, the living cell along with its hundreds of nanomachines is considered, today, to be the ultimate nanoscale fabrication system.

On the other hand, countless exciting questions in biology can be addressed in new ways by exploiting the rapidly growing capabilities of nanotechnological research approaches and tools. This research will form and shape the foundation for our understanding of how biological systems operate. We are exploiting nanofabrication to perform individual molecule analyses in biological systems, to study cellular responses to structured interfaces, and to explore dynamic life processes at reduced dimensions. Our research has advanced the ability to structure materials and pattern surface chemistry at subcellular and molecular dimensions.

The groundwork of each and every biological system is nanosized molecular building blocks and machinery that cooperate to produce living entities. These elements have ignited the imagination of nanotechnologists for many years and it is the combination of these two disciplines (nano and biotechnology) that has resulted in the birth of the new science of nanobiotechnology. Nanotechnology provides the tools and technology platforms for the investigation and transformation of biological systems, and biology offers inspirational models and bio-assembled components to nanotechnology. The difference between "nanobiology" to "nanobiotechnology" resides in the technology part of the term. Anything that is "man-made" falls into the technology section of nanobiotechnology. Nearly any molecular machinery that we can think of has its analog in biological systems and as for now, it appears that the first revolutionary application of nanobiotechnology will probably be in computer science and medicine. Nanobiotechnology will lead to the design of entirely new classes of micro- and nanofabricated devices and machines, the inspiration for which will be based on bio-structured machines, the use of biomolecules as building blocks, or the use of biosystems as the fabrication machinery.

3. NANOBIOTECHNOLOGY

Unlike nonbiological systems that are fabricated top-down, biological systems are built up from the molecular level (bottom-up). They do this via a collection of molecular tool kits of atomic resolution that are used to fabricate micro- and macrostructure architectures. Biological nanotechnology, or nanobiotechnology, can be viewed in many ways: one way is the incorporation of nanoscale machines into biological organisms for the ultimate purpose of improving the organism's quality of life. To date, there are a few methods for synthesizing nanodevices that have the potential to be used in an organism without risk of being rejected as antigens; another way is the use of biological "tool kits" to construct nano- to microstructures. However, the broad perspective is probably the one that will include both and will be defined as: **the engineering, construction, and manipulation of entities in the 1- to 100-nm range using biologically based approaches or for the benefit of biological systems**. The biological approaches can be either an inspired way of mimicking biological structures or the actual use of biological building blocks and building tools to assemble nanostructures. In a way, the first example of a nanobiotechnology system might be the production of recombinant proteins. Recombinant DNA technology can direct the ribosomal machinery to produce designed proteins both in vivo and in vitro that can serve as components of larger molecular structures.

As already mentioned, there are two basic fabrication approaches to creating nanostructures: bottom-up and top-down. The bottom-up approach exploits biological structures and processes to create novel functional materials, biosensors, and bioelectronics for different applications. This field encompasses many disciplines, including material science, organic chemistry, chemical engineering, biochemistry, and molecular biology. In the top-down approach, nanobiotechnology applies tools and processes of nano/microfabrication to build nanostructures and nanodevices. The tools that are used often involve optical and electron beam lithography and the processing of large materials into fine structures with defined surface features. One of the major differences between nanotechnology and nanobiotechnology is that in the former, the dominant approach is top-down, whereas in the latter, it is bottom-up.

An example of the bottom-up approach is the pioneering work of two leading groups on biomolecular motor proteins *(2–7)*. In these studies, naturally occurring motor proteins were engineered for compatibility with artificial interfaces to create new ways of joining proteins to synthetic nanomaterials. Biomolecular motors can provide chemically powered movement to micro- and nanodevices. Nanodevices utilizing motor proteins such as kinesin or F_1-ATPase can be used as nanoscale transporters, as probes for surface imaging, to control the movement of target substances, and to support the controlled assembly of nanostructures.

Structural properties that enable DNA to serve so effectively as genetic material can also be exploited to produce target materials with predictable three-dimensional (3D) structures in the bottom-up approach. Pioneering work using this approach is presented in studies performed by the group of Nadrian Seeman *(8–12)*. He uses DNA motifs with specific, structurally well defined, cohesive interactions involving hydrogen bonding or covalent interactions ("sticky ends") to produce target materials with predictable 2D and 3D structures. The complementarity that leads to the pairing of the DNA strands is the driving force for the complex assemblies with their branched structures. These efforts have generated a large number of individual species, including polyhedral catenanes, such as a cube and a truncated octahedron, a variety of single-stranded knots, and Borromean rings. The combination of these constructions with other chemical components is expected to contribute to the development of nanoelectronics, nanorobotics, and smart materials. Therefore, the organizational capabilities of structural DNA nanotechnology are just beginning to be explored, and the field is ultimately expected to be able to organize a variety of species in the material world.

Another fascinating example is the use of crystalline bacterial cell surface layer (S-layers) proteins as tools in nanofabrication and nanopatterning. The

S-layer is composed of identical protein or glycoprotein subunits that self-assemble into lattices, forming the outermost cell envelope component of many bacteria. As a result of their high degree of structural regularity, S-layers represent interesting model systems for studies on structural, functional, and dynamic aspects of supramolecular structure assembly. The nano-based approach of S-layer research was pioneered by Uwe B. Sleytr *(13–16)*. In one of those studies, lattices of cadmium sulfide quantum dots were synthesized by using self-assembled bacterial S-layers as templates. Au and CdSe nanoparticles were also deposited directly onto the protein lattice. Given that the macroscopic electronic or magnetic properties of nanoparticle arrays are influenced by interparticle distance and geometry, it should be possible to use various natural or engineered S-layer lattices as a "tuneable" system to obtain nanoparticle assemblies with designed properties for material science. In the future, engineered S-layer proteins might be used as tool kits for the positioning of proteins or nanoparticles in nanopatterned arrays. A faster route to nanopatterns might be a top-down approach wherein S-layer proteins are assembled on nanolithographically structured substrates. Metallic or semiconductor nanoparticle assemblies generated in this way will form the basis of materials with tailored electronic or magnetic properties. Several applications have been suggested for S-layers, such as their use as templates for the nanoscale patterning of inorganic materials or as immobilization matrices for biomedical applications. However, in particular, S-layer technologies provide new approaches for biotechnology, biomimetics, molecular nanotechnology, nanopatterning of surfaces, and formation of ordered arrays of metal clusters or nanoparticles as required for nanoelectronics.

4. REVIEWING MAJOR FIELDS IN NANOBIOTECHNOLOGY

In this section, we will very briefly review the major fields of nanobiotechnology. In fact, each and every one could stand by itself as a book title. However, here we only list the different fields along with a short compilation of recent reviews from the last 5 years published on the subject. Several recently published books are also listed here *(17–22)*.

4.1. Molecular Motors and Devices

A molecular machine can be defined as an assembly of a discrete number of molecular components designed to perform mechanical movement as a consequence of external stimulus. The concept of molecular motors is not new and in fact, every single cell contains several molecular motors as an integral part of its regular function. There are two basic types of natural molecular machines: the rotary motors, such as the F_1-ATPase of flagella, and the linear motors, such as myosin. The study of these molecular motors enables the use and design of new molecular motors based on biomolecules or other chemical components inspired by bio-motors *(23–26)*.

4.2. Self-Assembled Structures (Nano-Assemblies)

Self-assembly is the spontaneous organization of individual elements into ordered structures. Molecular self-assembly is a fabrication tool where engineering principles can be applied to design structures using basic principles adopted from naturally occuring self assemblies. Self-assembly's greatest advantage is that it is energetically efficient compared to direct assembly. In recent years, considerable advances have been made in the use of peptides and proteins as building blocks to produce a wide range of biological materials for diverse applications *(27–37)*.

4.3. Biomedical Application of Nanotechnology—Nanomedicine

Although major progress has been achieved in recent years, modern medicine is limited by both its knowledge and its treatment tools. It is only in the last 50 yr that medicine has started looking at diseases at the molecular level, and today's drugs are thus essentially single-effect molecules. The potential impact of nanotechnology on medicine stems directly from the dimension of the devices and materials that can interact directly with cells and tissues at a molecular level. Applied nanobiotechnology in medicine is in its infancy. However, the breadth of current nanomedicine research is extraordinary. It includes three major research areas: diagnostics, pharmaceuticals, and prosthesis and implants. Today, nanomedicine is one of the dominant and leading fields of nanobiotechnology *(38–51)*.

4.4. Biological Research at the Nanoscale

Living organisms and biomolecules are far more complex than engineered materials. In the last few decades, research has focused on the connection between structure, mechanical response, and biological function at the macro- and microlevels. The introduction of research tools at the nanolevel and nanomanipulation techniques stemming from the material world has launched a new paradigm of biomolecular research. Nanoresearch tools are capable of analyzing and visualizing properties of single molecules, thereby providing the opportunity to examine bio-processes of single cells and molecular motors *(52–55)*.

4.5. Biomimetics, Biotemplating, and De Novo-Designed Structures

One of the central goals of nanobiotechnology is the design and creation of novel materials on the nanoscale. Biomolecules, through their unique and specific interaction with other biomolecules and inorganic molecules, natively control complexed structures at the tissue and organ levels. With recent progress in nanoscale engineering and manipulation, along with developments in molecular biology and biomolecular structures, biomimetics and *de novo*-designed structures are entering the molecular level. The promise in biomimetics and biotemplating lies in the potential use of inorganic surface-specific proteins for controlled material assembly in vivo or in vitro *(56–67)*.

4.6. Nanocomputing

A comparison of biological systems to computers shows that both process information that is stored in a sequence of symbols taken from an unchanging alphabet, and both operate in a stepwise fashion. In recent years, great interest has arisen among researchers on developing new computers inspired from biological systems. Performing calculations employing biomolecules and using genetic engineering technology may soon find use as a tool for computation. The greatest promise of biological computers is that they can operate in biochemical environments *(68–71)*.

4.7. DNA-Based Nanotechnology and Nanoelectronics

DNA-based nanotechnology is intrinsic to all of the nanotechnological approaches mentioned thus far. An increasing number of scientists within nanoscience are using nucleic acids as building blocks in the bottom-up fabrication approach in order to produce novel structures and devices. The basic drive of this application is the well established Watson-Crick hybridization of complementary nucleic-acid strands. This force has been shown to be efficient in the construction of nanodevices, nanomachines, DNA-based nanoassemblies, DNA–protein conjugated structures, and DNA-based computation *(72–88)*.

5. CURRENT STATUS AND FUTURE TRENDS

Nanobiotechnology is still in the early stages of development; however, its development is multidirectional and fast-paced. Nanobiotechnology research centers are being founded and funded at a high frequency, and the numbers of papers and patent applications is also rising rapidly. In addition, the nanobiotechnology "tool box" is being rapidly filled with new and viable tools for bio-nanomanipulations that will speed up new applications. Finally, an analysis of the total investment in nanobiotechnology start-ups reveals that nearly 50% of the venture capital investments in nanotechnology is addressed to nanobiotechnology *(89)*.

One of the strongest driving forces in this research area is the semiconductor industry. Computer chips are rapidly shrinking according to Moore's law, i.e., by a factor of four every 3 yr. However, this simple shrinking law cannot continue for much longer, and computer scientists are therefore looking for solutions. One approach is moving to single-molecule transistors *(90–93)*. This shift is critically dependent on molecular nanomanipulations to form molecular computation that will write, process, store, and read information within the single molecule where proteins and DNA are some of the alternatives *(94–99)*.

As medical research and diagnostics steadily progresses based on the use of molecular biomarkers and specific therapies aimed at molecular markers and multiplexed analysis, the necessity for molecular-level devices increases. Technology platforms that are reliable, rapid, low-cost, portable, and that can handle large quantities are evolving and will provide the future foundation for personalized medicine. These new technologies are especially important in cases of early detection, such as in cancer. Future applications of nanobiotechnology will probably include nanosized devices and sensors that will be injected into, or ingested by, our bodies. These instruments could be used as indicators for the transmission of information outside of our bodies or they could actively perform repairs or maintenance. Nanotechnology-based platforms will secure the future realization of multiple goals in biomarker analysis. Examples for such platforms are the use of cantilevers, nanomechanical systems (NEMS), nanoelectronics (biologically gated nanowire), and nanoparticles in diagnostics imaging and therapy *(100–106)*.

The art of nanomanipulating materials and biosystems is converging with information technology, medicine, and computer sciences to create entirely new science and technology platforms. These technologies will include imaging diagnostics, genome pharmaceutics, biosystems on a chip, regenerative medicine, on-line multiplexed diagnostics, and food systems. It is clear that biology has much to offer the physical world in demonstrating how to recognize, organize, functionalize, and assemble new materials and devices. In fact, almost any device, tool, or active system known today can be either mimicked by biological systems or constructed using techniques originating in the bio-world. Therefore, it is plausible that in the future, biological systems will be used as building blocks for the construction of the material and mechanical fabric of our daily lives.

REFERENCES

1. http://www.zyvex.com/nanotech/feynman.html.
2. Hess H, Vogel V. Molecular shuttles based on motor proteins: active transport in synthetic environments. J Biotechnol 2001;82:67–85.
3. Hess H, Bachand G, Vogel V. Powering nanodevices with biomolecular motors. Chemistry 2004;10:2110–2116.
4. Hess H, Clemmens J, Brunner C, Doot R, Luna S, Ernst KH, Vogel V. Molecular self-assembly of "nanowires" and "nanospools" using active transport. Nano Lett 2005;5:629–633.

5. Montemagno C. Constructing nanomechanical devices powered by biomolecular motors. J Nanotechnol 1999;10:225–231.
6. Liu H, Schmidt JJ, Bachand GD, et al. Control of a biomolecular motor-powered nanodevice with an engineered chemical switch. Nat Mater 2002;1:173–177.
7. Xi J, Schmidt JJ, Montemagno CD. Self-assembled microdevices driven by muscle. Nat Mater 2005;4:180–184.
8. Seeman NC. From genes to machines: DNA nanomechanical devices. Trends Biochem Sci 2005;30:119–125.
9. Seeman NC. Structural DNA nanotechnology: an overview. Methods Mol Biol 2005;303:143–166.
10. Seeman NC. DNA enables nanoscale control of the structure of matter. Q Rev Biophys 2006;6:1–9.
11. Seeman NC. At the crossroads of chemistry, biology, and materials: structural DNA nanotechnology. Chem Biol 2003;10:1151–1159.
12. Seeman NC. DNA in a material world. Nature 2003;421:427–431.
13. Sara M, Pum D, Schuster B, Sleytr UB. S-layers as patterning elements for application in nanobiotechnology. J Nanosci Nanotechnol 2005;5:1939–1953.
14. Schuster B, Gyorvary E, Pum D, Sleytr UB. Nanotechnology with S-layer proteins. Methods Mol Biol 2005;300:101–123.
15. Shenton W, Pum D, Sleytr UB, Mann S. Synthesis of cadmium sulfide superlattices using self-assembled bacterial S-layers. Nature 1997;389: 585–587.
16. Gyorvary E, Schroedter A, Talapin DV, Weller H, Pum D, Sleytr UB. Formation of nanoparticle arrays on S-layer protein lattices. J Nanosci Nanotechnol 2004;4:115–120.
17. Niemeyer CM, Mirkin CA. Nanobiotechnology: Concepts, Applications and Perspectives. Weinheim: Wiley-VCH, 2004.
18. Rosenthal SJ, Wright DW. NanoBiotechnology Protocols. Methods in Molecular Biology, vol. 303. New Jersey: Humana Press, 2005.
19. Jain KK. Nanobiotechnology in Molecular Diagnosis: Current Techniques and Applications. Oxford: Taylor & Francis, 2005.
20. Nill KR. Glossary of Biotechnology and Nanobiotechnology Terms, 4th ed. London: CRC Press, 2005.
21. Nalwa HS. Handbook of Nanostructured Biomaterials and Their Applications in Nanobiotechnology. Valencia, CA: American Scientific Publishers, 2005.
22. Golden J. Nanobiotechnology. Oxford: Taylor & Francis, 2007.
23. Balzani VV, Credi A, Raymo FM, Stoddart JF. Artificial molecular machines. Angew Chem Int Ed Engl 2000;39:3348–3391.
24. Pennadam SS, Firman K, Alexander C, Gorecki DC. Protein-polymer nanomachines. Towards synthetic control of biological processes. J Nanobiotechnol 2004;2:8.
25. Astier Y, Bayley H, Howorka S. Protein components for nanodevices. Curr Opin Chem Biol 2005;9:576–584.
26. Paul N, Joyce GF. Minimal self-replicating systems. Curr Opin Chem Biol 2004;8:634–639.

27. Zhang S, Marini DM, Hwang W, Santoso S. Design of nanostructured biological materials through self-assembly of peptides and proteins. Curr Opin Chem Biol 2002;6:865–871.

28. Ghosh I, Chmielewski J. Peptide self-assembly as a model of proteins in the pre-genomic world. Curr Opin Chem Biol 2004;8:640–644.

29. Koltover I. Biomolecular self-assembly: stacks of viruses. Nat Mater 2004;3:584–586.

30. Boncheva M, Gracias DH, Jacobs HO, Whitesides GM. Biomimetic self-assembly of a functional asymmetrical electronic device. Proc Natl Acad Sci USA 2002;99:4937–4940.

31. Whitesides GM, Boncheva M. Beyond molecules: self-assembly of mesoscopic and macroscopic components. Proc Natl Acad Sci USA 2002;99:4769–4774.

32. Ghadiri MR, Tirrell DA. Chemistry at the crossroads. Curr Opin Chem Biol 2000;4:661–662.

33. Woolfson DN. The design of coiled-coil structures and assemblies. Adv Protein Chem 2005;70:79–112.

34. Wettig SD, Li CZ, Long YT, Kraatz HB, Lee JS. M-DNA: a self-assembling molecular wire for nanoelectronics and biosensing. Anal Sci 2003;19:23–26.

35. Yeates TO, Padilla JE. Designing supramolecular protein assemblies. Curr Opin Struct Biol 2002;12:464–470.

36. Wu LQ, Payne GF. Biofabrication: using biological materials and biocatalysts to construct nanostructured assemblies. Trends Biotechnol 2004;22:593–599.

37. Hamada D, Yanagihara I, Tsumoto K. Engineering amyloidogenicity towards the development of nanofibrillar materials. Trends Biotechnol 2004;22:93–97.

38. Emerich DF. Nanomedicine-prospective therapeutic and diagnostic applications. Expert Opin Biol Ther 2005;5:1–5.

39. Lutolf MP, Hubbell JA. Synthetic biomaterials as instructive extracellular microenvironments for morphogenesis in tissue engineering. Nat Biotechnol 2005;23:47–55.

40. Fortina P, Kricka LJ, Surrey S, Grodzinski P. Nanobiotechnology: the promise and reality of new approaches to molecular recognition. Trends Biotechnol 2005;23:168–173.

41. Jain KK. Role of nanobiotechnology in developing personalized medicine for cancer. Technol Cancer Res Treat 2005;4:645–650.

42. Ferrari M. Cancer nanotechnology: opportunities and challenges. Nat Rev Cancer 2005;5:161–171.

43. Labhasetwar V. Nanotechnology for drug and gene therapy: the importance of understanding molecular mechanisms of delivery. Curr Opin Biotechnol 2005;16:674–680.

44. Cheng MM, Cuda G, Bunimovich YL, et al. Nanotechnologies for biomolecular detection and medical diagnostics. Curr Opin Chem Biol 2006;10:11–19.

45. Jain KK. The role of nanobiotechnology in drug discovery. Drug Discov Today 2005;10:1435–1442.

46. Jain KK. Nanotechnology-based drug delivery for cancer. Technol Cancer Res Treat 2005;4:407–416.

47. Kubik T, Bogunia-Kubik K, Sugisaka M. Nanotechnology on duty in medical applications. Curr Pharm Biotechnol 2005;6:17–33.

48. Bogunia-Kubik K, Sugisaka M. From molecular biology to nanotechnology and nanomedicine. Biosystems 2002;65:123–138.
49. Silva GA. Neuroscience nanotechnology: progress, opportunities and challenges. Nat Rev Neurosci 2006;7:65–74.
50. Silva GA. Introduction to nanotechnology and its applications to medicine. Surg Neurol 2004;61:216–220.
51. Kubik T, Bogunia-Kubik K, Sugisaka M. Nanotechnology on duty in medical applications. Curr Pharm Biotechnol 2005;6:17–33.
52. Bao G, Suresh S. Cell and molecular mechanics of biological materials. Nat Mater 2003;2:715–725.
53. Haustein E, Schwille P. Single-molecule spectroscopic methods. Curr Opin Struct Biol 2004;14:531–540.
54. Curtis A, Wilkinson C. Nantotechniques and approaches in biotechnology. Trends Biotechnol 2001;19:97–101.
55. Ishii Y, Ishijima A, Yanagida T. Single molecule nanomanipulation of biomolecules. Trends Biotechnol 2001;19:211–216.
56. Mao C, Solis DJ, Reiss BD, et al. Virus-based toolkit for the directed synthesis of magnetic and semiconducting nanowires. Science 2004;303:213–217.
57. Ball P. Synthetic biology for nanotechnology. Nanotechnology 2005;16:R1–R8.
58. Sarikaya M, Tamerler C, Jen AK, Schulten K, Baneyx F. Molecular biomimetics: nanotechnology through biology. Nat Mater 2003;2:577–585.
59. Chen Y, Pepin A. Nanofabrication: conventional and nonconventional methods. Electrophoresis 2001;22:187–207.
60. Sarikaya M. Biomimetics: materials fabrication through biology. Proc Natl Acad Sci USA 1999;96:14,183–14,185.
61. Wilt FH. Developmental biology meets materials science: morphogenesis of biomineralized structures. Dev Biol 2005;280:15–25.
62. Penczek S, Pretula J, Kaluzynski K. Poly(alkylene phosphates): from synthetic models of biomacromolecules and biomembranes toward polymer-inorganic hybrids (mimicking biomineralization). Biomacromolecules 2005;6:547–551.
63. Vrieling EG, Sun Q, Beelen TP, et al. Controlled silica synthesis inspired by diatom silicon biomineralization. J Nanosci Nanotechnol 2005;5:68–78.
64. Bauerlein E. Biomineralization of unicellular organisms: an unusual membrane biochemistry for the production of inorganic nano- and microstructures. Angew Chem Int Ed Engl 2003;42:614–641.
65. Mastrobattista E, van der Aa MA, Hennink WE, Crommelin DJ. Artificial viruses: a nanotechnological approach to gene delivery. Nat Rev Drug Discov 2006;5:115–121.
66. Vriezema DM, Comellas Aragones M, Elemans JA, Cornelissen JJ, Rowan AE, Nolte RJ. Self-assembled nanoreactors. Chem Rev 2005;105:1445–1489.
67. Arora PS, Kirshenbaum K. Nano-tailoring; stitching alterations on viral coats. Chem Biol 2004;11:418–420.
68. Adar R, Benenson Y, Linshiz G, Rosner A, Tishby N, Shapiro E. Stochastic computing with biomolecular automata. Proc Natl Acad Sci USA 2004;101:9960–9965.
69. Benenson Y, Gil B, Ben-Dor U, Adar R, Shapiro E. An autonomous molecular computer for logical control of gene expression. Nature 2004;429:423–429.

70. Benenson Y, Paz-Elizur T, Adar R, Keinan E, Livneh Z, Shapiro E. Programmable and autonomous computing machine made of biomolecules. Nature 2001;414:430–434.

71. Hill RT, Lyon JL, Allen R, Stevenson KJ, Shear JB. Microfabrication of three-dimensional bioelectronic architectures. J Am Chem Soc 2005;127: 10,707–10,711.

72. Rothemund PW. Folding DNA to create nanoscale shapes and patterns. Nature 2006;440:297–302.

73. Rinaldi R, Maruccio G, Biasco A, Visconti P, Arima V, Cingolani R. A protein-based three terminal electronic device. Ann NY Acad Sci 2003;1006:187–197.

74. Wettig SD, Li CZ, Long YT, Kraatz HB, Lee JS. M-DNA: a self-assembling molecular wire for nanoelectronics and biosensing. Anal Sci 2003;19:23–26.

75. Davis JJ. Molecular bioelectronics. Philos Transact A Math Phys Eng Sci 2003;361:2807–2825.

76. Willner I, Katz E. Magnetic control of electrocatalytic and bioelectrocatalytic processesc. Angew Chem Int Ed Engl 2003;42:4576–4588.

77. Willner I, Willner B. Biomaterials integrated with electronic elements: en route to bioelectronics. Trends Biotechnol 2001;19:222–230.

78. Seeman NC. Structural DNA nanotechnology: an overview. Meth Mol Biol 2005;303:143–166.

79. Seeman NC. From genes to machines: DNA nanomechanical devices. Trends Biochem Sci 2005;30:119–125.

80. Seeman NC. At the crossroads of chemistry, biology, and materials: structural DNA nanotechnology. Chem Biol 2003;10:1151–1159.

81. Seeman NC. DNA in a material world. Nature 2003;421:427–431.

82. Seeman NC, Belcher AM. Emulating biology: building nanostructures from the bottom up. Proc Natl Acad Sci USA 2002;99:6451–6455.

83. Feldkamp U, Niemeyer CM. Rational design of DNA nanoarchitectures. Angew Chem Int Ed Engl 2006;45:1856–1876.

84. Niemeyer CM. Functional hybrid devices of proteins and inorganic nanoparticles. Angew Chem Int Ed Engl 2003;42:5796–5800.

85. Niemeyer CM, Adler M. Nanomechanical devices based on DNA. Angew Chem Int Ed Engl 2002;41:3779–3783.

86. Niemeyer CM. The developments of semisynthetic DNA-protein conjugates. Trends Biotechnol 2002;20:395–401.

87. Wngel J. Nucleic acid nanotechnology—towards Angstrom-scale engineering. Org Biomol Chem 2004;2:277–280.

88. Brucale M, Zuccheri G, Samori B. Mastering the complexity of DNA nanostructures. Trends Biotechnol 2006;24:235–243.

89. Paull R, Wolfe J, Hebert P, Sinkula M. Investing in nanotechnology. Nat Biotechnol 2003;21:1144–1147.

90. Jackel F, Watson MD, Mullen K, Rabe JP. Prototypical single-molecule chemical-field-effect transistor with nanometer-sized gates. Phys Rev Lett 2004;92:188,303.

91. Piva PG, DiLabio GA, Pitters JL, et al. Field regulation of single-molecule conductivity by a charged surface atom. Nature 2005;435:658–661.

92. D'Amico S, Maruccio G, Visconti P, D'Amone E, Bramanti A, Cingolani R, Rinaldi R. Ambipolar transistors based on azurin proteins. IEE Proc Nanobiotechnol 2004;151:173–175.

93. Brenning HT, Kubatkin SE, Erts D, Kafanov SG, Bauch T, Delsing P. A single electron transistor on an atomic force microscope probe. Nano Lett 2006;6:937–941.

94. Parker J. Computing with DNA. EMBO Rep 2003;4:7–10.

95. Soreni M, Yogev S, Kossoy E, Shoham Y, Keinan E. Parallel biomolecular computation on surfaces with advanced finite automata. J Am Chem Soc 2005;127:3935–3943.

96. Benenson Y, Adar R, Paz-Elizur T, Livneh Z, Shapiro E. DNA molecule provides a computing machine with both data and fuel. Proc Natl Acad Sci USA. 2003;100:2191–2196.

97. Cox JC, Ellington AD. DNA computation function. Curr Biol 2001;11:R336.

98. Cox JP. Long-term data storage in DNA. Trends Biotechnol 2001;19:247–250.

99. Unger R, Moult J. Towards computing with proteins. Proteins 2006;63:53–64.

100. Service RF. Materials and biology. Nanotechnology takes aim at cancer. Science 2005;310:1132–1134.

101. Cheng MM, Cuda G, Bunimovich YL, et al. Nanotechnologies for biomolecular detection and medical diagnostics. Curr Opin Chem Biol 2006;10:11–19.

102. Freitas RA, Jr. Nanomedicine, vol. I: Basic Capabilities. Georgetown, TX: Landes Bioscience, 1999.

103. Freitas RA, Jr. Nanomedicine, vol. Iia: Biocompatibility. Georgetown, TX: Landes Bioscience, 2003.

104. Roco MC. Nanotechnology: convergence with modern biology and medicine. Curr Opin Biotechnol 2003;14:337–346.

105. Kricka LJ, Park JY, Li SF, Fortina P. Miniaturized detection technology in molecular diagnostics. Expert Rev Mol Diagn 2005;5:549–559.

106. Fortina P, Kricka LJ, Surrey S, Grodzinski P. Nanobiotechnology: the promise and reality of new approaches to molecular recognition. Trends Biotechnol 2005;23:168–173.

II
BIOTEMPLATING

2

Experimental Strategies Toward the Use of the Porin MspA as a Nanotemplate and for Biosensors

Stefan H. Bossmann, Katharine Janik, Megh Raj Pokhrel, and Michael Niederweis

Summary

The porin MspA from *Mycobacterium smegmatis* has many unique properties, one being that it is the longest and the most stable porin identified to date. It is formed by supramolecular interaction of eight identical monomers of 184 amino acid residues (m = 20,000 Da). With dimensions of approx 10 nm in length and a diameter ranging from 1 nm (constriction zone) to 4.8 nm (opening of the MspA-goblet), it is ideal for bio-nanotechnological applications. The porin possesses a hydrophobic "docking zone," which enables it to reconstitute not only in lipid membranes, but also in numerous artificial (mono)membranes and hydrophobic, water-soluble polymer layers. Furthermore, we demonstrate here the design and proof-of-principle of an MspA porin-based biosensor for the TB-antibiotic isoniazid.

Key Words: Antibiotics; biosensor; HOPG; isoniazid; luminescence quenching; MspA; *Mycobacterium smegmatis*; photoinduced electron transfer; ruthenium-cathenane; sensitizer-relay assembly.

1. INTRODUCTION

1.1. The Mycobacterial Cell Envelope

The mycobacterial cell envelope forms an exceptionally strong barrier, rendering mycobacteria naturally impermeable to a wide variety of anti-microbial agents because of its unique structure *(1)*. In Fig. 1, the various layers of the mycobacterial cell envelope are shown schematically. The cytoplasmic membrane is the innermost layer of the envelope and has a thickness of approx 4 nm. Surrounding this membrane is the "cell-wall skeleton," a giant macromolecule consisting of peptidoglycan (a structure of oligosaccharides

From: *NanoBioTechonology: BioInspired Devices and Materials of the Future*
Edited by: Oded Shoseyov and Ilan Levy © Humana Press Inc., Totowa, NJ

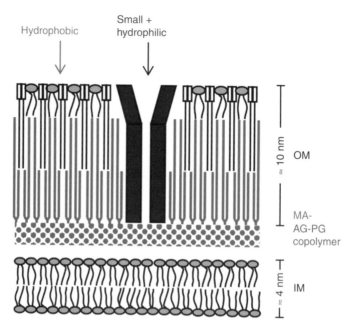

Fig. 1. Schematic representation of the mycobacterial cell envelope. (From ref. *6*, with permission.)

formed from disaccharide units of *N*-acetylglucosamine and *N*-glycolyl-muramic acid cross-linked by short peptides), arabinogalactan (a complex branched polysaccharide) and mycolic acids (long-chain, 2-alkyl-3-hydroxy fatty acids). Connected with the cell-wall skeleton, but not covalently attached to it, are a large variety of other lipids. Nikaido et al. have shown substantial evidence for the organization of the mycolic acids in a second lipid bilayer in addition to the cytoplasmic membrane *(2)*. Their X-ray diffraction measurements on purified mycobacterial envelopes, free of plasma membranes (<2% contamination or less), showed a strong reflection at 4.2 Å and a weaker, more diffuse one at 4.5 Å. These types of reflections are characteristic of ordered fatty acyl chains and were interpreted as indicating the presence of highly ordered and less ordered regions, respectively. By centrifuging a sample of cell walls onto a flat surface, measurements were obtained showing that the acyl chains were aligned perpendicular to the planes of the walls. The nature of the mycolic acids establishes the high-temperature phase change, which was discovered by studying purified walls and verifying that most of the associated lipids were previously removed with the detergent Triton X-114. Corynebacteria have a lower-temperature phase change and their mycolic acids are much shorter than those of mycobacteria. Chain length is another important factor, and the configuration of the double bond or cyclopropyl group proximal to the carboxyl group of the mycolate seems to be important as well, because a higher-temperature phase change correlates with a higher proportion of trans configuration. Some environmental mycobacteria can adjust the composition of their mycolates according to temperature, attempting to attain the required behavior of their outer permeability barrier *(3)*. Taking all of this into account, we conclude that *Mycobacterium smegmatis* becomes less permeable to lipophilic drugs when grown at higher temperatures.

Measurements by continuous-wave (CW)-electron paramagnetic resonance (EPR) of lipophilic probes—spin-labeled fatty acids—"dissolved" in purified walls or whole bacteria show that these enter only a less ordered and more fluid region *(4)*. This region may be that which is occupied by the alkyl chains of the associated lipids forming the exterior half of a bilayer, or that where these associated lipids intercalate into the part of the mycolate monolayer where the longer of the two alkyl chains of each mycolate is present. The insertion depth of the spin label determines the measured mobility, where the nitroxide type spin label is in the position of the carbon atom in the fatty acids. This effect has already been observed with conventional bilayers, but the change of mobility with depth was different in the case of mycobacterial walls, confirming the unusual nature of the mycobacterial outer permeability barrier. EPR spectra using whole cells were similar to those spectra using highly purified walls, telling us that the nitroxide-labeled fatty acids entered the outer part of the barrier only *(5)*.

The inner leaflet of the outer membrane (OM) is composed of mycolic acids (MA), which are covalently linked to the arabinogalactan (AG)-peptidoglycan (PG) copolymer. The outer leaflet is formed by a variety of extractable lipids such as trehalose-dimycolate ("cord factor"), lipo-oligosaccharides, sulfolipids, glycopeptidolipids, phenolic glycolipids, and glycerophospholipids. The diameters of the inner and outer membranes are rather poorly defined estimates from electron microscopic images of mycobacterial cell envelopes and are drawn to scale. Two general pathways through the mycobacterial OM exist: small and hydrophilic compounds diffuse through water-filled protein channels, the porins, whereas hydrophobic compounds use the lipid pathway by penetrating the OM directly (Fig. 1).

The mycolate monolayer can be formed even though the mycolate residues are covalently attached to the polysaccharide. This probably requires that the cross-linked glycan strands and the arabinogalactan strands run in a direction perpendicular to the cytoplasmic membrane *(5)*. The mycolates occur as esters of terminal arabinose units on the polysaccharide. The arabinosyl mycolate units are covalently linked to the galactan backbone, which is attached to the peptidoglycan. The whole polysaccharide is composed of sugars in their furanose form, giving additional flexibility to the chain and in all probability allowing the structure to accommodate itself to the close packing of the mycolate units through this characteristic-repeating motif *(5)*. These observations support the assumption that an asymmetric bilayer comprises the mycobacterial outer permeability barrier, with an inner leaflet of essentially "frozen" mycolate residues and an outer leaflet of more mobile lipids.

This outer membrane has unique properties: (1) it has a very low fluidity and will not melt at temperatures up to 70°C, in contrast to cytoplasmic membranes of other mesophilic organisms, which begin to disintegrate at 20°C; (2) it is thicker than all other known membranes, although it should be noted that the widely accepted thickness of about 10 nm is a rather poorly defined estimate from various electron microscopy pictures of mycobacteria and does not correspond to the length of the hydrophobic domain of MspA (3.7 nm; *7*); (3) it provides a very hydrophobic cell surface, which causes the bacteria to clump in a hydrophilic environment; and (4) its fluidity decreases toward the periplasmatic side of the membrane in contrast to that of the OM of Gram-negative bacteria *(8)*.

1.2. MspA from M. smegmatis is the Prototype of a New Family of Bacterial Porins

Hydrophilic molecules enter the mycobacteria by diffusing through channel-forming proteins, known as porins *(6)*. MspA is the major porin in the OM of *M. smegmatis* mediating the exchange of hydrophilic solutes between the environment and the periplasm *(9)*. Electron microscopy and crosslinking experiments indicated that MspA is a tetrameric protein with one central channel of 10 nm in length with a minimum inner diameter (constriction zone) of 1.0 nm *(7)*. The MspA crystal structure revealed a homo-octameric goblet-like conformation with a single channel and constitutes the first structure of a mycobacterial OM protein (Fig. 2B) *(7)*. MspA contains two consecutive 16-stranded β-barrels with nonpolar outer surfaces that confirm the very existence of an outer membrane in *M. smegmatis*. The length of the two membrane-spanning and pore-forming β-barrels is 3.7 nm, and the outer diameter of the 16-stranded β-barrel is 4.9 nm. The channel diameter varies between 4.8 nm and 1.0 nm at the pore eyelet, which is completely defined by two rings of aspartates. The β-sheet content is similar to that determined earlier by circular dichroism and infrared spectroscopy *(9)*. This makes MspA the membrane protein with the longest membrane-spanning domain known to date. These properties are drastically different from those of the trimeric porins of Gram-negative bacteria and classify MspA as the prototype of a new family of channel proteins.

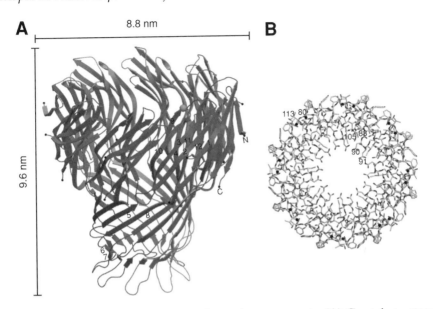

Fig. 2. Structure of MspA of *Mycobacterium smegmatis*. (**A**) Crystal structure (side view). (**B**) Crystal structure (top view). (From ref. 7, with permission.)

1.3. The Advantages of MspA in Nanotechnology Compared to Other Proteins

Proteins are macromolecules with dimensions in the nanometer range and can be tailored to specific needs by site-directed mutagenesis. Their use in nanotechnology has been severely hampered by the problem that most proteins lose their structural integrity in a nonnative environment, impeding their use in most technical processes. The MspA porin from *M. smegmatis* is an extremely stable protein, retaining its channel structure even after boiling in 2% sodium dodecyl sulfate (SDS) or extraction with organic solvents. This creates an extremely stable and adaptable environment and allows the use of MspA as a template for small molecules and nanoparticles in well defined arrangements on a nanometer scale. Ostwald processes, e.g., the coagulation of nanoparticles to bigger particles and, finally, precipitation, are prohibited or extremely decelerated when MspA is used as a template. The same principles are true for using the MspA channel for specific sensor functions. Longer template channels also allow the synthesis of longer nanowires, extending their application potential. The hydrophobic surface of MspA allows its assembly into biomimetic membranes. An additional useful feature is the tendency of MspA to self-assemble into ordered structures on

surfaces. This permits the synthesis of nanoarray lattices as well as nanoparticles and rods in a statistic distribution, embedded by a suitable hydrophobic
surface. Furthermore, MspA is the only mycobacterial porin to date that can
be purified in milligram quantities. It is selectively extracted by boiling cells
of *M. smegmatis* for 30 min in a buffer containing detergents and purified by
chromatography to apparent homogeneity *(10,11)*.

2. NANOSTRUCTURING BY DEPOSITION OF THE MSPA PORIN ON HIGHLY ORDERED PYROLYTIC GRAPHITE SURFACES

The generation of well defined nanostructures by protein or macromolecule
deposition on two-dimensional surfaces (areas of up to 1.0×10^{-4} m^2) represents a considerable advantage compared to the state-of-the-art technologies.
However, as already stated, most conventional proteins lose their structural
integrity in a nonnative environment, making their use in technical processes
highly improbable. Because of its extreme stability, MspA can be used in the
following simple and straightforward procedure for the nanopatterning of
extended surface areas: MspA, dissolved in a buffer solution, can be dispersed
into droplets employing a simple sonication procedure. The formed droplets
are deposited on a highly ordered pyrolytic graphite (HOPG) surface and,
depending on the exact deposition conditions (temperature, MspA concentration, sonication duration and intensity, and the length of the curing procedure after deposition), various nanostructures have been obtained. These
experimental results offer a simple and straightforward approach to the nanostructuring of surfaces by the deposition of protein containing buffer droplets.
At a deposition and curing temperature of 30°C, the generation of regular
nanostructures, which feature nanochannels, was unmistakably proven by
electron microscopy in combination with computer-assisted image analysis.
The formation of these highly desired nanochannels within the deposited layer
proceeds most likely by the reconstitution of biologically active MspA
nanopores (MW $\approx$ 160,000) within the codeposited MspA protein layer
formed by interaction of MspA monomers (MW $\approx$ 20,000). Three different
kinds of layer structures have been created by depositing protein/buffer
droplets on HOPG surfaces and independently analyzed by reflection electron
microscopy (REM) and transmission electron microscopy (TEM). The results
are summarized and compared in Fig. 3. It is clear that the temperature determined the formed nanostructures at the HOPG surface during the deposition
of droplets and during the curing process (the surface was allowed to [nano]
structure for at least 1 h at the chosen temperature) *(12)*.

Only isolated proteins, and no channel structures, are observed after
depositing MspA onto HOPG and curing at T = 20°C (Fig. 3, left image).

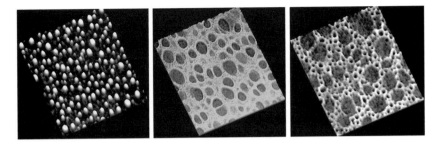

Fig. 3. Three types of layer structures created by depositing protein/buffer droplets onto highly ordered pyrolytic graphite surfaces (surface plots created by IMAGE). The extension of each image is 110 nm × 110 nm. (From ref. *12*, with permission.)

The main diameter of the units in these images is 2.15 ± 0.40 nm. Deposition of denatured MspA monomer results in similar structures (diameter 2.14 ± 0.70 nm), indicating that MspA deposits mainly in the form of its monomer (MW = 20,000). Be aware that denatured MspA monomer does not form any channels, which is in agreement with earlier findings wherein the MspA porin loses its ability to form nanochannels once it becomes MspA monomer *(8)*. Upon raising the deposition and curing temperature to T = 25°C, the MspA forms a supramolecular structure on the HOPG surface (Fig. 3, center image). The protein layer is 4.5 ± 0.25 nm thick. Indentations with a diameter of 9.2 ± 2.1 nm are found within the investigated structure. The residual thickness of the protein layer at the bottom of these dead-end channels is 1.4 ± 0.30 nm. It is evident that no open channels are present in this supramolecular structure.

The MspA pore dimensions were determined by electron microscopy, using negatively stained cell-wall preparations of *M. smegmatis* and of purified MspA. The MspA pore has an inner and outer diameter of 2.5 and 10 nm, respectively, and a length of about 10 nm *(9)*. In view of these dimensions, there are two possible interpretations of the supramolecular composition: (1) the MspA pore might be intact, but created at an angle or perpendicular to the HOPG surface, in which case it will elude detection, or (2) the observed indentations may have been made by larger aggregates of interacting MspA monomers. Please note that the TEM method will not completely show the whole configuration of these indentations. The filled bottoms remain elusive and the thickness of the protein structure can only be estimated. Real nanochannels possessing a diameter of 2.6 ± 0.50 nm have been identified upon increasing the deposition and the curing temperature to T = 30°C. The visible HOPG surface in the center region of the nanochannels confirms this finding (Fig. 3, right image). The diameter of these channels concurs exceptionally with the structure of MspA showing that the deposition procedure at T = 30°C did not denature the channel protein. It should be noted in this regard that the Tet repressor (TetR), which is a water-soluble DNA-binding protein and does not form channels, could not be deposited on a carbon surface using the sonication–deposition procedure described here. This indicated that this technique can only be used with stable proteins *(12)*.

3. MSPA-NANOCHANNELS GENERATED BY THE PORIN/POLYMER-TEMPLATE METHOD

3.1. MspA Reconstitution in a Poly-N-Isopropyl-Acrylamide Copolymer on HOPG (13)

Biologically active MspA octamers reconstitute naturally in a strongly hydrophobic environment. This is known from the structure of the mycobacterial cell envelope. Consequently, a poly-*N*-isopropyl-acrylamide (PNIPAM) copolymer was chosen to serve as the hydrophobic environment for MspA. The phenomenon of the "lower critical solution temperature" (LCST; whereas PNIPAM and many of its copolymers are soluble in cold water, the polymer becomes insoluble and precipitates out of solution when the temperature is increased) is of great value in this situation and thus PNIPAM and its copolymers with acrylic acid [P(NIPAM/AA)] were chosen as the medium to work with MspA *(14)*. The precipitation of PNIPAM occurs according to the following mechanism: upon approaching the LCST, individual coils of the macromolecule collapse and form so-called "globules" (coil-to-globule transition). During this process, the aqueous solvent is almost completely extruded as a result of the hydrophobic interaction of the NIPAM segments, which increases with rising temperature. Then the individual globules, having a diameter of several nanometers depending on the molecular weight of the PNIPAM (co)polymer, form clusters and precipitate out of solution. At this point, the solution becomes turbid ("cloud-point"). If comonomers possessing acid or base functions are present in the random PNIPAM copolymers, the occurrence of the LCST phenomenon becomes strongly influenced by the pH. In comparison to the pH, the concentration of salts or surfactants is of minor influence. Of much greater importance, with respect to the reconstitution of MspA, than the optical changes of a PNIPAM layer, is its remarkably increased hydrophobicity above the LCST. This formation of a strongly hydrophobic phase above the LCST permits the use of a P(NIPAM95.3/AA4.7) copolymer as a template for the reconstitution of MspA at the HOPG surface from an aqueous solution. In the first step, P(NIPAM95.3/AA4.7) was physisorbed from aqueous solution at HOPG. After treating the polymer-coated surface with ultrapure water,

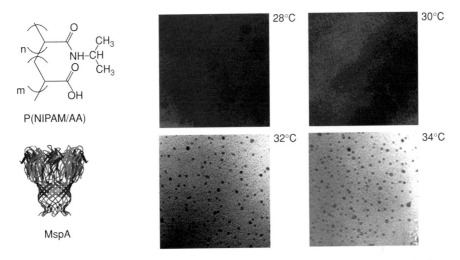

P(NIPAM/AA)

MspA

Fig. 4. MspA reconstitution within a P(NIPAM95.3/AA4.7) layer, physisorbed on highly ordered pyrolytic graphite at different temperatures. The dimensions of an image are 120 nm × 120 nm. (From ref. *13*, with permission.)

P(NIPAM95.3AA4.7)-covered HOPG was immersed in a PSO1 buffer solution containing biologically active MspA octamers *(13)*.

In Fig. 4, the building elements of the layer at HOPG, MspA and P(NIPAM95.3/AA4.7), and the effect of temperature at the HOPG surface are shown (TEM images). Note that water-soluble, but not hydrophobic, $Fe(CN)_6^{3-}$ has been added in order to enhance contrast.

Figure 5 shows the graphic result from an IMAGE-analysis of the MspA/P(NIPAM95.3/AA4.7) layer at HOPG. The nanochannels, randomly occurring within the P(NIPAM95.3/AA4.7) layer on HOPG, are clearly discernible. Note that in the absence of MspA, no channel structures were found, and MspA did not bind to the surface of HOPG in the absence of physisorbed P(NIPAM95.3/AA4.7). The results from the analysis of the nanopore diameters offer a surprising insight: there exist various types of pores within the P(NIPAM95.3/AA4.7) layer, as it appears by the apparent presence of four maxima in the histogram. In addition to the pores, which are typically found with diameters of approx 3 nm, there are at least three bigger pore structures present *(13)*.

3.2. MspA-Reconstitution Within the Cell-Wall Skeleton of M. tuberculosis (13)

The cell-wall skeleton of *M. tuberculosis* was chosen as a hydrophobic layer for the reconstitution of biologically active MspA octamers. This study is of special importance for future medical applications. The aim of this first

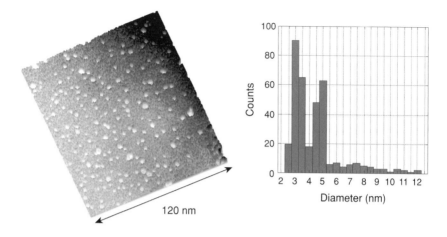

Fig. 5. Left: IMAGE analysis of MspA, embedded by P(NIPAM95.3/AA4.7) at the surface of highly ordered pyrolytic graphite (HOPG). The nanochannels reaching the surface of HOPG are clearly discernible. Right: size distribution of the MspA-nanochannels at HOPG. The main diameters found are (I) 3.25 ± 0.4 nm, (II) 4.9 ± 0.5 nm, (III) 6.3 ± 0.5 nm and (IV) 7.6 ± 0.7 nm. (From ref. *13*, with permission.)

and trailblazing experiment was to investigate whether MspA nanopores can also be formed in the OM of one of its relatives, the pathogenic *M. tuberculosis*. Alongside the general scientific interest in the reconstitution behavior of mycobacterial porins, a possible therapeutic approach against tuberculosis might arise from the incorporation of a porin, like MspA, into cell-wall fragments of *M. tuberculosis* (received as part of National Institutes of Health [NIH], National Institute of Allergy and Infectious Disease [NIAID] Contract No. HHSN266200400091C, entitled "Tuberculosis Vaccine Testing and Research Materials"). This experiment was based on the observation that incorporated MspA porin increases the sensitivity of *M. tuberculosis* to small and hydrophilic antibiotics *(15)*. The result is shown in Fig. 6. The experimental approach consisted of three steps: the cell wall of *M. tuberculosis* was physisorbed at HOPG by immersion in a solution containing PS01-buffer and 2.00 mg/mL T = 310 K for at least 12 h. After washing with H_2O and the removal of most of the remaining H_2O in vacuum, the first TEM image was recorded (Fig. 6A). Figure 6B–D were obtained by treating the HOPG/ mycobacterial membrane surface with PS01-buffer containing 1.25 µg/mL purified MspA porin at T = 310 K for 60 s, 300 s, and 3000 s *(13)*.

The cell wall of *M. tuberculosis* at HOPG appears to be smooth. Unfortunately, pore structures are not discernible using these experimental conditions. These findings are similar to our previous findings with cell-wall fragments of *Mycobacterium bovis* BCG *(15)*. Figure 6B,C show that dramatic

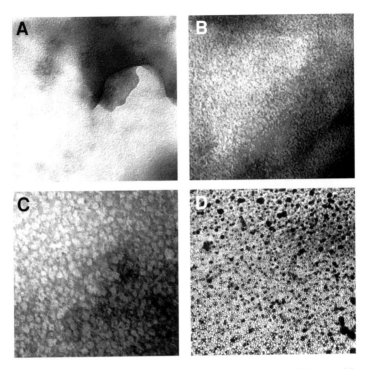

Fig. 6. (A) Cell-wall fragment of *Mycobacterium tuberculosis* (90 nm × 90 nm). **(B)** Cell-wall fragment of *M. tuberculosis* exposed to PS01-buffer containing 1.25 mg/mL purified MspA porin at T = 310 K for 60 s (90 nm × 90 nm). **(C)** Exposure time 300 s, conditions identical to **B** (90 nm × 90 nm). **(D)** Exposure time 3000 s, conditions identical to **B** and **C** (500 nm × 500 nm). (From ref. *13*, with permission.)

changes of the surface structure occur upon contact with MspA. After a contact time of 300 s, pore structures are clearly discernible (*see* Fig. 7). The presence of MspA appears to be the only change in these systems. Thus, the subsequent reconstitution of MspA within the cell wall of *M. tuberculosis* appears to be responsible for the observed changes. After a contact time of 3000 s, the membrane breaks down completely. This result can be regarded as experimental evidence for the use of the MspA porin as a transport vector through the outer mycobacterial membrane of *M. tuberculosis*.

4. PORIN-TRANSPORT ASSAY

The ability of pores to allow the diffusion of solutes can be analyzed in vitro using the liposome swelling assay *(16)*. However, the theoretical basis of this assay is only poorly understood and it is not amenable to a large number of solutes. Here, we present here the porin-transport assay (PTA) as an alternative

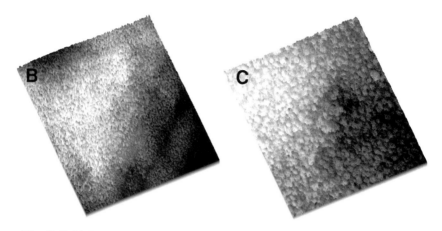

Fig. 7. IMAGE analysis of Fig. 6**B**,**C** (both: 90 nm × 90 nm). The appearance of porin channels is clearly discernible. (From ref. *13*, with permission.)

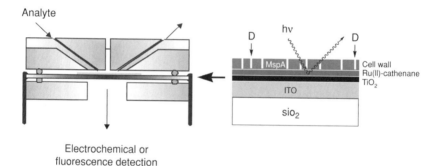

Scheme. 1. Right: working principle of a porin-transport assay. Left: construction of a sensor for solutes that pass through protein pores (D: solutes [e.g. antibiotics] diffusing through the mycobacterial channels, acting as sacrificial donors).

approach that would circumvent these problems. Mycobacterial cell-wall fragments containing the pores of interest are attached to a Ru(II)-cathenane layer on a TiO_2 layer. In the analyte, various solutes are offered to the PTA. Molecules that are able to diffuse through the pores can be detected by means of electrochemical detection (impedance analysis) or fluorescence detection using either fluorescent target molecules or fluorophores, which compete for binding sites at MspA (Scheme 1). Such an assay offers the potential to rapidly screen thousands of modifications of existing antibiotics or large libraries in a high-throughput format to identify compounds with improved transport efficiencies through porins of *M. tuberculosis*.

4.1. A Sensor Prototype for the Detection of Possible Antibiotics Against Mycobacteria

The fight against tuberculosis (and other pathogenic mycobacteria) critically depends on the development of new experimental strategies and the corresponding technology. It is only by identifying chemical (sub)structures, which enable molecules to rapidly overcome the OM permeability barrier of *M. tuberculosis*, for example by diffusing through mycobacterial porin channels, that novel, more efficient TB drugs can be developed. However, the research on these so-called enabling technologies should be performed with nonpathogenic organisms, because of serious time constraints and the delay that protective measures against pathogenic organisms naturally cause. Thus, our first prototype of a sensor for prospective antibiotics that can transgress the mycobacterial membranes through the embedded porin channels was realized using the membrane of *M. smegmatis* and additional MspA isolated from the same mycobacterium.

4.2. Construction of a Porin-Based Sensor

This light-absorption sensor is built in several stages:

1. A silicium dioxide chip ($2.0 \times 1.0 \times 0.1$ cm), covered with a thin layer (200 µm) of indium-tin-oxide (ITO), serves as the base plate of the light-absorption sensor.

2. A microlayer of titanium-dioxide is chemically deposited on top of the ITO by slow hydrolysis of titanium(IV)-tetra-n-butoxide [$Ti(OC_4H_9)_4$] (1.0 *M* in methanol/acetone [1:1, v/v] at T = 40°C for 24 h, followed by immersion in H_2O for 1 h and drying for 24 h at 120°C in an air-atmosphere). This procedure is included in the sensor preparation. The roughness of the interface designed for the physisorption of the sensitizer-relay assembly (vide infra) had to be increased in order to permit the detection of absorption spectra. An REM image of the ITO surface, covered with TiO_2, is shown in Fig. 8A.

3. A ruthenium(II)-cathenane (*Rucath*) (*17,18*) is used as a light-absorbing metal complex. It combines the very suitable photophysical properties of a ruthenium(II)-polypyridyl complex with a mechanically connected electron-bis-relay of the viologen-type. Photoinduced electron transfer between the electronically excited metal complex and the mechanically attached electron relay occurs at a reasonable rate and quantum efficiency ($k_{ET} = 2.6 \times 10^7$ s^{-1}, $\Phi \approx 0.55$). In the absence of a sacrificial donor, rapid back-electron transfer and only a transient chemical reaction are observed (depicted in Scheme 2A).

   Although the photoinduced electron transfer from the excited Ru(II)-sensitizer proceeds with very high quantum efficiency (vide supra), a minor fraction of this excited state deactivates by means of luminescence ($\Phi \approx 0.002$ at 664 nm in H_2O). Within the nanosecond time window, a bis-exponential luminescence decay pattern is observed ($\tau_1 = 0.24$ ns, [85%], $\tau_2 = 428$ ns [15%] in H_2O), which originates in the very complex motion characteristics of the mechanically linked cathenane structure.

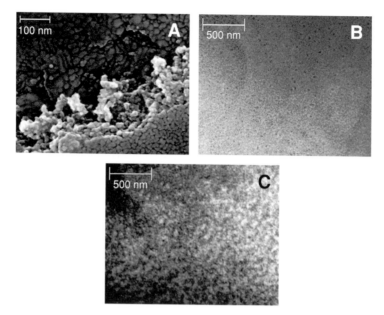

Fig. 8. Three stages in sensor preparation (reflection electron microscopy images): (**A**) TiO$_2$-deposition on indium-tin-oxide. (**B**) Cell wall of *M. smegmatis* deposited onto the ruthenium(II)-cathenanes on TiO$_2$. (**C**) Cell wall of *M. smegmatis* after reconstitution of additional MspA.

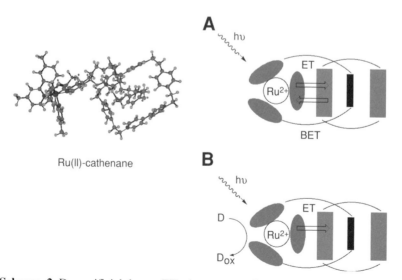

Ru(II)-cathenane

Scheme. 2. D, sacrificial donor; ET, electron transfer; BET, back-electron transfer.

Table 1
Eight Structurally Related "Sacrificial Donors"

	1, INZ	3	5	7
	—N(H)—NH$_2$	—N$_3$	—O—	—N(H)—
	2	4	6	8
	—N(H)—NH$_2$	—N$_3$	—O—	—N(H)—

INZ, isoniazid

As becomes apparent from Table 1 and Fig. 9A, three isonicotinic acid derivatives [1 (isoniazid),3,5] and three nicotinic derivatives (2,4,6), quench the luminescence from the ^{3}MLCT state of the *Rucath*, with very similar but only modest efficiency. When 1–6 were employed as quencher, linear Stern-Volmer kinetics were found *(19)*. In Fig. 9B, one example of this linear quenching behavior is shown. However, when *N*-isopropylisonicotinamide *(7)* and *N*-isopropylnicotinamide *(8)* were used, distinctly nonlinear, downward-sloping quenching curves were obtained. We estimated quenching constants on the order of 1×10^7 $M^{-1}s^{-1}$ from the first three recorded measurements.

It is worth noting that the quenching behavior of *Rucath* in solution differs remarkably from its photochemical reactivity in the presence of exactly the same molecules when embedded within a mycobacterial membrane (discussed later). We discovered at least one reason for the observed differences between the (photoinduced) reactivity of *Rucath* in solution and when bound to MspA: binding to MspA changes the photophysical properties of the *Rucath* considerably, because both the chemical environment and the main conformations of the cathenane change. As becomes apparent from Fig. 10, the observed luminescence lifetime increases dramatically. Upon binding within MspA, a bis-exponential luminescence decay was detected in the ns/µs-domain {τ_1 = 810 ns [82%], τ_2 = 27,550 ns [27.5 µs (!), 18%]}. Note that the shorter component increased upon binding by a factor of 3375,

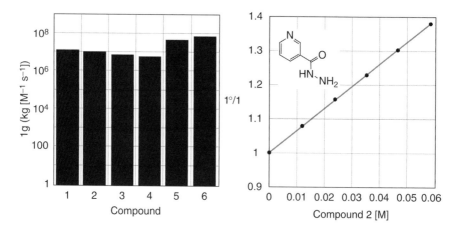

Fig. 9. (A) (left) Stationary quenching constants of *Rucath* (5.0×10^{-5} *M* in acetonitrile/H$_2$O [1/1, v/v] according to Stern-Volmer kinetics). The longest luminescence lifetime in this solvent composition has been measured at 635 ns (18%) using single-photon counting. This value was used for the calculation of the quenching constants summarized in Fig. 9A. **(B)** (right) Stern-Volmer plot of the luminescence quenching of *Rucath* by nicotinohydrazide *(2)*.

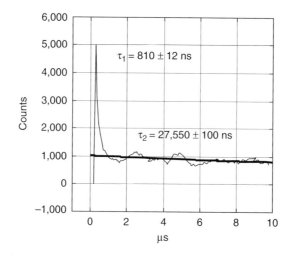

Fig. 10. Single-photon-counting measurement of the luminescence decay of *Rucath* (1×10^{-7} *M*) bound to MspA (1×10^{-6} *M*) in aqueous phosphate buffer (0.05 *M*) (λ_{ex} = 460 nm, λ_{em} = 680 nm).

whereas the longer component increased 43 times. Because such an extreme enhancement of the luminescence lifetime of any ruthenium(II)-polypyridyl complex has never, to the best of our knowledge, been reported, we excluded the possibility of luminescence arising from a ^{3}MLCT or similar state. Furthermore, we attributed the observed light emission to chemoluminescence

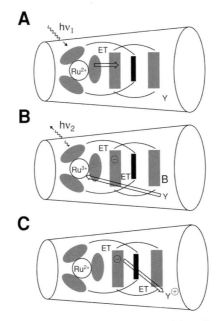

Scheme. 3. Paradigm for chemoluminescence arising from *Rucath* bound to MspA. **(A)** Photoelectrontransfer between the Ru(II)-complex and the mechanically linked bis-viologen acceptor. **(B)** Thermal electron transfer from a tyrosine residue within MspA and chemoluminescence. **(C)** Thermal electron transfer between the previously reduced bis-viologen acceptor and the oxidized tyrosine-unit.

from the *Rucath*. Ruthenium(II)-polypyridyl complexes are well known to exhibit the phenomenon of chemoluminescence. MspA contains four tyrosine residues (Y48, Y66, Y82, Y177). Three of these tyrosine residues (Y48, Y66, Y82) are exposed to the exterior of MspA. However, one tyrosine residue (Y177) is accessible from the interior (water-filled porin channel). Because tyrosine is the amino-acid residue possessing the lowest redox potential, it is likely that it can participate in photoinduced electron-transfer events. Scheme 3 summarizes the proposed electron-transfer pathways.

Unfortunately, in the presence of MspA, we have been unable to perform the above-described quenching experiments. We could not achieve stable solutions, either in acetonitrile/H_2O or in a phosphate buffer (0.05 M). In all six cases investigated, a precipitate of *Rucath*@MspA was formed at quencher concentrations of approx 1×10^{-3} M, which prevented the recording of meaningful luminescence intensity data.

4. In Table 1, isoniazid (INZ), which is a potent drug against tuberculosis, and seven other structurally related model compounds, are presented. All eight molecules were employed as "sacrificial donors" in our study. In principle, each molecule possessing an oxidation potential <1.20 V (vs SHE) can be used. When present, the donor undergoes sacrificial oxidation by means of a thermal electron transfer to ruthenium(III), which has been previously generated by photoelectron transfer from ruthenium(II) to one of the mechanically linked

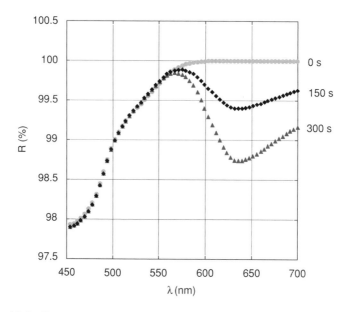

Fig. 11. Reflectance spectra of the *Rucath*, physisorbed on TiO_2/indium-tin-oxide, immersed in an aqueous phosphate buffer (0.05 *M*, pH 7.0) containing 1×10^{-3} *M* of *N*-isopropylisonicotinamide (compound 7) as sacrificial donor, under continuous irradiation employing a TQ 150 light source (further explanations are provided in the text).

viologen acceptors. Following this electron-transfer step, the sacrificial donor decomposes irreversibly. Note that the *Rucath* and the sacrificial donor have to be in close proximity (van der Waals contact) so that the sacrificial electron-transfer reaction can compete with the back-electron transfer (*see* Scheme 2B). This sacrificial photoinduced reaction works within several minutes (t < 300 s) for all substrates 1–8, if the donor concentration is sufficient (c = 1×10^{-3} *M*). The result obtained when using *N*-isopropylisonicotinamide (7) as sacrificial donor is shown in Fig. 11.

5. The next step in the design of the sensor for prospective antibiotics consists of the deposition of patches (up to approx 1200 nm × 1200 nm) of cell wall from *M. smegmatis* on top of the layer of *Rucath* adsorbed to TiO_2 (17,18). The deposition is achieved by immersion of the layered quartz/ITO/TiO_2/cathenane-assembly in a phosphate buffer solution (0.05 *M*, pH 7.0) containing 20 μg of cell wall/mL for 24 h. An REM characterization of the obtained cell-wall-covered sensor surface is shown in Fig. 8B. Relatively few porin channels, which correspond to the natural abundance of MspA (and other mycobacterial porins) in the cell wall of *M. smegmatis*, appear as black dots (approx 15 channels per 100 nm × 100 nm). The low density of the hydrophilic porin channels, which provide the only known pathways through the thick and hydrophobic mycobacterial membrane, is most likely responsible for the failure of this sensing array. Irradiation of this layered

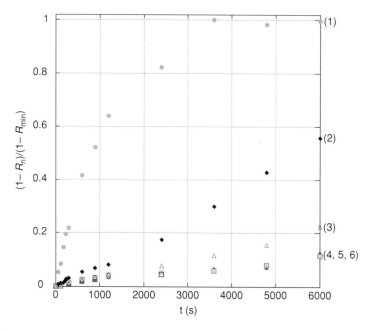

Fig. 12. Normalized visible absorption at 635 nm of the layered sensor assembly, caused by eight structurally related "sacrificial donors" (1×10^{-3} *M*): (1)–(6), dissolved in an aqueous phosphate buffer (0.05 *M*, pH 7.0), under continuous irradiation employing a TQ 150 light source (further explanations are provided in the text). (7) (*see* Fig. 11) and (8) led to very similar results as (4) (6), considering the experimental error from three repetitions (± 0.05 units).

    sensor assembly for 6000 s using a medium-pressure mercury lamp (TQ 150, Heraeus) with a pyrex socket (d = 30 cm) did lead to a very small increase in absorption in the red region of the visible spectrum ($\lambda_{max} \approx 635$ nm), causing a slight decrease of the measured reflectance to R ≈ 0.985–0.99 for all sacrificial donors 1–8. However, these spectral changes most likely originate from the ridges between the cell-wall patches absorbed on the *Rucath* layer.

6. Immersion of the layered quartz/ITO/TiO$_2$/cathenane/cell wall assembly in an aqueous phosphate buffer (0.05 *M*, pH 7.0) containing 0.55 µg purified MspA/mL for 24 h leads to the reconstitution of additional MspA within the previously adsorbed cell wall of *M. smegmatis*. We estimate from REM images (*see* Fig. 8C for a typical result) that the number of porin channels increases by roughly a factor of 40 or 50 during this procedure. As it becomes apparent from Figs. 11 and 12, these modified bacterial cell walls are able to function in layered sensor assemblies for the detection of possible candidates for therapy of various mycobacteria. Isoniazid, which is a known antibiotic against tuberculosis *(20)*, clearly shows the best result, and reaches a reflectance of 0.928 after 3600 s of continuous irradiation (TQ 150). In Fig. 12, the normalized results obtained with compounds 1–8 (c = 1.0×10^{-3} *M* in an aqueous phosphate buffer [0.05 *M*, pH 7.0]), which have been administered at the exterior of the modified

cell wall of *M. smegmatis*, are summarized. Isoniazid is clearly able to cause the steepest increase in adsorption. Its constitutional isomer (compound 2) shows the second fastest incline. Note that this difference is not observed in the absence of the modified mycobacterial membrane. This is a clear indication of enhanced transgression of isoniazid (compound 1) through the employed membrane. The constriction zone of MspA features two hydrophilic rings, formed by aspartates, and two hydrophobic rings, formed by leucines and isoleucines, which are believed to play an important role in this selectivity *(7)*. The biomolecular reasons for this behavior are the target of further investigation.

5. CONCLUSION

We demonstrate that the porin MspA from *M. smegmatis* can be successfully employed in the fields of bio-nanotechnology and medicinal chemistry. As a result of its extraordinary stability against thermal and chemical decomposition, the presence of a hydrophobic docking region on its exterior, and the availability of its water-filled, hydrophilic interior, MspA is a very versatile channel protein. MspA can be easily deposited on surfaces using microscopic buffer-droplets, which are formed by standardized sonication procedures. It is able to reconstitute in water-soluble hydrophobic polymer layers [P(NIPAM/AA) copolymers] and forms microscopic "letters" and "molds." The geometric dimensions of MspA and the presence of a constriction zone, which is formed by a double-ring of 2×8 aspartates, permit its application in a porin-based biosensor for the anti-TB-antibiotic isoniazid. For this purpose, a layered sensor design is chosen. A *Rucath* is bound within the porin channel and a change in visible absorption due to *Rucath* photoreduction in the presence of isoniazid and seven related compounds is observed. Isoniazid causes the strongest response in this sensor.

REFERENCES

1. Brennan PJ, Nikaido H. The envelope of mycobacteria. Annu Rev Biochem 1995;64:29–63.
2. Nikaido H, Kim SH, Rosenberg EY. Physical organization of lipids in the cell wall of *Mycobacterium chelonae*. Mol Microbiol 1993;8:1025–1030.
3. Liu J, Barry CE III, Besra GS, Nikaido H. Mycolic acid structure determines the fluidity of the mycobacterial cell wall. J Biol Chem 1996;271:29,545–29,551.
4. Trias J, Jarlier V, Benz R. Porins in the cell wall of mycobacteria. Science 1992;258:1479–1481.
5. Dmitriev BA, Ehlers S, Rietschel ET, Brennan PJ. Molecular mechanics of the mycobacterial cell wall: from horizontal layers to vertical scaffolds. Int J Med Microbiol 2000;290:251–258.
6. Niederweis M. Mycobacterial porins—new channel proteins in unique outer membranes. Mol Microbiol 2003;49:1167–1177.

7. Faller M, Niederweis M, Schulz GE. The structure of a mycobacterial outer-membrane channel. Science 2004;303:1189–1192.

8. Stahl C, Kubetzko S, Kaps I, Seeber S, Engelhardt H, Niederweis M. MspA provides the main hydrophilic pathway through the cell wall of *Mycobacterium smegmatis*. Mol Microbiol 2001;40:451–464.

9. Engelhardt H, Heinz C, Niederweis M. A tetrameric porin limits the cell wall permeability of *Mycobacterium smegmatis*. J Biol Chem 2002;277:37,567–37,572.

10. Heinz C, Niederweis M. Selective extraction and purification of a mycobacterial outer membrane protein. Anal Biochem 2000;285:113–120.

11. Heinz C, Roth E, Niederweis M. Purification of porins from *Mycobacterium smegmatis*. Meth Mol Biol 2003;228:139–150.

12. Niederweis M, Heinz C, Janik K, Bossmann SH. Nanostructuring by deposition of protein channels formed on carbon surfaces. Nano Lett 2002;2;1263–1268.

13. Bossmann SH, Janik K, Pokhrel MR, Heinz C, Niederweis M. Reconstitution of a porin from *Mycobacterium smegmatis* at HOPG covered with hydrophobic host layers. Surface and Interface Analysis 2004;36:127–134.

14. Ottaviani MF, Winnik FM, Bossmann SH, Turro NJ. Phase separation of poly(Nisopropylacrylamide) in mixtures of water and methanol: a spectroscopic study of the phase-transition process with a polymer tagged with a fluorescent dye and a spin label. Helv Chim Acta 2001;84:2476–2492.

15. Mailaender C, Reiling N, Engelhardt H, Bossmann S, Ehlers S, Niederweis M. The MspA porin promotes growth and increases antibiotic susceptibility of both *Mycobacterium bovis* BCG and *Mycobacterium tuberculosis*. Microbiology 2004;150:853–864.

16. Nikaido H, Nikaido K, Harayama S. Identification and characterization of porins in Pseudomonas aeruginosa. J Biol Chem 1991;266:770–779.

17. Hu YZ, van Loyen D, Schwarz O, et al. Intramolecular electron transfer between noncovalently linked donor and acceptor in a [2]catenane. J Am Chem Soc 1998;120:5822–5823.

18. Hu YZ, Bossmann SH, Van Loyen D, Schwarz O, Duerr H. A novel 2,2′ bipyridine[2]catenane and its ruthenium complex: synthesis, structure, and intramolecular electron transfer—a model for the photosynthetic reaction center. Chemistry—A European Journal 1999;5:1267–1277.

19. Duerr H, Bossmann S. Ruthenium polypyridine complexes. On the route to biomimetic assemblies as models for the photosynthetic reaction center. Acc Chem Res 2001;34:905–917.

20. O'Neil MJ. The Merck Index: An Encyclopedia of Chemicals, Drugs and Biologicals, 12th ed. Whithouse Station, New Jersey: Merck Research Laboratories, 2000.

3

Bionanotechnology and Bionanoscience of Artificial Bioassemblies

Steven S. Smith and Katarzyna Lamparska-Kupsik

Summary

Bionanotechnology is now creating an entire class of new devices that will improve and augment existing approaches to biology and medicine. Information from physics, chemistry, and molecular biology is being used to reassemble biological and nonbiological molecules into useful and informative devices. These artificial bioassemblies are generally built on nanoscale scaffolds. This flexible construction principle allows them to serve as models of cellular assemblies, models of possible intermediates in molecular evolution, reporters for studying intracellular dynamics, and tools for detecting and tagging different cell types, including cancer cells. This chapter assesses recent progress in this area.

Key Words: Bioassemblies; bionanotechnology; DNA methyltransferases; DNA scaffolds; DNA tethers; nanoscale devices; nanoscale scaffolds; quantum dots.

1. INTRODUCTION

There are four general types of molecular scaffold now in use in bionanotechnology. In each case, the scaffold serves as a template or connector that links functional groups in a predetermined arrangement. Crystalline organic scaffolds, organic chains and polymers, nucleic-acid scaffolds, and protein scaffolds comprise the set of scaffolds that have been most carefully studied.

2. QUANTUM DOTS AS SCAFFOLDS

Quantum confinement produces a number of interesting properties in small three-dimensional (3D) clusters of atoms called quantum dots. Molecular orbital theory permits the clusters to be viewed as single molecules. For noble metal clusters, the Hartree-Fock level of theory permits the calculation of HOMO-LUMO energy differences (band gaps) when the cluster contains

From: *NanoBioTechonology: BioInspired Devices and Materials of the Future*
Edited by: Oded Shoseyov and Ilan Levy © Humana Press Inc., Totowa, NJ

fewer than 100 atoms *(1)*. The band gap is generally inversely related to the size of the cluster. Larger clusters ranging in size up to 10,000 atoms can be produced by semiconductor methods. These are also amenable to electronic structure calculations for band-gap prediction, but here the calculations rely on the Effective Mass Approximation *(2)* or other semi-empirical methods *(3,4)*.

In each case, experimental observation and molecular theory confirm the broad absorption energies and sharp emission energies (often in the visible spectrum) that are associated with quantum dots. Moreover, because these are fundamentally molecular absorption and emission properties, quantum dots are very stable to continuous excitation at high intensity. Although they can undergo cycles of photoinduced ionization and neutralization, they are much more stable than organic fluorophores under continuous excitation. Even so, under prolonged excitation, quantum dots can shrink as a result of the photo-oxidation process, causing their emission spectrum to shift to the blue *(5)*.

To date, most applications have involved the semiconductor-based quantum dots. These systems are intrinsically insoluble in water. Coating the dot with a layer of ZnS *(6)* is generally used to solubilize them. When the crystals are prepared, trioctylphosphine (TOPO) *(7,8)* is used to terminate growth. A ZnS coating is produced when the phosphine is exchanged in a reaction with diethylzinc and hexamethylsilathiane. Mercaptoacetic acid is then reacted with the surface ZnS. The resulting carboxylic acid groups solubilize the dot and can be used to couple other moieties to the surface.

A variety of molecular probes are now being studied with quantum dots. This has been made possible by the rich bioconjugate chemistry available for surface linking to quantum dot scaffolds. For example, avidin-tagged chimeric proteins have been detected with biotinylated quantum dots *(9)*. Growth factors *(10)*, antibodies *(11–13)*, and peptides *(14,15)* linked to quantum dots have all been employed in detecting surface markers. In addition, fluorescent *in situ* hybridization (FISH) with quantum dots linked to nucleic-acid probes has been used to detect specific chromosomes *(16)* and specific genes on chromosomes *(17)*. In this latter application, the resistance of quantum dots to photobleaching suggests that they may permit quantitative and perhaps even archival FISH methods *(17)*.

3. ORGANIC CHAINS AND POLYMERS AS SCAFFOLDS

By initiating synthesis with a core compound that carries multiple linking functionalities, branched polymerization can be continued for multiple rounds to obtain extended scaffolds called dendrimers. In general, the first round of branches brings the system to what is called G0, the second round to G1, the third round to G2, etc. Free rotation around multiple bonds in the branched system allows the polymer to display a spherically distributed set of surface functionalities that permit the bioconjugation of almost any desired biomolecule. Interesting biomolecules are then tethered at multiple sites along this roughly spherical surface. Scaffolds of this type are now commercially available with generation levels of 10 or less and linked at any of several core molecules to one of several branch systems so as to display one of several surface chemistries. One example would be a diethylamine core linked to poly(amido)amine branches displaying an amine for surface coupling.

Asymmetric or half dendrimers called dendrons carrying fluorophores bound to termini on each branch terminus have been attached to nucleic-acid probes to yield enhanced detection of herpes virus *(18)*. Radiolabeled dendrimers have also been linked to antibodies for the enhanced detection of antigens *(19)*. Generation 3 *(20)* and generation 5 *(21)* polyamidoamide dendrimers linked to folate ligands have been used to target folate receptors on tumor cells. As noted above, the larger dendrimers (i.e., those with diameters as large as 20 nm) are roughly spherical in solution *(22)*. Even so, their flexibility allows them to flatten out when they contact a substrate with multiple targets on its surface *(23)*.

Short organic linkers have also been used as scaffolds. For example, early attempts to increase antibody avidities used short synthetic chains to tether antibodies together to produce enhanced binding *(24–26)*. The observed enhanced forward binding rate appears to result from the increased probability of initial binding coupled with the tendency for cooperative binding to adjacent sites, whereas the decreased off rate appears to be a result of the requirement for coordinate release at multiple binding sites.

4. DNA STRUCTURES AS SCAFFOLDS

4.1. Tethers of Single-Stranded DNA

Molecular beacons and TaqMan® quantitative PCR probes are essentially DNA-tethered fluorophores. As such, they are currently the most widely used nanoscale bioassemblies. The initial implementation of the TaqMan concept *(27)* utilized ^{32}P as the end-label and showed that its release by the 5′ exonuclease activity of the Taq polymerase was proportional to the generation of the PCR product as cycle number increased. Adding a phosphate to its 3′ end prevents extension of the probe. Because the sequence is homologous to the target amplicon, additional specificity is achieved in the PCR. Soon after the invention of the TaqMan concept, tethered fluorophores were introduced *(28)*. Quenching by Förster resonance energy transfer *(29)* is achieved by choosing a fluorophore with an emission spectrum that overlaps the absorption spectrum of a second fluorophore. This is a nonradiative energy transfer mediated by induced-dipole interaction that requires close proximity of the two fluorophores. The rate of energy transfer is 50% at the Förster radius (generally between 2 and 6 nm) and falls off with the sixth power of the distance between the fluorophores. At 24 nucleotides, the dyes used in TaqMan quantitative PCR are about 8 nm apart where they are efficiently quenched. Once 5′ exonucleolytic activity of the Taq polymerase degrades the DNA tether, the 5′ fluorophore is released to solution where it is no longer quenched by the tethered quencher. Fluorescence data can be acquired in real time using a number of commercially available instruments. This permits initial target concentrations to be estimated from a threshold cycle number in the PCR. The successful applications of this test are too numerous to mention *(30)*, and it may be anticipated that an even larger list of applications will appear in the near future.

Molecular beacons operate on roughly the same principle except that the presence of a short tandem repeat at each end of the probe oligodeoxynucleotide causes it to form a hairpin loop that juxtaposes the fluorophore and the quencher. The region of the loop is complementary to one strand of the amplicon so that it hybridizes with the amplified DNA during the PCR, causing the short terminal repeats to denature, which moves the fluorophores apart so as to diminish quenching *(31,32)*. Because the short terminal repeats are not designed to hybridize with the probe, they form 5' and 3' flaps at the ends of the probe. Oddly enough, the 5' flap endonuclease activity of Taq polymerase *(33)* apparently does not degrade the molecular beacon during amplification when Taq polymerase *(31)* or its variant Taq Gold® *(34)* are used. The probe strand appears to be displaced by the polymerase as it moves through the region occupied by the probe sequence, whereupon it snaps back into the hairpin conformation, quenching the fluorophores. As with the TaqMan approach, molecular beacon PCR methodology has yielded a large number of applications *(35)*.

A similar concept has been used in another kind of nanotechnological device. Here, the fluorophore tetrafluoro-flourescein (TET) and the quencher carboxytetramethylrhodamine (TAMRA) are initially positioned in a 40-bp duplex containing a central two-nucleotide gap. In this duplex, quenching by resonance energy transfer is minimal because the fluorophores are about 13 nm apart. The duplex is designed so that each end carries a single-strand overhang of 24 nucleotides that is available for hybridization. The gapped duplex is forced to fold when a 56-nucleotide strand that is complementary to the two 24-nucleotide overhangs is added to the system. This forces the two fluorophores into apposition, extinguishing their fluorescence. In this state, the 56-mer presents an eight-nucleotide single-strand overhang at one end. To reestablish fluorescence, a 56-nucleotide strand is added that is fully complementary to the 56-nucleotide strand now holding the gapped duplex in the hairpin conformation. Strand displacement releases a 56-bp duplex from the system and the gapped 40-bp duplex adopts the extended unquenched conformation present at the beginning of the process. In this manner, the system can be cycled as many as seven times between an open and closed conformation *(36)*. Cycling beyond this point was hindered by bleaching of the fluorophore and the accumulation of unusual DNA structures *(36)*. Other DNA structural interconversions have also been used to construct similar nanoscale devices *(37,38)*.

Systems for tethering metal ions to DNA have also been developed for a variety of applications. For example, a complex was formed by dimerization of an IDA-modified oligonucleotide with a DNA duplex in the presence of metal ions *(39)*. The IDA was postsynthetically conjugated to the 5'-end of the DNA through an amino-linker by the reaction with N,N-bis(ethoxycarbonyl-methyl)glycine p-nitrophenyl ester *(40)*. In these experiments, a 14-mer poly T modified by linking IDA to its 5'-end was designed so that it was complementary to both ends of a palindromic 44-mer DNA duplex. When the DNA duplex was annealed to the 14-mer poly T-IDA conjugate, the melting temperature (T_m) of the complex in high salt was observed to increase from 21°C to 31°C in the presence of Lu^{3+} ions *(39)*. The authors suggested that the complex comprises a linear triple helix with both parallel and antiparallel pyrimidine strands, permitting chelation of the Lu^{3+} ions in the center of a linear structure. Other possibilities for the structure of the complex can be imagined; however, the linkage of the two 14-mers by Lu^{3+} appears to have been established.

Oligonucleotides modified by IDA under the influence of metal ions (Gd^{3+}) have also been used to enhance duplex stability *(41)*. The two 9-mer oligonucleotides with IDA were hybridized to the same 18-mer target to form the helix with the IDAs in the center of the duplex. The T_m of the duplex increased by 15°C in the presence of Gd^{3+}. This provided a significant enhancement of duplex stability under the influence of appropriate metal ions. In this case, the IDA was joined to the 3'- and 5'-ends of the oligomer through the amino-linker. The synthesis was based on a protocol described by Endo and Komiyama *(42)*, in which IDA was appended to the DNA as a sequence dependent T-IDA-T phosphoramidite.

The lanthanide complex with DNA has also been used to selectively hydrolyze RNAs *(40)*. In this case, a 15-mer of DNA with IDA on the 5'-end was hybridized to the RNA. In the presence of Lu^{3+}, the IDA-oligodeoxynucleotide selectively hydrolyzed the RNA at the 3'-end, closest to the Lu^{3+} ion. The IDA was postsynthetically attached to the 5'-end of the DNA through an amino-linker *(40)*.

Dervan and co-workers used Fe^{2+} to specifically cleave a double-stranded DNA to analyze the groove location and orientation of DNA binding ligands *(43–47)*. The Fe^{2+} ions were incorporated into the DNA helix by formation a triple helix through the short complementary oligonucleotide carrying an EDTA moiety. The EDTA was postsynthetically attached to the oligonucleotides through the modified thymidine containing an amino linker at the C5 position *(44)* or as a phosphoramidite of modified thymidine with the triethyl ester of EDTA *(45)*.

Another chelator, nitrilotriacetic acid (NTA), was used as a guide for the creation of photocrosslinked protein–DNA conjugates *(48)*. A trifunctional molecule called NBzM containing an NTA His-tag targeting group, a Benzophenone photoconjugating group, was linked through a Maleimide functional group to 5'-thio-modified oligonucleotides to create the protein targeting conjugate. The NBz-oligonucleotide conjugate was noncovalently coordinated to the His-tag on the protein through its Ni:NTA group. Ultraviolet (UV) light was then used to create a covalent bond between the photoreactive group and the protein surface to create protein–DNA conjugates.

4.2. Immobile DNA Junctions as Scaffolds

Under the appropriate conditions, DNA will spontaneously adopt numerous 2D and 3D structures that can serve as scaffolds *(49,50)*. The methods used to prepare immobile junctions with up to six arms emanating from a single point *(51,52)* have been extended to the construction of complex structures with the connectivity of several of the platonic solids *(53–55)* and to Borromean Rings *(56)*. Among the most useful of these are the analogs of the double-crossover recombination intermediates. This assembly is fixed in a planar conformation that has permitted its use as a nanoscale tile for the production of extended periodic patterns *(50)*. Moreover, in another implementation of the resonance energy transfer concept, DNA double-crossover molecules have been adapted to sense changes in salt concentration. Here, the B-Z transition in a short region of duplex DNA *(57)* was used to change the distance between fluorophores that are otherwise rigidly constrained by the double-crossover architecture on either side of the short duplex. As the DNA region twists into the left-handed conformation at high salt concentration, the dye molecules at the ends of each double crossover move beyond the Förster radius and fluoresce.

4.3. Order in DNA Scaffolds

Prebiotic metabolosomes are postulated to have been among the first use of ordered components in evolution *(58)*. This evolutionary principle has been used to order as many as four proteins coupled to single-stranded DNAs. In this concept, the coupled single strands are designed to be complementary to distinct regions of a longer guide RNA or DNA strand *(59,60)*. This targets the coupled proteins to preselected sites along the guide RNA or DNA. In general, the streptavidin–biotin coupling system has been used to link proteins to the single-stranded DNAs. However, it has been suggested that His$_6$-tagged proteins might also be used in this application by employing NTA-derivatized oligodeoxynucleotides *(61)*.

Proteins can also be ordered on DNA as fusions with DNA (cytosine-5) methyltransferases, because most cytosine methyltransferases can form a covalent link to duplex DNA structures containing 5-fluorocytosine. These bacterial enzymes occur with many DNA recognition sequences so that fusion proteins of many types can be ordered. This technology has been used to construct bioassemblies that are based on linear and branched DNA scaffolds *(62–64)*.

Directed positioning has also been demonstrated by using distinct methyltransferases targeted to unique recognition sites in the DNA scaffold *(62–64)*. Each of the DNA scaffolds can be modified for stability and easy detection. The methyltransferase technology has the advantage that any fusion protein that can be expressed in bacteria can be used. Of particular interest are peptides, which bind to receptors of different types *(65)*.

5. PROTEINS AS SCAFFOLDS

Like nucleic-acid complementarity, protein–protein interaction can provide a mechanism for the self-assembly of nanoscale scaffolds. Scaffolds of this type generally form either interlocked extended systems or closed shells. Systems of this type have been extensively studied in molecular biology and the basic rules of assembly are known.

5.1. Open Protein Scaffolds

Protein–protein interaction can be used to form arrays and filaments *(66,67)*. However, systems with high interaction stabilities are best in this application *(67)*. One such system is the streptavidin tetramer. The core protein–protein interaction system in the tetramer displays the biotin binding sites in a roughly tetrahedral array permitting the system to link biotin-conjugated proteins together *(68,69)*. Biotin-conjugated antibodies linked in this fashion exhibit a multivalency effect, generating an almost 35-fold enhancement in antigen-binding avidity *(70)*. The RNAse barnase and its proteinaceous inhibitor barstar also interact tightly enough to serve as a linking system. In this case, bioassemblies have been constructed that display antibodies on flexible dimers or trimers, yielding increased antigen-binding avidity even though the antibodies are displayed on a flexible tether *(66)*.

5.2. Closed Protein Scaffolds

The studies of Caspar and Klug *(71)* yielded much of what we now know about viral capsid assembly. Recent molecular simulations *(72,73)* of viral assembly confirm one of Caspar and Klug's basic proposals, namely, that the two different geometric subunit forms (pentagons and hexagons) needed for the assembly of icosahedral capsids are composed of the same set of protein subunits assembled to form components that are approximately triangular. That is to say that the capsids are deltahedra. When hexagons are formed, they are roughly planar. When pentagons are formed, their vertex is out of the plane because one of the triangular elements is missing. The simulations suggest that the pentagons and hexagons must be in equilibrium for self-assembly to occur *(72,73)*.

Icosahedral symmetry requires two-, three-, and fivefold rotational axes. Each of these is present in icosahedral virus capsids, but platonic geometry is not generally found. The pentagonal subunits are present at each of the 12 vertices that are expected to exhibit fivefold rotational symmetry, but the region between these vertices is tiled with hexagonal subunits displaying the same protein–protein interaction surfaces. These natural properties—a limited number of capsid proteins forming delta that are displayed in predictable ways on the surface of an icosahedral deltahedron—make viral capsids excellent nanoscale protein scaffolds.

To date, Cowpea Mosaic Virus capsids are the most carefully studied system of this type. The capsids of this virus are formed by two proteins, a small subunit containing one of the interacting domains in the triangular facet, and a large subunit that contains two of the interacting domains. As expected, the capsid is formed from 120 protein subunits assembled into 60 triangular facets, with 12 pentagonal elements. A reactive lysine residue is exposed on each of the 60 facets, allowing for surface modification of the assembled capsid. This reactive lysine is selectively modified by fluorescein isothiocyanate or fluorescein *N*-hydroxysuccinimide ester when the input ratio of dye molecules to viral particles is roughly stoichiometric (i.e., 60 to 70 dye molecules per capsid). Above this level, the dyes react with other amino acids on the surface and on the interior of the capsid *(74)*. In vitro mutagenesis has also been used to display a reactive cysteine-containing loop on the surface of each of the 60 triangular facets of the capsid *(75)*. Fluorescein, rhodamine, biotin, and 900-nm diameter gold particles can be linked to these cysteine-modified capsids. Moreover, precise patterns of these capsids have been laid down on gold surfaces using scanning probe nanolithography *(76)*.

These examples exploit the symmetry of the cage-like properties of a purified virion. However, it may also be possible to design cages *de novo*. For example, Padilla et al. *(67)* fused two protein–protein interaction domains, one that normally dimerizes (M1 matrix protein of influenza virus) with one that normally trimerizes (bromoperoxidase) by using a short alpha helical linker.

This fusion produced a curved delta that can spontaneously assemble to form a 12-subunit spherical shell with tetrahedral symmetry. Shells with octahedral symmetry composed of 24 subunits or icosahedral symmetry composed of 60 subunits may be possible if the interaction domains can be placed at appropriate angles by choosing appropriate linkers *(67)*.

6. TAKE-HOME LESSON

From the foregoing, it is clear that our understanding of the materials science of nanoscale assemblies and biomolecules is contributing to the development of useful devices on the nanoscale. Bionanotechnology is rapidly capitalizing on that understanding to produce a rich set of tools for probing and detecting normal and diseased cells. The fundamental physical chemistry of the nanoscale scaffold has already been fruitfully exploited as a flexible construction principle, and is rapidly yielding a bionanoscience of its own that is contributing to our understanding of molecular evolution, intracellular dynamics and supramolecular structure.

ACKNOWLEDGMENT

This work was supported by grant R01 CA10252-01A1 PI from the National Institutes of Health/National Cancer Institute.

REFERENCES

1. Lammers U, Borstel G. Electronic and atomic structure of copper clusters. Phys Rev B. Condensed Matter 1994;49:17,360–17,377.
2. Thoai DB, Hu YZ, Koch SW. Influence of the confinement potential on the electron-hole-pair states in semiconductor microcrystallites. Phys Rev B. Condensed Matter 1990;42:11,261–11,266.
3. Franceschetti A, Zunger A. Quantum-confinement-induced Gamma – –>X transition in GaAs/AlGaAs quantum films, wires, and dots. Phys Rev B. Condensed Matter 1995;52:14,664–14,670.
4. Wang LW, Zunger A. Pseudopotential calculations of nanoscale CdSe quantum dots. Phys Rev B. Condensed Matter 1996;53:9579–9582.
5. Nirmal M, Dabbousi BO, Bawendi MG, et al. Fluorescence intermittency in single cadmium selenide nanocrystals. Nature 1996;383:802–804.
6. Chan WC, Nie S. Quantum dot bioconjugates for ultrasensitive nonisotopic detection. Science 1998;281:2016–2018.
7. Dabbousi BO, Rodriguez-Viejo J, Mikulec FV, et al. (CdSe)ZnS core-shell quantum dots: synthesis and characterization of a size series of highly luminescent nanocrystallites. J Phys Chem B 1997;101:9463–9475.
8. Weber MH, Lynn KG, Barbiellini B, Sterne PA, Denison AB. Direct observation of energy-gap scaling law in CdSe quantum dots with positrons. Phys Rev B 2002;66:041301–041305.

9. Pinaud F, King D, Moore HP, Weiss S. Bioactivation and cell targeting of semiconductor CdSe/ZnS nanocrystals with phytochelatin-related peptides. J Am Chem Soc 2004;126:6115–6123.

10. Lidke DS, Nagy P, Heintzmann R, et al. Quantum dot ligands provide new insights into erbB/HER receptor-mediated signal transduction. Nat Biotechnol 2004;22:198–203.

11. Gao X, Cui Y, Levenson RM, Chung LW, Nie S. In vivo cancer targeting and imaging with semiconductor quantum dots. Nat Biotechnol 2004;22:969–976.

12. Wu X, Liu H, Liu J, et al. Immunofluorescent labeling of cancer marker Her2 and other cellular targets with semiconductor quantum dots. Nat Biotechnol 2003;21:41–46.

13. Jaiswal JK, Mattoussi H, Mauro JM, Simon SM. Long-term multiple color imaging of live cells using quantum dot bioconjugates. Nat Biotechnol 2003;21:47–51.

14. Akerman ME, Chan WC, Laakkonen P, Bhatia SN, Ruoslahti E. Nanocrystal targeting in vivo. Proc Natl Acad Sci USA 2002;99:12,617–12,621.

15. Winter JOL, Korgel BA, Schmidt CE. Recognition molecule directed interfacing between semiconductor quantum dots and nerve cells. Advanced Materials 2001;13:1673–1677.

16. Pathak S, Choi SK, Arnheim N, Thompson ME. Hydroxylated quantum dots as luminescent probes for in situ hybridization. J Am Chem Soc 2001;123: 4103–4104.

17. Xiao Y, Barker PE. Semiconductor nanocrystal probes for human metaphase chromosomes. Nucl Acids Res 2004;32:e28.

18. Striebel HM, Birch-Hirschfeld E, Egerer R, Foldes-Papp Z, Tilz GP, Stelzner A. Enhancing sensitivity of human herpes virus diagnosis with DNA microarrays using dendrimers. Exp Mol Pathol 2004;77:89–97.

19. Woller EK, Cloninger MJ. Mannose functionalization of a sixth generation dendrimer. Biomacromolecules 2001;2:1052–1054.

20. Shukla S, Wu G, Chatterjee M, et al. Synthesis and biological evaluation of folate receptor-targeted boronated PAMAM dendrimers as potential agents for neutron capture therapy. Bioconjug Chem 2003;14:158–167.

21. Choi Y, Thomas T, Kotlyar A, Islam MT, Baker JR Jr. Synthesis and functional evaluation of DNA-assembled polyamidoamine dendrimer clusters for cancer cell-specific targeting. Chem Biol 2005;12:35–43.

22. Ballauff M, Likos CN. Dendrimers in solution: insight from theory and simulation. Angew Chem Int Ed Engl 2004;43:2998–3020.

23. Mecke A, Lee I Jr, Holl MM, Orr BG. Deformability of poly(amidoamine) dendrimers. Eur Phys J E Soft Matter 2004;14:7–16.

24. Kiessling LL, Gestwicki JE, Strong LE. Synthetic multivalent ligands in the exploration of cell-surface interactions. Curr Opin Chem Biol 2000;4:696–703.

25. Dower SK, DeLisi C, Titus JA, Segal DM. Mechanism of binding of multivalent immune complexes to Fc receptors. 1. Equilibrium binding. Biochemistry 1981;20:6326–6334.

26. Dower SK, Titus JA, DeLisi C, Segal DM. Mechanism of binding of multivalent immune complexes to Fc receptors. 2. Kinetics of binding. Biochemistry 1981;20: 6335–6340.

27. Holland PM, Abramson RD, Watson R, Gelfand DH. Detection of specific polymerase chain reaction product by utilizing the 5'-3' exonuclease activity of *Thermus aquaticus* DNA polymerase. Proc Natl Acad Sci USA 1991;88: 7276–7280.
28. Lee LG, Connell CR, Bloch W. Allelic discrimination by nick-translation PCR with fluorogenic probes. Nucl Acids Res 1993;21:3761–3766.
29. Förster T. Delocalized excitation and excitation transfer. In: Sinanoglu O, ed. Modern Quantum Chemistry, vol. 3. New York: Academic Press, 1965:93–137.
30. Bonetta L. Prime time for real-time PCR. Nat Meth 2005;2:305–312.
31. Tyagi S, Kramer FR. Molecular beacons: probes that fluoresce upon hybridization. Nat Biotechnol 1996;14:303–308.
32. Piatek AS, Tyagi S, Pol AC, Telenti A, Miller LP, Kramer FR, Alland D. Molecular beacon sequence analysis for detecting drug resistance in *Mycobacterium tuberculosis.* Nat Biotechnol 1998;16:359–363.
33. Lyamichev V, Brow MA, Varvel VE, Dahlberg JE. Comparison of the 5' nuclease activities of taq DNA polymerase and its isolated nuclease domain. Proc Natl Acad Sci USA 1999;96:6143–6148.
34. Tapp I, Malmberg L, Rennel E, Wik M, Syvanen AC. Homogeneous scoring of single-nucleotide polymorphisms: comparison of the 5'-nuclease TaqMan assay and Molecular Beacon probes. Biotechniques 2000;28:732–738.
35. Tan W, Wang K, Drake TJ. Molecular beacons. Curr Opin Chem Biol 2004;8:547–553.
36. Yurke B, Turberfield AJ, Mills AP Jr, Simmel FC, Neumann JL. A DNA-fuelled molecular machine made of DNA. Nature 2000;406:605–608.
37. Yan H, Zhang X, Shen Z, Seeman NC. A robust DNA mechanical device controlled by hybridization topology. Nature 2002;415:62–65.
38. Alberti P, Mergny JL. DNA duplex-quadruplex exchange as the basis for a nanomolecular machine. Proc Natl Acad Sci USA 2003;100:1569–1573.
39. Sueda S, Ihara T, Takagi M. Metallo-regulation of DNA triple helix formation through cooperative dimerization of two oligonucleotides. Chem Lett 1997;26:1085–1086.
40. Matsumura K, Endo M, Komiyama M. Lanthanide complex-oligo-DNA hybrid for sequence-selective hydrolysis of RNA. J Chem Soc, Chem Commun 1994;1994:2019–2020.
41. Horsey I, Krishnan-Ghosh Y, Balasubramanian S. Enhanced cooperative binding of oligonucleotides to form DNA duplexes mediated by metal ion chelation. Chem Commun 2002;2002:1950–1961.
42. Endo M, Komiyama M. Novel phosphoramidite monomer for the site-selective incorporation of a diastereochemically pure phosphoramidate to oligonucleotide. J Org Chem 1996;61:1994–2000.
43. Baliga R, Singleton JW, Dervan PB. RecA oligonucleotide filaments bind in the minor groove of double-stranded DNA. Proc Natl Acad Sci USA 1995;92:10,393–10,397.
44. Han H, Dervan PB. Different conformational families of pyrimidine purine pyrimidine triple helices depending on backbone composition. Nucl Acids Res 1994;22:2837–2844.

45. Dreyer GB, Dervan PB. Sequence-specific cleavage of single-stranded DNA: oligodeoxynucleotide-EDTA Fe(II). Proc Natl Acad Sci USA 1985; 82:968–972.

46. Moser HE, Dervan PB. Sequence-specific cleavage of double helical DNA by triple helix formation. Science 1987;238:645–650.

47. Strobel SA, Dervan PB. Site-specific cleavage of a yeast chromosome by oligonucleotide-directed triple-helix formation. Science 1990;249:73–75.

48. Meredith GD, Wu HY, Allbritton NL. Targeted protein functionalization using His-tags. Bioconjug Chem 2004;15:969–982.

49. Churchill ME, Tullius TD, Kallenbach NR, Seeman NC. A Holliday recombination intermediate is twofold symmetric. Proc Natl Acad Sci USA 1988; 85:4653–4656.

50. Winfree E, Liu F, Wenzler LA, Seeman NC. Design and self-assembly of two-dimensional DNA crystals. Nature 1998;394:539–544.

51. Marky LA, Kallenbach NR, McDonough KA, Seeman NC, Breslauer KJ. The melting behavior of a DNA junction structure: a calorimetric and spectroscopic study. Biopolymers 1987;26:1621–1634.

52. Wang YL, Mueller JE, Kemper B, Seeman NC. Assembly and characterization of five-arm and six-arm DNA branched junctions. Biochemistry 1991;30:5667–5674.

53. Chen JH, Seeman NC. Synthesis from DNA of a molecule with the connectivity of a cube. Nature 1991;350:631–633.

54. Zhang Y, Seeman NC. The construction of a DNA truncated octahedron. J Am Chem Soc 1994;116:1661–1669.

55. Shih WM, Quispe JD, Joyce GF. A 1.7-kilobase single-stranded DNA that folds into a nanoscale octahedron. Nature 2004;427:618–621.

56. Mao C, Sun W, Seeman NC. Assembly of Borromean rings from DNA. Nature 1997;386:137–138.

57. Mao C, Sun W, Shen Z, Seeman NC. A nanomechanical device based on the B-Z transition of DNA. Nature 1999;397:144–146.

58. Gibson TJ, Lamond AI. Metabolic complexity in the RNA world and implications for the origin of protein synthesis. J Mol Evol 1990;30:7–15.

59. Niemeyer CM, Sano T, Smith CL, Cantor CR. Oligonucleotide-directed self-assembly of proteins: semisynthetic DNA—streptavidin hybrid molecules as connectors for the generation of macroscopic arrays and the construction of supramolecular bioconjugates. Nucl Acids Res 1994;22:5530–5539.

60. Niemeyer CM, Koehler J, Wuerdemann C. DNA-directed assembly of bienzymic complexes from in vivo biotinylated NAD(P)H:FMN oxidoreductase and luciferase. Chembiochem 2002;3:242–245.

61. Meredith GD, Wu HY, Allbritton NL. Targeted protein functionalization using his-tags. Bioconjug Chem 2004;15:969–982.

62. Smith SS, Niu L, Baker DJ, Wendel JA, Kane SE, Joy DS. Nucleoprotein-based nanoscale assembly. Proc Natl Acad Sci USA 1997;94:2162–2167.

63. Smith SS. A self-assembling nanoscale camshaft: implications for nanoscale materials and devices constructed from proteins and nucleic acids. Nano Lett 2001;1:51–56.

64. Clark J, Shevchuk T, Swiderski PM, et al. Mobility-shift analysis with microfluidics chips. Biotechniques 2003;35:548–554.
65. Smith SS. Designs for the self-assembly of open and closed macromolecular structures and a molecular switch using DNa methyltransferase to order proteins on nucleic acid scaffolds. Nanotechnology 2002;13:413–419.
66. Deyev SM, Waibel R, Lebedenko EN, Schubiger AP, Pluckthun A. Design of multivalent complexes using the barnase*barstar module. Nat Biotechnol 2003;21:1486–1492.
67. Padilla JE, Colovos C, Yeates TO. Nanohedra: using symmetry to design self assembling protein cages, layers, crystals, and filaments. Proc Natl Acad Sci USA 2001;98:2217–2221.
68. Ringler P, Schulz GE. Self-assembly of proteins into designed networks. Science 2003;302:106–109.
69. Moll D, Huber C, Schlegel B, Pum D, Sleytr UB, Sara M. S-layer-streptavidin fusion proteins as template for nanopatterned molecular arrays. Proc Natl Acad Sci USA 2002;99:14,646–14,651.
70. Kipriyanov SM, Little M, Kropshofer H, Breitling F, Gotter S, Dubel S. Affinity enhancement of a recombinant antibody: formation of complexes with multiple valency by a single-chain Fv fragment-core streptavidin fusion. Protein Eng 1996;9:203–211.
71. Caspar DL, Klug A. Physical principles in the construction of regular viruses. Cold Spring Harbor Symp Quant Biol 1962;27:1–24.
72. Bruinsma RF, Gelbart WM, Reguera D, Rudnick J, Zandi R. Viral self-assembly as a thermodynamic process. Phys Rev Lett 2003;90:248101.
73. Zandi R, Reguera D, Bruinsma RF, Gelbart WM, Rudnick J. Origin of icosahedral symmetry in viruses. Proc Natl Acad Sci USA 2004;101:15,556–15,560.
74. Wang Q, Kaltgrad E, Lin T, Johnson JE, Finn MG. Natural supramolecular building blocks. Wild-type cowpea mosaic virus. Chem Biol 2002;9:805–811.
75. Wang Q, Lin T, Johnson JE, Finn MG. Natural supramolecular building blocks. Cysteine-added mutants of cowpea mosaic virus. Chem Biol 2002;9:813–819.
76. Cheung CL, Camarero JA, Woods BW, Lin T, Johnson JE, De Yoreo JJ. Fabrication of assembled virus nanostructures on templates of chemoselective linkers formed by scanning probe nanolithography. J Am Chem Soc 2003;125:6848–6849.

4

Genetically Engineered S-Layer Proteins and S-Layer-Specific Heteropolysaccharides as Components of a Versatile Molecular Construction Kit for Applications in Nanobiotechnology

Eva-M. Egelseer, Margit Sára, Dietmar Pum, Bernhard Schuster, and Uwe B. Sleytr

Summary

One of the key challenges in material sciences is the technological utilization of self-assembly systems, wherein molecules spontaneously associate under equilibrium conditions into supramolecular structures joined by noncovalent bonds. Although molecular self-assembly is the governing principle in morphogenesis of biological systems, so far only a few molecular species have been exploited for controlled self-assembly into defined nanostructures. Crystalline bacterial cell surface layer (S-layer) proteins represent a first-order self-assembly system that has been optimized in the course of evolution. S-layers are composed of single protein or glycoprotein species which self-assemble into lattices with oblique, square, or hexagonal symmetry. Self-assembly into highly ordered monomolecular protein lattices occurs not only on the bacterial cell surface but also on artificial supports, such as polymers, silicon wafers, noble metals, lipid films, liposomes, lipid-plasmid particles, or on hollow polyelectrolyte nanoparticles. For recrystallization in an oriented manner, S-layer-specific polysaccharides as the natural anchoring molecules for S-layer proteins in the bacterial cell wall have been exploited as biomimetic linkers. To generate oriented functional monomolecular protein lattices, S-layer fusion proteins have been constructed, which incorporated either IgG-binding sequences, streptavidin, hypervariable regions of heavy chain camel antibodies, allergens, green fluorescent protein, metal-binding sequences, or a single cysteine residue. The fusion sites were selected such that after recrystallization on artificial supports precoated with S-layer-specific polysaccharides, the functional sequence remains exposed on the outermost surface of the protein lattice. Arrangement of specific functions in ordered fashion and their controlled confinement to defined areas of subnanometer

From: *NanoBioTechonology: BioInspired Devices and Materials of the Future*
Edited by: Oded Shoseyov and Ilan Levy © Humana Press Inc., Totowa, NJ

dimensions are key requirements for many applications in nanobiotechnology, including the development of label-free detection systems, biocompatible surfaces, signal processing between cells and integrated circuits, or even non-life science applications, such as molecular electronics and data storage.

Key Words: Nanobiotechnology; protein lattices; self-assemsly; S-layer proteins.

1. INTRODUCTION

Nanobiotechnology uses concepts from molecular biology, biochemistry, and chemistry to identify components, processes, and principles for the construction of self-assembling materials and devices. In particular, biological systems provide an enormous diversity of higher-order functional structures and patterns arising from molecular self-assembly. Most frequently, the inital step of molecular organization into functional units and complex supramolecular structures requires arrangement of molecules into ordered arrays. Recently, considerable effort has been devoted to exploiting natural self-assembly systems and to introducing variations into natural molecules (e.g., proteins and DNA) to achieve basic building blocks for specific structures and applications *(1–8)*.

This chapter is intended to provide a survey of the unique general principles of S-layer proteins and S-layer fusion proteins, as well as of the exploitation of S-layer-specifc heteropolysaccharides as biomimetic linkers to solid supports and liposomes. It describes how they are used as building blocks and templates for the generation of functional nanostructures at the meso- and macroscopic scale for both life and non-life science applications.

2. GENERAL ASPECTS OF S-LAYER PROTEINS

Ultrastructural analyses in combination with chemical and genetic studies have revealed that in the course of billions of years of evolution, prokaryotic organisms have developed a broad spectrum of cell-envelope structures *(9)*. Despite this diversity, one of the most commonly observed cell-surface structures are monomolecular arrays composed of identical species of protein or glycoprotein subunits *(10)*. Such surface layers, or S-layers, have been identified on organisms of nearly every taxonomic group of walled bacteria and they represent an almost universal feature of archaea *(11)*.

In archaea lacking a rigid cell-wall layer, S-layers represent the only wall component external to the plasma membrane. Being composed of a single species of constituent protein or glycoprotein subunits, S-layers can be considered the simplest type of biological membrane developed during evolution *(9–11)*. Interestingly, monomolecular arrays of proteinaceous subunits have also been observed in the bacterial sheath *(12)*, in exosporial membranes, and on eukaryotic algae *(13)*. As S-layers are present in Gram-positive and Gram-negative bacteria and archaea, they can be associated with quite different supporting structures that must provide suitable templates for maintaining a monomolecular protein lattice during all stages of cell growth and division *(5)*. In most organisms, S-layers must be considered nonconservative structures with the potential to fulfill a broad spectrum of functions *(14,15)*. For a great variety of applications in nanobiotechnology, it is of considerable importance that isolated S-layer subunits have the capability to recrystallize into coherent monomolecular protein lattices in suspension, at liquid–surface interfaces, on lipid films, and on liposomes, as well as on a great variety of solid supports, such as silicon wafers, noble metals, and polymers. As S-layers are periodic structures, they exhibit identical physicochemical properties on each constituent subunit down to the subnanometer scale. Moreover, S-layers are isoporous lattices, with pores of identical size and morphology.

S-layers have proven to be particularly well suited as building blocks in a biomolecular construction kit for "bottom-up" strategies involving all major classes of biological molecules (proteins, lipids, nucleic acids, glycans, and their combinations). It is now evident that native and genetically modified S-layer proteins fulfill most requirements for generating novel nanostructured materials and devices, as is required in nanobiotechnology, molecular nanotechnology, and biomimetics *(4–7,16–18)*.

2.1. Structural Analysis of S-Layer Lattices

High-resolution transmission electron microscopy is widely used to image and characterize S-layers that may be used as S-layer fragments, self-assembly products, or S-layer monolayers. The latter are usually generated by recrystallization of isolated S-layer proteins on solid supports *(4)*. Freeze-etching and freeze-drying in combination with heavy metal shadowing are the most straightforward approaches to obtaining information on the lattice type (Fig. 1) and surface structure of S-layers *(19)*. Freeze-etching provided detailed information on the in vivo self-assembly and continuous recrystallization of S-layer proteins on intact growing and dividing cells. This dynamic self-assembly process guarantees the complete coverage of the cell surface with a monomolecular protein lattice *(10,18)*. High-resolution unidirectional or rotary shadowing can depict characteristic topographical details of S-layers, such as lattice symmetry and lattice parameters. A structural feature of many S-layers is a smooth outer and a more corrugated inner face *(20–22)*. This difference is of particular importance when the orientation (sidedness due to attachment via the inner or outer surface) of S-layers on artificial substrates must be determined. Negative staining is an easy preparation technique for

A

B   Lattice types

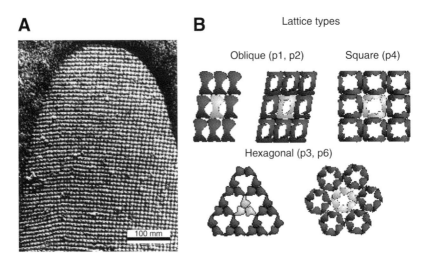

Fig. 1. (A) Electron micrograph of a freeze-etched and Pt/C-shadowed preparation of a gram-positive organism exhibiting a square (p4) S-layer lattice. **(B)** Schematic drawing illustrating the various S-layer lattice types. In the oblique lattice, one morphological unit (grey) consists of one (p1) or two (p2) identical subunits. Four subunits constitute one morphological unit in the square (p4) lattice type, whereas the hexagonal lattice type is either composed of three (p3) or six (p6) subunits. The center-to-center spacing of the morphological units is a strain-specific feature and may range from 5 to 35 nm.

electron microscopic investigation. Particularly in combination with two-dimensional (2D) and 3D image-reconstruction techniques, it allows high-resolution studies of the ultrastructure of S-layer lattices *(20–22)*. However, the highest resolution (3.5 Å to 5 Å) is obtained by cryo-electron microscopy, where the S-layer is embedded in a thin layer of amorphous ice and imaged with extremely low electron doses (~1–10 e/Å^2) *(21)*. Such low electron doses are forced by the high sensitivity of the protein to electron irradiation. As quantum noise dominates image formation under such conditions, image-processing methods are necessary to enhance the signal-to-noise ratio in the recorded micrographs *(20)*.

Contrary to the electron microscopic preparation techniques, scanning force microscopy allows an investigation of S-layer monolayers in their native environment *(23–25)*. In fact, contact mode microscopy in liquid is most frequently used to investigate S-layer monolayers at subnanometer resolution. S-layer proteins are highly susceptible to the applied tip loading forces which should not exceed 0.5 to 1 nN. To obtain the best resolution, the ionic content and strength of the buffer solution in the liquid cell must be carefully adjusted in order to minimize electrostatic interactions between tip and sample. As a result of their stiffness and hardness, silicon wafers and mica are the most commonly used substrates for scanning force microscopic investigations. S-layer proteins recrystallize into large-scale monomolecular protein lattices on silicon, whereas S-layer fragments or self-assembly products are preferably deposited on mica. If the S-layer protein is recrystallized on flat solid supports such as silicon wafers, lattice formation can be followed in real time *(23)*. By this approach, it could be demonstrated that crystal growth starts at several distant nucleation points and proceeds in-plane until a closed layer consisting of a mosaic of crystalline domains is formed. The scanning force microscope has also been used as a nanotool for inducing conformational changes in S-layer proteins *(26,27)*. Furthermore, the capability of scanning force microscopy to resolve molecular details on biological samples together with its force detection sensitivity has led to the development of the so-called "topography and recognition mode," a method suitable for visualizing the chemical composition of a sample while mapping its topography *(28)*. It is anticipated that the simultaneous investigation of both topography and recognition will enable the unsurpassed elucidation of the structure–function relationship of a broad spectrum of biological samples.

2.2. Self-Assembly and Recrystallization Properties of S-Layer Proteins

In contrast to many archaeal S-layer proteins, those of bacteria are non-covalently linked to each other and to the supporting cell-wall component. Thus, complete solubilization of S-layers into their constituent subunits and release from the bacterial cell envelope can be achieved by treatment with high concentrations of hydrogen-bond-breaking agents (e.g., guanidinium hydrochloride), by dramatic changes in the pH value, or in the salt concentration *(4,18)*. During removal of the disrupting agent, e.g., by dialysis, the S-layer subunits frequently show the ability to self-assemble into 2D arrays (Fig. 2). Such self-assembly products may have the form of flat sheets or open-ended cylinders, and they represent monolayers or double layers *(4,10,11,18)*. With some S-layer proteins, the self-assembly process has been shown to depend on the presence of bivalent cations.

Before recrystallization on artificial supports (Fig. 2), S-layer proteins must be kept in a water-soluble state. Depending on the type of S-layer protein, this can be achieved (a) in the absence of bivalent cations *(29)*, (b) by adjusting to concentrations that are subcritical for self-assembly *(30)*, or (c) in the presence of S-layer-specific secondary cell-wall polymers (SCWP) *(29,31)*. The latter have been found to inhibit the self-assembly in suspension, but they promote the recrystallization of soluble S-layer proteins on artificial supports *(29,31)*. The formation of coherent crystalline arrays strongly depends

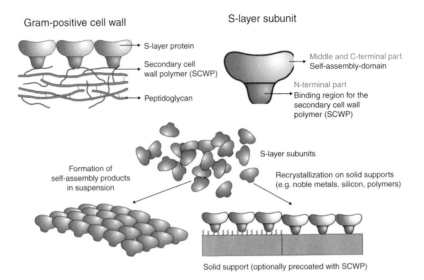

Fig. 2. Schematic drawing showing (i) that S-layer proteins from Bacillaceae are bound to the peptidoglycan-containing layer via the secondary cell-wall polymer (SCWP); (ii) the structural organization of S-layer proteins consisting of the N-terminal cell-wall-anchoring domain and the self-assembly domain; (iii) the formation of self-assembly products in suspension; (iv) the recrystallization of isolated S-layer subunits on supports optionally pre-coated with SCWP. On supports not pre-coated with SCWP, the S-layer subunits may either attach with their inner or outer surface.

on the S-layer protein species, the environmental conditions of the bulk phase (e.g., temperature, pH, ion composition, and ionic strength), and, in particular, on the surface properties of the substrate. Recrystallization of isolated S-layer subunits at the air/water interface and on Langmuir films has proven to be an easy and reproducible way of generating coherent S-layer lattices on a large scale. In accordance with S-layer proteins recrystallized on solid substrates, the orientation of the protein arrays (sidedness due to the attachment via the inner or outer surface of the S-layer subunits) at liquid interfaces is determined by the anisotropy in the physicochemical surface properties of the protein lattice, as well as by the physicochemical properties of the support used for recrystallization *(4,16,18)*.

2.3. Chemical Properties and Molecular Biology

Chemical analyses and genetic studies revealed that the monomolecular S-layer is the result of the secretion and subsequent crystallization of a single homogeneous protein or glycoprotein species with a molecular mass ranging from 40 to 200 kDa *(4,15,16,18)*. Most S-layer proteins are weakly acidic, with isoelectric points (pI) in the range of 4 to 6 *(15)*. Exceptions have been reported for the S-layer proteins of lactobacilli *(32)* and those of *Methanothermus fervidus (33)*, which possess pIs of 9 to 11 and 8.4, respectively. Typically, S-layer proteins have a large proportion of hydrophobic amino acids (40–60 mol%), possess little or no sulfur-containing amino acids, and consist of about 25 mol% charged amino acids. The most frequent posttranslational modification of S-layer proteins is glycosylation *(34–36)*. The glycan chains of S-layer proteins from Gram-positive bacteria more closely resemble the architecture of lipopolysaccharides (LPS) occurring in the outer membrane of Gram-negative bacteria than those of eukaryotic glycan structures.

Information regarding the secondary structure of S-layer proteins is either derived from the amino-acid sequence or from circular dichroism (CD) measurements, indicating that approx 20% of the amino acids are organized as α-helices and about 40% occur as β-sheets. Aperiodic folding and β-turn content may vary between 5 and 45%. Secondary-structure predictions based on protein sequence data revealed that most α-helical segments are arranged at the N-terminal end.

During the last two decades, numerous S-layer genes from bacteria and archaea of quite different taxonomical position have been sequenced and cloned *(4–7,15,18)*. To elucidate the structure–function relationship of distinct segments of S-layer proteins, N- and/or C-terminally truncated forms were produced and their self-assembly and recrystallization properties investigated *(37–39)*. Another approach was seen in performing a cysteine scanning mutagenesis and screening the accessibility of the introduced cysteine residue in the soluble, self-assembled, and recrystallized S-layer proteins *(30)*. The final aim of this study was to find out which amino-acid positions in the primary sequence are located on the outer surface of the subunits, inside the pores, at the subunit-to-subunit interface, or on the inner S-layer surface.

In the case of Bacillaceae, sequencing and cloning of S-layer genes further revealed that identities are limited to the N-terminal region. This part was found to be responsible for anchoring the S-layer subunits to the underlying rigid cell envelope layer by binding to a heteropolysaccharide, termed SCWP. The polymer chains are covalently linked to the peptidoglycan backbone; this occurs most probably via phosphodiester bonds *(40)*. Basically, two types of binding mechanisms between the N-teminal part of S-layer proteins and SCWP have been described. The first type of binding mechanism, which involves so-called S-layer-homologous (SLH) domains and pyruvylated SCWP *(29,37,41–46)*, has been found to be widespread among prokaryotes and is considered as having been conserved in the course of evolution *(46)*. The second type of binding mechanism has been described for *Geobacillus stearothermophilus* PV72/p6 and ATCC 12980 *(15,47,48)* and a temperature-derived strain variant from the latter *(49)*. This binding mechanism involves an SCWP that consists of N-acetyl glucosamine, glucose and 2,3-dideoxy-diacetamido mannosamine uronic acid in the molar ratio of 1:1:2 *(50)*, and a highly conserved N-terminal region that does not possess an SLH domain *(47–49)*. Concerning the first binding mechanism, the construction of knock-out mutants in *Bacillus anthracis* and *Thermus thermophilus* in which the gene encoding a putative pyruvyl transferase was deleted demonstrated that the addition of pyruvic acid residues to the peptidoglycan-associated cell-wall polymer was a necessary modification to bind SLH-domain-containing proteins *(41,46)*. This observation was also supported by surface plasmon resonance (SPR) spectroscopy measurements for which the S-layer protein SbsB of *G. stearothermophilus* PV72/p2 and native and chemically modified SCWP devoid of pyruvic acid residues were used for interaction studies *(44)*. By applying the dissection approach, the SLH domain (rSbsB$_{32–208}$) of SbsB was found to be exclusively responsible for SCWP-binding. This protein–carbohydrate interaction proved to be highly specific *(15,16)*, and evidence that the pyruvyate ketals are a necessary modification for the binding process was provided *(44)*. The residual part of SbsB devoid of the SLH domain (rSbsB$_{209–920}$) did not show any affinity for the SCWP, but retained the ability to self-assemble, thereby forming the oblique (p1) lattice typical of full-length SbsB. As these findings were supported by applying optical spectroscopic methods and electron microscopy *(51)*, rSbsB is considered to consist of two functionally and structurally independent domains, namely the SLH domain, which is responsible for "cell-wall targeting" by recognizing the SCWP, and the larger C-terminal part, which corresponds to the self-assembly domain *(44,51)*. Furthermore, thermal as well as guanidinium hydrochloride-induced equilibrium unfolding profiles monitored by intrinsic fluorescence and CD spectroscopy allowed the characterization of rSbsB$_{32–208}$ as an α-helical protein with a single cooperative unfolding transition. The C-terminal rSbsB$_{209–920}$ could be characterized as a β-sheet protein with typical multi-domain unfolding and a lower stability as a stand-alone protein *(51)*. Thus, the dissection approach might be useful for the whole group of S-layer proteins, leading to truncated forms that would be accessible to direct structure determination. The fact that no structural model at atomic resolution of an S-layer protein has been available until now may be explained by the molecular mass of the subunits being too large for nuclear magnetic resonance (NMR) analysis, as well as by the intrinsic property of S-layer proteins to self-assemble into 2D lattices, thereby hindering the formation of the isotropic 3D crystals required for X-ray crystallography. In addition, the low solubility of S-layer proteins is a general hindrance to both methods.

In the case of the S-layer protein SbsC of *G. stearothermophilus* ATCC 12980, water-soluble N- or C-terminally truncated forms were used for first

3D crystallization studies. For the C-terminally truncated form, rSbsC$_{31-844}$ crystals, which diffracted to a resolution of 3 Å using synchrotron radiation, could be obtained by sitting-drop vapor-diffusion with polyethylene glycol 6000 as precipitating agent *(52)*. Native and heavy atom derivative data confirmed the results of the secondary structure prediction, which indicated that the N-terminal region comprising the first 257 amino acids is mainly organized as α-helices, whereas the middle and C-terminal part of SbsC consist of loops and β-sheets *(52)*. Information on the 3D structure of S-layer proteins would contribute to the possibility of rationally designing S-layer fusion proteins that incorporate functional domains, for example, within the pore areas of the protein lattice.

2.4. Functional Aspects

The high physiological expense and the widespread occurrence of S-layers raise the question of what selection advantage S-layer-carrying organisms have in their natural habitats over their nonlayered counterparts. So far, no general, all-encompassing natural function has been found and many of the functions assigned to S-layers remain hypothetical. However, it is now recognized that S-layers can act as (a) a framework to determine and maintain cell shape and cell division in archaea that possess S-layers as exclusive wall component; (b) protective coats, molecular sieves, as well as molecule and ion traps; (c) structures involved in cell adhesion and surface recognition; (d) templates for fine grain mineralization; (e) adhesion sites for cell-associated exoenzymes, and (f) virulence factors in pathogenic organisms with an important role in invasion and survival within the host *(9,12,15)*. In this context, S-layer variation, which might have developed as a pivotal mechanism to respond to changing environmental conditions in the course of evolution, must be mentioned. S-layer variation leads to the synthesis of alternate S-layer proteins, either by the expression of different S-layer genes or by recombination of partial coding sequences, and has been described as occurring in pathogens as well as in nonpathogens *(53–56)*. In the latter, S-layer variation is frequently induced in response to environmental stress factors, such as increased oxygen supply, whereas in pathogens, altered cell-surface properties most probably protect the cells from the lytic activity of the immune system.

3. A MOLECULAR CONSTRUCTION KIT BASED ON S-LAYER FUSION PROTEINS AND S-LAYER-SPECIFIC HETEROPOLYSACCHARIDES

S-layer proteins, SCWPs, and the specific interactions between them have played an important role in the development of a biomolecular construction kit. As a result of their excellent recrystallization properties, the S-layer proteins

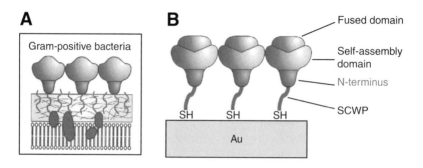

Fig. 3. Schematic drawing of (**A**) the cell envelope of a Gram-positive S-layer-carrying organism. (**B**) Exploitation of the thiolated secondary cell-wall polymer (SCWP) as biomimetic linker to a gold chip for an oriented binding of S-layer fusion proteins. In S-layer fusion proteins, the C-terminal part is replaced by the fused foreign functional sequence or domain.

SbpA of *Bacillus sphaericus* CCM 2177 *(38)* and SbsB of *G. stearothermophilus* PV72/p2 *(31)* have been used for most developments relevant to nanobiotechnology. Both S-layer proteins self-assemble in suspension and recrystallize on various types of solid supports, as well as on lipid layers and liposomes. However, depending on the physicochemical properties of the underlying support, the S-layer subunits attach either with the outer or via the inner surface, so that the complementary surface remains exposed to the ambient environment. To guarantee that the S-layer subunits attach in uniform orientation, which is extremely important if S-layer fusion proteins are used for building up functional monomolecular protein lattices, SCWP have been exploited as biomimetic linkers for pre-coating solid supports (Fig. 3). In this case, the S-layer subunits bind with the inner surface comprising the SCWP-binding SLH domain *(16)*. For some specific applications, the S-layer protein SbsC of *G. stearothermophilus* ATCC 12980 has been used *(48)*. SbsC shows excellent self-assembly properties, but, as this S-layer protein cannot be kept in the water-soluble state, it is not suited to recrystallization on artificial supports.

3.1. The S-Layer Proteins SbpA, SbsB and SbsC and the Corresponding S-Layer-Specific Heteropolysaccharides

The S-layer protein SbpA of *B. sphaericus* CCM 2177 consists of a total (including a 30-amino-acid-long signal peptide) of 1268 amino acids *(38)*. SbpA self-assembles into a square (p4) lattice structure with a center-to-center spacing of the morphological units of 13.1 nm. The self-assembly process is strongly dependent on the presence of bivalent cations, such as calcium ions *(29)*. In the absence of bivalent cations, this S-layer protein stays in the water-soluble state. The N-terminal part of SbpA comprises three typical SLH motifs, but for reconstituting the functional SCWP-binding domain, an additional 58-amino-acid-long SLH-like motif is required *(37)*. The SCWP of *B. sphaericus* CCM 2177 consists of disaccharide repeating units that are composed of *N*-acetyl glucosamine (GlcNAc) and *N*-acetyl mannosamine (ManNAc). The ManNAc residues carry a pyruvate ketal, which endows the polymer chains with a negative net charge *(29)*. Studies on the structure–function relationship revealed that up to 237 C-terminal amino acids can be deleted without influencing the formation of the p4 lattice structure *(29,37)*. On the other hand, the deletion of 350 C-terminal acids was linked to a change from square (p4) to oblique (p1) lattice symmetry *(37)*. By producing various C-terminally truncated forms and performing surface-accessibility screens, it became apparent that amino-acid position 1068 is located on the outer surface of the square lattice *(38)*. This was the reason that the C-terminally truncated form rSbpA$_{31-1068}$ was used as the base form for the construction of several S-layer fusion proteins *(57–61)*. An N- and C-terminally truncated form (rSbpA$_{203-1031}$) was capable of self-assembling into the square (p4) lattice structure with a center-to-center spacing of the morphological units and a protein mass distribution similar to that formed by full-length rSbpA *(37)*. These findings indicated that the segment between amino acids 203 and 1031 is responsible for the self-assembly process and for pore formation.

The S-layer protein SbsB of *G. stearothermophilus* PV72/p2 consists of a total (including a 31-amino-acid-long signal peptide) of 920 amino acids. By applying the dissection approach, it could be demonstrated that SbsB is composed of the N-terminal SCWP-binding domain, which corresponds to the SLH domain, and the C-terminal self-assembly domain *(44,51)*. As the removal of fewer than 10 C-terminal amino acids led to water-soluble rSbsB forms, the C-terminal part can be considered extremely sensitive to deletions. When the C-terminal end of full-length SbsB was exploited for linking foreign functional sequences, water-soluble S-layer fusion proteins were obtained *(62)*, which recrystallized into the oblique (p1) lattice only on solid supports pre-coated with SCWP of *G. stearothermophilus* PV72/p2. As demonstrated by SPR spectroscopy, the SLH domain comprising the three SLH motifs specifically recognizes the SCWP of *G. stearothermophilus* PV72/p2 as a binding site *(44)*. The polymer chains are composed of GlcNAc and ManNAc in the molar ratio of approx 2:1 *(31)* and contain pyruvate ketals, which provide a net negative charge. The use of chemically modified SCWP in SPR interaction studies clearly showed that the pyruvic acid residues, and not the N-acetyl groups of the amino sugars, play a crucial role in the recognition and binding process *(44)*.

Recently, chimeric S-layer proteins comprising either the N-terminal part of SbpA and the C-terminal part of SbsB (rSbpA-SbsB), or vice versa (rSbsB-SbpA), have been constructed *(37)*. The aim was to create a spectrum of

S-layer proteins that bind to the same type of SCWP, but assemble into different lattice types. Accordingly, rSbsB and rSbsb-SbpA specifically recognized the SCWP of *G. stearothermophilus* PV72/p2 as binding site, but in contrast to rSbsB, which assembles into an oblique (p1) lattice, the chimeric protein rSbsb-SbpA formed a square (p4) lattice structure. In addition, in rSbsB-SbpA, the surface-located amino-acid position 1068 of the SbpA primary sequence can be exploited as a fusion site for foreign functional sequences *(57–61)*. On the other hand, rSbpA and the chimeric protein rSbpA-SbsB recognized the SCWP from *B. sphaericus* CCM 2177 as a binding site, but they self-assembled into different lattice types, namely, square (p4) for rSbpA and oblique (p1) for rSbpA-SbsB *(37)*.

SbsC is the S-layer protein of *G. stearothermophilus* ATCC 12980 and consists of a total (including a 30-amino-acid-long signal peptide) 1099 amino acids. Isolated SbsC self-assembles into an oblique (p2) lattice. SbsC does not possess an SLH domain on the N-terminal part, but reveals a conserved N-terminus region, which has also been found in S-layer proteins of other *G. stearothermophilus* wild-type strains, such as PV72/p6, DSM 2358 *(47,48)*, and NRS 2004/3a *(63)*, as well as in a temperature-derived strain variant of ATCC 12980 *(49)*. This type of N-terminal part, which comprises amino acids 31 to 257, specifically recognizes the SCWP that contains 2,3-dideoxy-diacetamido mannosamine uronic acid as a negatively charged component *(39,48)*. Studies on the structure–function relationship of SbsC revealed that the N-terminal part is responsible for cell-wall anchoring via the specific SCWP. In the C-terminal part, up to 179 amino acids leading to rSbsC$_{31-920}$ could be deleted without interfering with the formation of the oblique lattice structure *(39)*. Further deletion of C-terminal amino acids led to SbsC forms that were still capable of self-assembling but such self-assembly products did not show a regular lattice structure (rSbsC$_{31-880}$), whereas rSbsC$_{31-844}$ even represented a completely water-soluble form that was used for 3D crystallization experiments *(52)*.

3.2. Functionalization of Solid Supports with S-Layer-Specific Heteropolysaccharides

As the S-layer protein SbsB binds with its outer surface to positively charged liposomes and to silicon wafers, N-terminal-core-streptavidin fusion proteins have been constructed, mixed with separately expressed free core-streptavidin in a molar ratio of 1:3, and folded into functional heterotetramers (HT) *(62)*. Such HT recrystallized in the expected orientation and left the streptavidin moiety exposed to the external environment *(62)*. However, when the composition of the liposomes was changed, or solid supports other than silicon were used, binding of SbsB or fusion proteins thereof via the

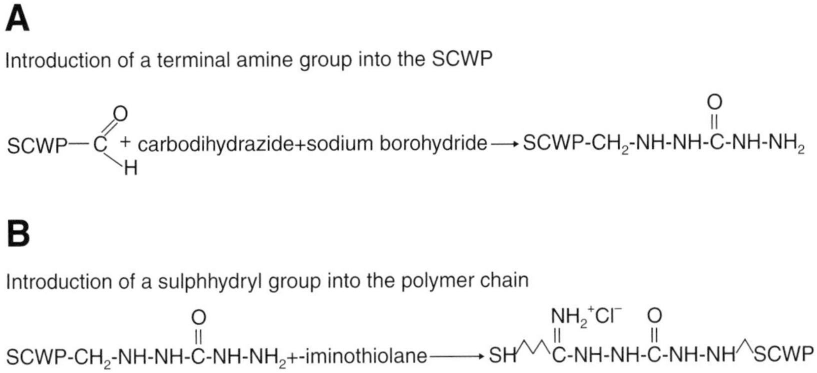

A

Introduction of a terminal amine group into the SCWP

SCWP—C(=O)H + carbodihydrazide+sodium borohydride → SCWP-CH$_2$-NH-NH-C(=O)-NH-NH$_2$

B

Introduction of a sulphhydryl group into the polymer chain

SCWP-CH$_2$-NH-NH-C(=O)-NH-NH$_2$ + iminothiolane ⟶ SH~~C(=NH$_2^+$Cl$^-$)-NH-NH-C(=O)-NH-NH~~SCWP

Fig. 4. Chemical modification of the secondary cell-wall polymer (SCWP) for binding to solid supports or preparing conjugates with lipid molecules for integration into lipid layers or liposomes. **(A)** The terminal amine group is introduced by linking carbodihydrazide to the latent aldehyde groups of the polymer chain which is followed by reduction of the Schiff' base with sodium borohydride. **(B)** Sulfhydryl groups are subsequently introduced by reaction of the terminal amine group with 2-iminothiolane.

outer or inner surface strongly depended on the specific physicochemical properties of the supporting material and the final orientation (sidedness) could not be predicted. As the same behavior was observed for SbpA, the natural anchoring molecules for these S-layer proteins in the bacterial cell wall, the specific SCWP were exploited as biomimetic linkers for pre-coating solid supports in order to guarantee that the fusion proteins attach via the inner surface carrying the SCWP-binding region (Fig. 3) *(59–61)*. In the case of liposomes, the synthesis of amphiphilic molecules consisting of the smallest functional unit of the respective SCWP and a lipid anchor is currently being performed. Such functionalized lipid molecules shall be incorporated into liposomes, lipid-plasmid particles, or flat lipid layers.

In general, SCWPs are isolated from bacterial cell-wall fragments by applying a hydrofluoric acid (HF) extraction procedure *(31,64)*. After purification by size-exclusion chromatography, the latent aldehyde group of the reducing end of the polymer chain is modified with carbodihydrazide and the Schiff' base is reduced with sodium borohydride *(44)*. Sulfhydryl groups are introduced by reaction of the free amine group with 2-methyl mercaptobutyrimidate (2-iminothiolane) (Fig. 4). Such modified SCWP carrying a free terminal sulfhydryl group can be used for direct adsorption to gold substrates (Fig. 3), as required for SPR *(59–61)* or surface plasmon field enhanced spectroscopy *(57)*. For covalent binding to supports carrying free amine groups, the sulfhydryl group of the polymer chain (termed thiolated

SCWP) is activated with m-maleimidobenzoyl-*N*-hydroxysuccinimide ester
(MBS) which—as a heterobifunctional crosslinker—can then react with
amine groups, e.g. of amine-modified cellulose microbeads *(61)*. Another
possible way to activate thiolated SCWP can be seen in the use of the heterobifunctional crosslinker sulfosuccinimidyl 6-[3´-2(-pyridyldithio)-
propionamido] hexanoate (Sulfo-LC-SPDP) *(44)*.

3.3. S-Layer Fusion Proteins

During the last few years, S-layer technology has advanced by the con-
struction of S-layer fusion proteins that comprise (a) an accessible N-terminal
SCWP-binding domain, which can be exploited for oriented binding and
recrystallization on artificial supports pre-coated with SCWP, (b) the self-
assembly domain, and (c) a C-terminally fused functional sequence (Fig. 3).
Chimeric proteins following this construction principle have been termed
C-terminal S-layer fusion proteins. However, for some selected applications,
N-terminal S-layer fusion proteins that were mainly based on SbsB, the S-layer
proteins of *G. stearothermophilus* PV72/p2 were constructed. Such N-terminal
fusion proteins attached with the outer surface to positively charged liposomes
and silicon wafers, thereby leaving the N-terminal region with the fused
functional sequence exposed to the external environment *(62)*.

To gain sufficient knowledge about which amino-acid positions of the
S-layer proteins foreign peptide sequences could be fused to without interfering with the self-assembly and recrystallization properties, the structure–
function relationship of distinct segments of different S-layer proteins had to
be elucidated. In the case of the S-layer protein SbpA, surface-located amino-
acid positions were identified by fusing *Strep*-Tag I (Fig. 5), a short affinity
peptide specific for streptavidin or strep-Tactin, to the C-terminal end *(38)*.
As it could be demonstrated that the deletion of 200 C-terminal amino acids
had no influence on the self-assembly properties of this S-layer protein and
that *Strep*-Tag I was accessible to a significantly higher extent in the C-terminally truncated form rSbpA$_{31-1068}$ than in full-length rSbpA (Fig. 5), the C-terminal truncation was used as the base form for the construction of various
S-layer fusion proteins *(38,57–61)*. All chimeric proteins that were heterolo-
gously expressed in *Escherichia coli* retained the ability to self-assemble and
to recrystallize into a monomolecular protein lattice on peptidoglycan-containing sacculi of *B. sphaericus* CCM 2177, as well as on gold chips or glass substrates pre-coated with SCWP.

Owing to the versatile applications of the streptavidin-biotin interaction
as a biomolecular coupling system, minimum-sized core-streptavidin (118
amino acids) was fused either to N- or C-terminal positions of SbsB or
linked to the C-terminal end of rSbpA$_{31-1068}$ *(57,62)*. The fusion proteins and
core-streptavidin were produced independently in *E. coli*, isolated, and

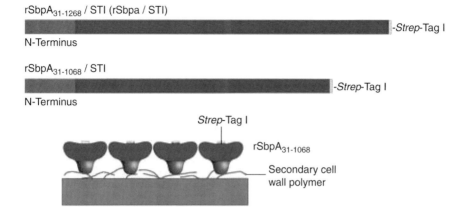

Fig. 5. By linking the short affinity tag *Strep*-Tag I (STI) to the C-terminal end of full-length rSbpA (rSbpA$_{31-1268}$) and the C-terminally truncated form rSbpA$_{31-1068}$, and subsequent binding of strep-Tactin with a molecular weight of 66,000, it could be demonstrated that only amino-acid position 1068 is located on the outer surface of the S-layer subunits and is therefore exploited as a fusion site for various S-layer fusion proteins.

refolded to obtain HT (Fig. 6) that consist of one chain fusion protein and three chains core-streptavidin. In the case of SbsB, HT were used for recrystallization on liposomes and on silicon wafers *(62)*. For rSbpA-streptavidin HT, fluorescence quenching of the tryptophan residues in the binding pockets of streptavidin demonstrated that in comparison to free streptavidin, HT had a higher binding capacity for D-biotin, biotinylated insulin (MW 5800), and biotinylated horseradish peroxidase (MW 44,000). As a first application approach, a monolayer generated by recrystallization of such HT on gold chips pre-coated with thiolated SCWP (Fig. 7) was used as a matrix for hybridization experiments *(57)*. Surface plasmon field enhanced spectroscopy demonstrated that biotinylated oligonucleotides (30-mers) could bind to the streptavidin moiety of the HT with high specificity and that the hybridization reaction of the complementary fluorescently labeled oligonucleotides (15-mers) followed the Langmuir isotherm. These new tools, which combine the recrystallization ability of the S-layer protein moiety with the biotin-binding property of streptavidin, reveal a promising application potential to create a functional sensor surface of broad interest (Fig. 8).

In a first approach, an S-layer fusion protein incorporating the hypervariable region of a heavy chain camel antibody directed against lysozyme was produced *(59)*. As proof of principle could be provided by this model system, an S-layer fusion protein incorporating the camel antibody sequence recognizing prostate-specific antigen (PSA), termed rSbpA$_{31-1068}$/cAb-PSA-N7, was produced, recrystallized on gold chips pre-coated with thiolated SCWP,

Egelseer et al.

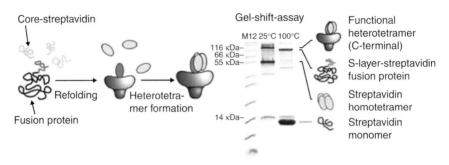

Fig. 6. Refolding of heterotetramers (HT) by applying the rapid dilution protocol. The S-layer-streptavidin fusion protein is mixed with core-streptavidin in a molar ratio of 1:3. During the refolding procedure, functional HT are formed which are separated from other components by size-exclusion chromatography and further purified by affinity chromatography using 2-imino biotin agarose *(62)*. To confirm the formation of HT, the gel-shift assay is carried out. In this figure, a C-terminal S-layer-streptavidin fusion protein is shown.

Fig. 7. Scanning force microscopic image of a monolayer consisting of heterotetramers of the rSbpA-streptavidin fusion protein, which were recrystallized on gold chips pre-coated with the secondary cell-wall polymer. The image was obtained in contact mode under water. Bar, 20 nm.

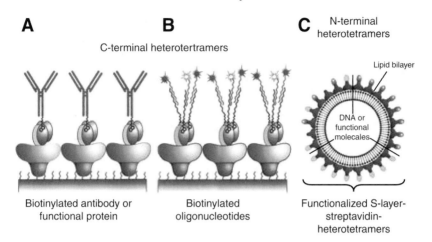

Fig. 8. HT of C-terminal fusion proteins are recrystallized on solid supports pre-coated with secondary cell-wall polymer (SCWP) to generate sensing layers for the development of protein or DNA chips (**A, B**). In the case of protein chips, the catching antibody is biotinylated (**A**), whereas biotinylated oligonucleotides are bound to the monolayer of HT in the case of DNA chips (**B**). HT of N-terminal fusion proteins are recrystallized on liposomes to which the S-layer subunits bind with their outer surface (**C**).

and exploited as a nanopatterned sensing layer in SPR to detect PSA *(60)*. As derived from response levels measured for binding of PSA to a monolayer consisting of $rSbpA_{31-1068}$/cAb-PSA-N7, the molar ratio between bound PSA and the S-layer fusion protein was 0.78, which means that at least three PSA molecules were bound per morphological unit of the square lattice consisting of four identical subunits of S-layer fusion protein. The advantage of this monomolecular S-layer protein lattice can be seen in the constant and short distance of the ligands to the optically active gold layer at its constant height of 10–15 nm, which implies that binding events occur at a defined distance from the gold layer. This chimeric S-layer can be considered a key element for the development of sensing layers for label-free detection systems such as SPR, surface acoustic wave (SAW), or quartz crystal microbalance (QCM-D), in which the binding event can be measured directly by the mass increase without the need of any labeled molecule (Fig. 8).

The S-layer fusion protein carrying two copies of the 58-amino-acid-long Fc-binding Z-domain which is a synthetic analogue of the B-domain of Protein A on the C-terminal end ($rSbpA_{31-1068}$/ZZ) was recrystallized on gold chips pre-coated with thiolated SCWP *(61)*. The binding capacity of the native or cross-linked monolayer for human IgG was determined by SPR

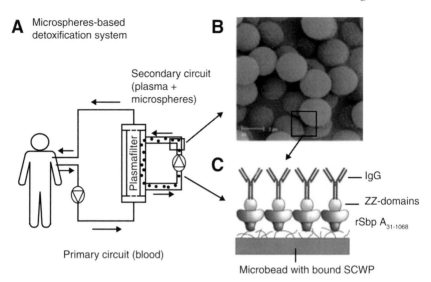

A Microspheres-based detoxification system

Secondary circuit (plasma + microspheres)

Plasmafilter

Primary circuit (blood)

B

C

— IgG
— ZZ-domains
rSbp A$_{31\text{-}1068}$

Microbead with bound SCWP

Fig. 9. (A) Schematic drawing of the working principle of the microsphere-based detoxification system (MDS) and **(B)** amine-modified cellulose microparticles. In **C**, the rSbpA$_{31\text{-}1068}$/ZZ fusion protein was recrystallized on amine-modified cellulose microparticles to which thiolated secondary cell-wall polymer (SCWP) had been covalently bound.

spectroscopy. In the case of the native monolayer, the binding capacity for human IgG was 5.1 ng/mm^2, whereas after crosslinking with dimethyl pimelimidate, 4.4 ng IgG/mm^2 were bound. These values corresponded to 78% or 65% of the theoretical saturation capacity of a planar surface for IgG aligned in an upright position, with condensed state of the Fab regions. For batch-adsorption experiments using human serum, 3-μm large, biocompatible, cellulose-based, amine-modified, and SCWP-coated microbeads were used for recrystallization of the S-layer fusion protein *(61)*. This novel type of microbead shall find application in the microsphere-based detoxification system (MDS; *65*) (Fig. 9).

The S-layer fusion protein incorporating the sequence of enhanced green fluorescent protein (EGFP) (rSbpA$_{31\text{-}1068}$/EGFP) retained the ability to self-assemble in suspension and to recrystallize on peptidoglycan-containing sacculi and on positively charged liposomes *(58)*. Investigation of self-assembly products formed by rSbpA$_{31\text{-}1068}$/EGFP revealed that the fluorescence activity of the EGFP portion in the fusion protein was not affected by the self-assembly process *(38)*. Because of the ability of the EGFP moiety to fluoresce, the rSbpA$_{31\text{-}1068}$/EGFP fusion protein represents a useful tool to visualize the uptake of liposomes by human cells *(58)*.

In the chimeric S-layer proteins $rSbsC_{31-920}$/Bet v1 and $rSbpA_{31-1068}$/Bet v1 carrying the major birch pollen allergen Bet v1 at the C-terminal end, the surface location and functionality of the fused allergen was demonstrated by binding Bet v1-specific IgE or a Bet v1-specific monoclonal mouse antibody *(38,66)*. These fusion proteins can be used to generate arrays for diagnostic test systems to determine the concentration of Bet v1-specific IgE in patients' whole blood, plasma, or serum samples *(67)*. In a recent study, the feasibility of the fusion protein $rSbsC_{31-920}$/Bet v1 as a novel approach to designing vaccines with reduced allergenicity and immunomodulating capacity for specific immunotherapy of type I allergy was described *(67)*. The fusion protein $rSbsC_{31-920}$/Bet v1 contained all relevant Bet v1-specific B and T cell epitopes, but was significantly less efficient at releasing histamine than free Bet v1. In cells of birch pollen-allergic individuals, $rSbsC_{31-920}$/Bet v1 induced interferon (IFN)-γ along with interleukin (IL)-10, but no Th2-like response, as observed after stimulation with free Bet v1 *(67)*.

So far, all of the above described S-layer fusion proteins have been expressed in *E. coli*. Although isolation and purification procedures have been optimized, the use of *E. coli* as an expression host is particularly disadvantageous when S-layer fusion proteins absolutely free of LPS are required, as for vaccine development or for application as an immunoadsorbent in the MDS. To circumvent these problems, the development of a homologous expression system in a Gram-positive host is currently under way for the production of S-layer fusion proteins that are self-assembled on the cell surface. In the case of knock-out mutants that have lost the ability to pyruvylate the SCWP, the S-layer fusion proteins will be secreted into the culture fluid *(46)*.

3.4. Recrystallization of S-Layer Fusion Proteins on Liposomes

Biomolecular self-assembly can be used as a powerful tool for nanoscale engineering. Thus, the development of building blocks for nanobiotechnology, which are based on the fusion of a functional domain to an S-layer protein, is of paramount importance *(4–7,16–18)*.

Liposomes are colloidal, vesicular structures based on (phospho)lipid bilayers or on tetraetherlipid monolayers *(68)*, and they are widely used as delivery systems for enhancing the efficiency of various biologically active molecules and for the transport of therapeutic agents to the site of disease in vivo *(69,70)*. Liposomes can encapsulate water-soluble agents in their aqueous compartment and lipid-soluble substances within the lipid bilayer itself *(71)*. These agents include small molecular drugs used in cancer chemotherapy and genetic drugs as plasmids encoding therapeutic genes *(72)*. Generally, liposomes release their contents by interaction with target cells, either by adsorption, endocytosis, lipid exchange, or fusion *(70,73)*.

In previous studies, the wild-type S-layer proteins of *Bacillus coagulans* E38-66 and *G. stearothermophilus* PV72/p2 were recrystallized on positively charged liposomes composed of dipalmitoylphosphatidylcholine, cholesterol, and hexadecylamine in a molar ratio of 10:5:1 *(74–77)*. Such S-layer-coated liposomes (S-liposomes) with a diameter of 50 to 200 nm represent simple model systems resembling the architecture of artificial virus envelopes. For that reason, S-liposomes could reveal a broad application potential, particularly as drug-delivery systems or in gene therapy *(4)*.

S-liposomes possess significantly enhanced stability towards thermal and mechanical stress factors *(76)*. For generating targeted S-liposomes, the S-layer lattice on liposomes was crosslinked, biotinylated, and exploited for covalent binding of functional macromolecules, like biotinylated antibodies, via the streptavidin-biotin bridge *(77)*. These immuno-S-liposomes comprise several components with specific functions: the liposome as drug carrier, the antibody as homing device, and the S-layer lattice as stabilizing structure for the liposome, as anchoring layer for the antibodies, and, most probably, as protective sheath for prolonged blood circulation times.

To avoid chemical modification reactions and to prevent diffusion of potentially toxic agents through the lipid bilayer into the interior of the vesicles, S-layer fusion proteins incorporating the sequence of core-streptavidin have been constructed. Functional HT were prepared according to the procedure described by Moll et al. *(62)*. In the case of HT that were based on the S-layer protein SbsB, three of the four binding pockets remained accessible for binding biotinylated molecules. After recrystallization of HT on positively charged liposomes, the protein lattice was further functionalized by binding biotinylated peroxidase or biotinylated ferritin *(62)*. Binding of biotinylated ligands to liposomes coated with a monolayer of HT can be used to enable receptor-mediated uptake into human cells. A further promising application potential can be seen in the development of drug targeting and delivery systems based on lipid–plasmid complexes coated with functional HT for transfection of human cells.

Another interesting approach is the generation of a functional chimeric S-layer EGFP fusion protein based on the C-terminally truncated form of SbpA (rSbpA$_{31-1068}$/EGFP) to follow the uptake of S-liposomes into human cells *(58)*. Liposomes coated with a monolayer of rSbpA$_{31-1068}$/EGFP were applied to HeLa cells. After incubation, the cells were fixed and the cell membrane was stained with a transferrin-tetramethylrhodamine conjugate. For evaluation of the cellular localization and the intracellular fate of S-liposomes, confocal laser scanning microscopy (CLSM) was applied and the ongoing interaction between the cell membrane and the green fluorescent S-liposomes could be demonstrated. The CLSM images revealed that most of the S-liposomes were internalized within 2 h of incubation and that the major part entered the HeLa cells by endocytosis *(58)*. To our knowledge, rSbpA$_{31-1068}$/EGFP is the first fusion protein to maintain the ability to fluoresce and to recrystallize into a monomolecular protein lattice. Because of its ability to fluoresce, liposomes coated with rSbpA$_{31-1068}$/EGFP represent a useful tool to visualize the uptake of S-liposomes into eukaryotic cells. With regard to further experiments, the most interesting advantage can be seen in the recrystallization of fusion proteins incorporating EGFP in combination with HT on the same liposome surface. In that case, it would be possible to simultaneously investigate the uptake of these specially coated S-liposomes by target cells and the functionality of transported drugs without the necessity of additional labeling procedures.

3.5. S-Layers as Templates in the Formation of Nanoparticle Arrays

Currently, there is great interest in developing strategies for the deposition of functional biomolecules and nanoparticles in an ordered manner in solution, and on surfaces and interfaces. Their controlled confinement to defined areas of nanometer dimensions is of particularly great interest. Such technologies are seen as key requirements for many applications, including the development of biosensors, biochips and diagnostic systems, biocompatible surfaces and controlled biomineralization, signal processing between cells and integrated circuits, and drug and gene targeting and delivery, as well as molecular electronics and data storage and catalytic processes. This section focuses on the use of S-layers as patterning elements in the formation of ordered arrays of nanoparticles.

3.5.1. In Situ *Synthesis of Nanoparticles on S-Layers*

The use of an S-layer as template for the formation of perfectly ordered nanoparticle arrays was originally reported by Douglas et al. *(78)*. In this approach, S-layer fragments of the archaean *Sulfolobus acidocaldarius* were attached to a solid substrate, shadowed by Ta/W by evaporation, and ion-milled in order to reduce the thickness of the metal film in such a way that nanometric metal clusters remained in the pores of the hexagonally ordered protein lattice. Later on, nanostructured metal layers were prepared in a basically similar approach *(79–82)*.

Recently, it has been demonstrated that self-assembly products or monolayers obtained by recrystallization of S-layer proteins on solid supports induce the formation of CdS particles *(83)*, or gold nanoparticles *(83–84)*. Inorganic CdS super-lattices with either oblique or square lattice symmetry of approx 10-nm repeat distance were fabricated by exposing self-assembly products to Cd(II) solutions, which was followed by slow reaction with H$_2$S.

Precipitation of the inorganic phase was confined to the pores with the result that CdS super-lattices with prescribed symmetries were obtained *(83)*. In addition, two-tier stacks of nanoparticles were formed in the presence of double-layered self-assembly products. The two associated back-to-back S-layers were an accurate register of every fourth row of the double-layered microstructure. This particulate arrangement of organized CdS nanocrystals led to the characteristic stripe pattern, with stripes 16 nm in width and 32 nm apart.

In a similar procedure, a square super-lattice of uniformly sized 4- to 5-nm gold nanoparticles at a 13.1-nm repeat distance was fabricated by exposing the square S-layer lattice formed by SbpA in which sulfhydryl groups had been introduced to a tetrachloroauric(Ill) acid solution *(84)*. Transmission electron microscopic analysis showed that the gold nanoparticles were formed in the pore region during electron irradiation of an initially grainy gold coating covering the whole S-layer lattice. The shape of the gold particles resembled the morphology of the pore region of the square lattice structure (Fig. 10). By electron diffraction and energy dispersive X-ray analysis, the crystallites were identified as gold [Au(0)]. Electron diffraction patterns revealed that the gold nanoparticles were crystalline but in the long range not crystallographically aligned. Similar experiments were performed with a broad range of different metal salt solutions such as $PdCl_2$, $NiSO_4$, $KPtCl_6$, $Pb(NO_3)_2$, and KFe $[Fe(CN)_6]$. Wet chemistry was also applied for producing platinum nanoparticles on the S-layer of *Sporosarcina ureae* *(85–87)*. One morphological unit of the S-layer lattice of *S. ureae* revealed seven platinum cluster sites with a diameter of approx 1.9 nm.

3.5.2. Binding of Preformed Nanoparticles on S-Layers in Ordered Arrays

An alternative approach to the direct chemical synthesis of nanoparticle arrays on S-layer lattices can be seen in the binding of pre-formed nanoparticles to crystalline arrays. Nanoparticles may be either bound by making use of exposed charged groups on the S-layer, such as carboxylic acid or amine groups, or by introducing specific functionalities into the S-layer subunits by genetic engineering. The pattern of bound nanoparticles resembles the spacing and symmetry of the underlying S-layer lattice.

Binding of nanoparticles by making use of the surface chemistry of S-layers involves either noncovalent or covalent bonds. The controlled binding of polycationic ferritin (PCF), which is a positively charged topographical marker with a diameter of 12 nm, on the hexagonal S-layer lattice of *Thermoproteus tenax* was demonstrated nearly two decades ago *(88)*. PCF was bound electrostatically by carboxylic acid groups localized in the center of the hexagonal unit cells. Contrary to the noncovalent binding of PCF, carbodiimide-activation of free carboxylic acid groups that had been intro-

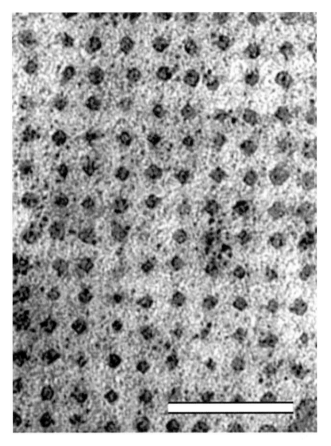

Fig. 10. Transmission electron microscopic image of 4- to 5-nm gold nanoparticles (formed by wet chemistry) regularly arranged on the S-layer of *Bacillus sphaericus* CCM 2177. The superlattice of the gold nanoparticles resembles the square lattice symmetry and the lattice constant (13.1 nm) of the underlying S-layer. Bar, 50 nm.

duced by succinylation of the carbohydrate moiety of an S-layer glycoprotein led to the densest possible packing order and to a hexagonally ordered super-latttice of the 12-nm large ferritin molecules *(89)*. Both approaches were originally developed for binding biologically active molecules, such as enzymes or antibodies, as monomolecular sensing layers for the development of biosensors or affinity matrices *(4–7,15–18)*.

Recently, citrate-stabilized gold nanoparticles (~5 nm in diameter) have been bound as large regular arrays on the inner surface of the square lattice formed by the S-layer proteins from *B. sphaericus* strains *(90,91)*, whereas amine-functionalized nanoparticles were immobilized on the outer surface

(91). The design and expression of S-layer fusion proteins has opened a new area for the functionalization of S-layer lattices, as it has become possible to fuse or insert a defined number of foreign peptide sequences per S-layer subunit. Another approach to introducing a defined number of functionalities can be seen in the use of mutated S-layer proteins having only a single cysteine residue inserted and by coupling of a functional sequence to the sulfhydryl group of the cysteine residue *(30)*. Concerning specific deposition of metals or nanoparticles on S-layer lattices, chimeric proteins with a defined number of appropriate inserted or fused peptide sequences *(92–95)* are currently being constructed. It should be stressed that the structural diversity of S-layer lattices, the possibility of inserting or fusing metal-binding peptides with different specificities to a single S-layer subunit at different amino-acid positions of the primary sequence, the ease of recrystallization of such chimeric S-layer proteins on a broad spectrum of substrates, and their potential for further chemical modifications offer the possibility of creating a wide range of ordered nanoparticle arrays.

4. CONCLUSIONS AND PERSPECTIVES

It is now evident that there are only a few examples in nature where proteins reveal the intrinsic capability to self-assemble into monomolecular arrays in suspension and on surfaces and interfaces. Until now, extensive studies on the use of such systems for nanotechnological applications have only been performed with S-layers. A broad spectrum of basic and applied research on S-layer systems has clearly demonstrated that these unique self-assembly systems represent an ideal patterning element for nanobiotechnological and biomimetic nanotechnology applications. In particular, the repetitive physicochemical properties and isoporosity of S-layer lattices down to the sub-nanometer scale make them unique building blocks and matrices for generating complex and multilayered supramolecular assemblies. The prime attractiveness of such "bottom-up" strategies lies in both their capability of generating uniform nanostructures and the possibility of exploiting such structures at the meso- and macroscopic scale *(6,16,18)*. Moreover, it has been demonstrated that S-layers represent versatile and adaptable self-assembly systems, which can be combined with all major species of biological or synthesized (macro)molecules. More recently, an astonishingly broad spectrum of genetic modifications of S-layer proteins has been achieved, as the incorporation of single or multifunctional domains was made possible without loss of their self-assembly capabilities.

Regarding the physicochemical properties of the inner and outer surface, S-layers are highly anisotropic structures. Thus, it was essential to ensure that recrystallization of mainly genetically engineered S-layer proteins occurred in defined orientation on solid supports, lipid membranes, and liposomes. Biomimetic approaches copying the physicochemical properties of cell-envelope structures supporting native S-layer proteins provide an elegant solution for this problem *(59–61)*.

In particular, S-layer-stabilized lipid membranes resembling the supramolecular construction principle of archaea dwelling under extreme environmental conditions or virus envelopes will enable the exploitation of specific membrane functions under different formats while considerably increasing the lifetime of the membranes as required for biosensors (e.g., lipid chips, drug screening, diagnostics, targeting, and drug-delivery systems) *(16,96–99)*. S-layer fragments or self-assembly products have been demonstrated to be well suited for a geometrically defined covalent attachment of haptens and immunogenic or immuno-stimulating substances. Moreover, haptenated S-layer structures act as strong immuno-potentiators *(100)*. Recently, a fusion protein comprising a C-terminally truncated form of the S-layer protein SbsC and Bet v1 has revealed remarkable immunomodulating capacity *(67)*. It is expected that highly specific immunogenic components with intrinsic targeting and delivery functionalities can be developed by combining recombinant S-layer proteins with the supramolecular construction principle of virus envelopes.

Another line of development is directed to the use of S-layer lattices for non-life science applications. It has been demonstrated that S-layers can be exploited for biological templating and the formation of arrays of metal clusters and nanoparticles as required in molecular electronics, biocatalysis and nonlinear optics. Most recently, it has been shown that spatially well-defined, ordered S-layer arrays at silicon supports can be fabricated by exploiting the simple soft lithography technique, micromolding in capillaries *(91)*.

Despite the fact that a broad spectrum of applications for S-layers has been developed, many other areas may emerge from life and non-life science research. We assume that in particular, data on specific biological functions and biophysical properties of S-layers of organisms growing and surviving under extreme environmental conditions and genetic modifications of such S-layer (glyco)proteins will strongly stimulate applied S-layer research and broaden the application potential *(4–7,16–18)*.

ACKNOWLEDGMENTS

This work was supported by the Austrian Science Fund (FWF), projects P16295-B10, P17170-B10, and P18510-B12, by the EU project NAS-SAP, by the Austrian Federal Ministry of Transport, Innovation and Technology (MNA-Network), by the Erwin Schrödinger Society for Nanosciences, and by the US Air Force Office of Scientific Research (AFOSR), projects F49620-03-1-0222 and "Biocat."

Egelseer et al.

REFERENCES

1. Clark J, Singer EM, Korns DR, Smith SS. Design and analysis of nanoscale bioassemblies. BioTechniques 2004;36:992–1001.
2. Niemeyer CM, Mirkin CA. Nanobiotechnology. Concepts, Applications and Perspectives. Weinheim: Wiley-VCH, 2004.
3. Seeman NC, Belcher AM. Emulating biology: building nanostructures from the bottom up. Proc Natl Acad Sci USA 2002;99:6451–6455.
4. Sleytr UB, Messner P, Pum D, Sára M. Crystalline bacterial cell surface layers (S-layers): from supramolecular cell structure to biomimetics and nanotechnology. Angew Chem Int Ed 1999;38:1034–1054.
5. Sleytr UB, Sára M, Pum D. Crystalline bacterial cell surface layers (S-layers): a versatile self-assembly system. In: Ciferri A, ed. Supramolecular Polymerization. New York, Basle: Marcel Dekker, 2000:177–213.
6. Sleytr UB, Sára M, Pum D, Schuster B, Messner P, Schäffer C. Self-assembly protein systems: microbial S-layers. In: Steinbüchel A, Fahnestock S, eds. Biopolymers, vol. 7. Weinheim: Wiley-VCH, 2003:285–338.
7. Sleytr UB, Pum D, Sára M, Schuster B. Molecular nanotechnology with 2-D protein crystals. In: Nalwa HS, ed. Encyclopedia of Nanoscience and Nanotechnology, vol. 5. San Diego: Academic Press, 2004:693–702.
8. Yeates TO, Padilla JE. Designing supramolecular protein assemblies. Curr Opin Struct Biol 2002;12:464–470.
9. Sleytr UB, Beveridge TJ. Bacterial S-layers. Trends Microbiol 1999;7:253–260.
10. Sleytr UB. Regular arrays of macromolecules on bacterial cell walls: structure, chemistry, assembly and function. Int Rev Cytol 1978;53:1–64.
11. Sleytr UB, Messner P, Pum D, Sára M. Occurrence, location, ultrastructure and morphogenesis of S-layers. In: Sleytr UB, Messner P, Pum D, Sára M, eds. Crystalline Bacterial Cell Surface Proteins. Austin: Landes Company, Academic Press, 1996:5–33.
12. Beveridge TJ, Graham LL. Surface layers of bacteria. Microbiol Rev 1991; 55:684–705.
13. Roberts K, Hills GJ, Shaw, PJ. The structure of algael cell walls. In: Harris JR, ed. Electron Microscopy of Proteins, vol. 3. London: Academic Press, 1982:1–40.
14. Sára M, Egelseer EM. Functional aspects of S-layers. In: Sleytr UB, Messner P, Pum D, Sára M, eds. Crystalline bacterial cell surface proteins. Austin: Academic Press, 1996:103–131.
15. Sára M, Sleytr, UB. S-layer proteins. J Bacteriol 2000;182:859–868.
16. Sára M, Pum D, Schuster B, Sleytr UB. S-layers as patterning elements for application in nanobiotechnology. J Nanosci Nanotechnol 2005;5:1939–1953..
17. Sleytr UB, Pum D, Schuster B, Sára M. Molecular nanotechnology and nanobiotechnology with two-dimensional protein crystals (S-layers). In: Rosoff M. ed. Nano-Surface Chemistry. New York, Basle: Marcel Dekker, 2001:333–389.
18. Sleytr UB, Sára M, Pum D, Schuster B. Crystalline bacterial cell surface layers (S-layers):a versatile self-assembly system. In: Ciferri A, ed. Supramolecular Polymers. Boca Raton: CRC Press, 2005: 583–612.

19. Robards AW, Sleytr UB. Low temperature methods in biological electron microscopy. In: Glauert AM, ed. Practical Methods in Electron Microscopy, vol. 10. Amsterdam: Elsevier, 1985.

20. Amos LA, Henderson R, Unwin PNT. Three-dimensional structure determination by electron microscopy of two-dimensional crystals. Progr Biophys Mol Biol 1982;39:183–231.

21. Baumeister W, Engelhardt H. Three-dimensional structure of bacterial surface layers. In: Harris JR, Horne RW, eds. Electron Microscopy of Proteins, vol. 6. London: Academic Press, 1987:109–154.

22. Hovmöller S, Sjögren A, Wang DN. The structure of crystalline bacterial surface layers. Prog Biophys Mol Biol 1988;51:131–163.

23. Györvary ES, Stein O, Pum D, Sleytr UB. Self-assembly and recrystallization of bacterial S-layer proteins at silicon supports imaged in real time by atomic force. J Microscopy 2003;212:300–306.

24. Karrasch S, Hegerl R, Hoh J, Baumeister W, Engel A. Atomic force microscopy produces faithful high-resolution images of protein surfaces in an aqueous environment. Proc Natl Acad Sci USA 1994;91:836–838.

25. Pum D, Sleytr UB. Monomolecular reassembly of a crystalline bacterial cell surface layer (S-layer) on untreated and modified silicon surfaces. Supramol Sci 1995;2:193–197.

26. Müller DJ, Baumeister W, Engel A. Conformational change of the hexagonally packed intermediate layer of *Deinococcus radiodurans* monitored by atomic force microscopy. J Bacteriol 1996;178:3025–3030.

27. Scheuring S, Stahlberg H, Chami M, Houssin C, Rigaud JL, Engel A. Charting and unzipping the surface layer of *Corynebacterium glutamicum* with the atomic force microscope. Mol Microbiol 2002;44:675–684.

28. Stroh CM, Ebner A, Geretschläger M, et al. Simultaneous topography and recognition imaging using force microscopy. Biophys J 2004;87:1981–1990.

29. Ilk N, Kosma P, Puchberger M, et al. Structural and functional analyses of the secondary cell wall polymer of *Bacillus sphaericus* CCM 2177 that serves as an S-layer-specific anchor. J Bacteriol 1999;181:7643–7646.

30. Howorka S, Sára M, Wang Y, et al. Surface-accessible residues in the monomeric and assembled forms of a bacterial surface layer protein. J Biol Chem 2000;275:37,876–37,886.

31. Sára M, Dekitsch C, Mayer HF, Egelseer EM, Sleytr, UB. Influence of the secondary cell wall polymer on the reassembly, recrystallization and stability properties of the S-layer protein from *Bacillus stearothermophilus* PV72/p2. J Bacteriol 1998;180:4146–4153.

32. Boot HJ, Kolen CP, Pouwels PH. Identification, cloning, and nucleotide sequence of a silent S-layer protein gene of *Lactobacillus acidophilus* ATCC 4356 which has extensive similarity with the S-layer protein gene of this species. J Bacteriol 1995;177:7222–7230.

33. Bröckl G, Behr M, Fabry S, et al. Analysis and nucleotide sequence of the genes encoding the surface-layer glycoproteins of the hyperthermophilic methanogens *Methanothermus fervidus* and *Methanothermus sociabilis*. Eur J Biochem 1991;199:147–152.

34. Messner P, Schäffer C. Surface layer glycoproteins of bacteria and archaea. In: Doyle RJ, ed. Glycomicrobiology. New York: Kluwer Academic/Plenum Publishers, 2000:93–125.
35. Schäffer C, Messner P. Glycobiology of surface layer proteins. Biochimie 2001;83:591–599.
36. Schäffer C, Messner, P. Prokaryotic glycoproteins. In: Herz W, Falk H, Kirby GW, Moore RE, Tamm C, eds. Progress in the Chemistry of Organic Natural Products, vol. 85. Wien, New York: Springer, 2003:51–124.
37. Huber C, Ilk N, Rünzler D, et al. The three S-layer-like homology motifs of the S-layer protein SbpA of *Bacillus sphaericus* CCM 2177 are not sufficient for binding to the pyruvylated secondary cell wall polymer. Mol Microbiol 2005;55:197–205.
38. Ilk N, Völlenkle C, Egelseer EM, Breitwieser A, Sleytr UB, Sára M. Molecular characterization of the S-layer gene, *sbpA*, of *Bacillus sphaericus* CCM 2177 and production of a functional S-layer fusion protein with the ability to recrystallize in a defined orientation while presenting the fused allergen. Appl Environ Microbiol 2002;68:3251–3260.
39. Jarosch M, Egelseer EM, Huber C, et al. Analysis of the structure-function relationship of the S-layer protein SbsC of *Bacillus stearothermophilus* ATCC 12980 by producing truncated forms. Microbiology 2001;147:1353–1363.
40. Steindl C, Schäffer C, Wugeditsch T, et al. The first biantennary bacterial secondary cell wall polymer and its influence on S-layer glycoprotein assembly. Biochem J 2002;368:483–494.
41. Cava F, de Pedro MA, Schwarz H, Henne A, Berenguer J. Binding to pyruvylated compounds as an ancestral mechanism to anchor the outer envelope in primitive bacteria. Mol Microbiol 2004;52:677–690.
42. Chauvaux S, Matuschek M, Beguin P. Distinct affinity of binding sites for S-layer homologous domains in *Clostridium thermocellum* and *Bacillus anthracis* cell envelopes. J Bacteriol 1999;181:2455–2458.
43. Lemaire M, Miras I, Gounon P, Beguin P. Identification of a region responsible for binding to a cell wall within the S-layer protein of *Clostridium thermocellum*. Microbiology 1998;144:211–217.
44. Mader C, Huber C, Moll D, Sleytr UB, Sára M. Interaction of the crystalline bacterial cell surface layer protein SbsB and the secondary cell wall polymer of Geobacillus stearothermophilus PV72 assessed by real-time surface plasmon resonance biosensor technology. J Bacteriol 2004;186:1758–1768.
45. Mesnage S, Tosi-Couture E, Mock M, Fouet A. The S-layer homology domain as a means for anchoring heterologous proteins on the cell surface of *Bacillus anthracis*. J Appl Microbiol 1999;87:256–260.
46. Mesnage S, Fontaine S, Mignot T, Delepierre M, Mock M, Fouet A. Bacterial SLH domain proteins are non-covalently anchored to the cell surface via a conserved mechanism involving wall polysaccharide pyruvylation. EMBO J 2000;19:4473–4484.
47. Egelseer EM, Leitner K, Jarosch M, et al. The S-layer proteins of two *Bacillus stearothermophilus* wild-type strains are bound via their N-terminal region to a secondary cell wall polymer of identical chemical composition. J Bacteriol 1998;180:1488–1495.

48. Jarosch M, Egelseer EM, Mattanovich D, Sleytr UB, Sára M. S-layer gene *sbsC* of *Bacillus stearothermophilus* ATCC 12980: molecular characterization and heterologous expression in *Escherichia coli*. Microbiology 2000;146:273–281.

49. Egelseer, EM, Danhorn T, Pleschberger M, Hotzy C, Sleytr UB, Sára M. Characterization of an S-layer glycoprotein produced in the course of S-layer variation of *Bacillus stearothermophilus* ATCC 12980 and sequencing and cloning of the *sbsD* gene encoding the protein moiety. Arch Microbiol 2001;177:70–80.

50. Schäffer C, Kählig H, Christian R, Schulz G, Zayni S, Messner P. The diacetamido-dideoxyuronic-acid-containing glycan chain of *Bacillus stearothermophilus* NRS 2004/3a represents the secondary cell wall polymer of wild-type *B. stearothermophilus* strains. Microbiology 1999;145:1575–1583.

51. Rünzler D, Huber C, Moll D, Köhler G, Sára M. Biophysical characterization of the entire bacterial surface layer protein SbsB and its two distinct functional domains. J Biol Chem 2004;279:5207–5215.

52. Pavkov T, Oberer M, Egelseer EM, Sára M, Sleytr UB, Keller W. Crystallization and preliminary structure determination of the C-terminal truncated domain of the S-layer protein SbsC. Acta Crystallogr D Biol Crystallogr 2003;59: 1466–1468.

53. Dworkin J, Blaser MJ. Molecular mechanisms of *Campylobacter fetus* surface layer protein expression. Mol Microbiol 1997;26:433–440.

54. Sára M, Kuen B, Mayer HF, Mandl F, Schuster KJ, Sleytr UB. Dynamics in oxygen-induced changes in S-layer protein synthesis from *Bacillus stearothermophilus* PV72 and the S-layer-deficient variant T5 in continuous culture and studies of the cell wall composition. J Bacteriol 1996;178:2108–2117.

55. Scholz H, Riedmann E, Witte A, Lubitz W, Kuen B. S-layer variation in *Bacillus stearothermophilus* PV72 is based on DNA rearrangement between the chromosome and the naturally occurring megaplasmids. J Bacteriol 2001; 183:1672–1679.

56. Jakava-Viljanen M, Avall-Jääskeläinen S, Messner P, Sleytr UB, Palva A. Isolation of three new surface layer protein genes (*slp*) from *Lactobacillus brevis* ATCC 14869 and characterization of the change in their expression under aerated and anaerobic conditions. J Bacteriol 2002;184:6786–6795.

57. Huber C, Liu J, Egelseer EM, et al. Heterotetramers formed by an S-layer-streptavidin fusion protein and core streptavidin as nanoarrayed template for biochip development. Small 2006;2:142–150.

58. Ilk N, Küpcü S, Moncayo G, et al. A functional chimaeric S-layer-enhanced green fluorescent protein to follow the uptake of S-layer-coated liposomes into eukaryotic cells. Biochem J 2004;379:441–448.

59. Pleschberger M, Neubauer A, Egelseer EM, et al. Generation of a functional monomolecular protein lattice consisting of an S-layer fusion protein comprising the variable domain of a camel heavy chain antibody. Bioconjug Chem 2003;14:440–448.

60. Pleschberger M, Saerens D, Weigert D, et al. An S-layer heavy chain camel antibody fusion protein for generation of a nanopatterned sensing layer to detect the prostate-specific antigen by surface plasmon resonance technology. Bioconjug Chem 2004;15:664–671.

61. Völlenkle C, Weigert S, Ilk N, et al. Construction of a functional S-layer fusion protein comprising an immunoglobulin G-binding domain for development of specific adsorbents for extracorporeal blood purification. Appl Environ Microbiol 2004;70:1514–1521.

62. Moll D, Huber C, Schlegel B, Pum D, Sleytr UB, Sára M. S-layer-streptavidin fusion proteins as template for nanopatterned molecular arrays. Proc Natl Acad Sci USA 2002;99:14,646–14,551.

63. Schäffer C, Wugeditsch T, Kählig HP, Scheberl A, Zayni S, Messner P. The surface layer (S-layer) glycoprotein of *Geobacillus stearothermophilus* NRS 2004/3a. Analysis of its glycosylation. J Biol Chem 2002;277:6230–6239.

64. Ries W, Hotzy C, Schocher I, Sleytr UB, Sára M. Evidence that the N-terminal part of the S-layer protein from *Bacillus stearothermophilus* PV72/p2 recognizes a secondary cell wall polymer. J Bacteriol 1997;179:3892–3898.

65. Weber V, Weigert S, Sára M, Sleytr UB, Falkenhagen D. Development of affinity microparticles for extracorporeal blood purification based on crystalline bacterial cell surface proteins. Ther Apher 2001;5:433–438.

66. Breitwieser A, Egelseer EM, Moll D, et al. A recombinant bacterial cell surface (S-layer)-major birch pollen allergen-fusion protein (rSbsC/Bet v1) maintains the ability to self-assemble into regularly structured monomolecular lattices and the functionality of the allergen. Protein Eng 2002;15:243–249.

67. Bohle B, Breitwieser A, Zwölfer B, et al. A novel approach to specific allergy treatment: the recombinant fusion protein of a bacterial cell surface (S-layer) protein and the major birch pollen allergen Bet v1 (rSbsC-Bet v1) combines reduced allergenicity with immunomodulating capacity. J Immunol 2004;172: 6642–6648.

68. Crommelin D, Storm G. Liposomes: from the bench to the bed. J Liposome Res 2003;13:33–36.

69. Lasic DD, Papahadjopoulos D. Liposomes revisited. Science 1995;267: 1275–1276.

70. Torchilin VP. Recent advances with liposomes as pharmaceutical carriers. Nat Rev Drug Discov 2005;4:145–160.

71. Lasic DD. Novel applications of liposomes. Trends Biotechnol 1998;16: 307–321.

72. Templeton NS, Lasic DD. New directions in liposome gene delivery. Mol Biotechnol 1999;11:175–180.

73. Ostro MJ, Cullis PR. Use of liposomes as injectable-drug delivery systems. Am J Hosp Pharm 1989;46:1576–1587.

74. Küpcü S, Sára M, Sleytr UB. Liposomes coated with crystalline bacterial cell surface protein (S-layer) as immobilization structures for macromolecules. Biochim Biophys Acta 1995;1235:263–269.

75. Küpcü S, Lohner K, Mader C, Sleytr UB. Microcalorimetric study on the phase behaviour of S-layer coated liposomes. Mol Membr Biol 1998;15:69–74.

76. Mader C, Küpcü S, Sára M, Sleytr UB. Stabilizing effect of an S-layer on liposomes towards thermal or mechanical stress. Biochim Biophys Acta 1999; 1418:106–116.

77. Mader C, Küpcü S, Sleytr UB, Sára M. S-layer-coated liposomes as a versatile system for entrapping and binding target molecules. Biochim Biophys Acta 2000;1463:142–150.

78. Douglas K, Clark NA, Rothschild KJ. Nanometer molecular lithography. Appl Phys Lett 1986;48:676–678.
79. Douglas K, Devaud G, Clark NA. Transfer of biologically derived nanometer-scale patterns to smooth substrates. Science 1992;257:642–644.
80. Winningham TA, Gillis HP, Choutov DA, Martin KP, Moore JT, Douglas K. Formation of ordered nanocluster arrays by self-assembly on nanopatterned Si(100) surfaces. Surf Sci 1998;406:221–228.
81. Panhorst M, Brückl H, Kiefer B, Reiss G, Santarius U, Guckenberger R. Formation of metallic surface structures by ion etching using a S-layer template. J Vac Sci Technol B 2001;19:722–724.
82. Malkinski L, Camley RE, Celinski Z, Winningham TA, Whipple SG, Douglas K. Hexagonal lattice of 10-nm magnetic dots. J Appl Phys 2003; 93:7325–7327.
83. Shenton W, Pum D, Sleytr UB, Mann S. Biocrystal templating of CdS super-lattices using self-assembled bacterial S-layers. Nature 1997;389:585–587.
84. Dieluweit S, Pum D, Sleytr UB. Formation of a gold superlattice on an S-layer with square lattice symmetry. Supramol Sci 1998;5:15–19.
85. Mertig M, Kirsch R, Pompe W, Engelhardt H. Fabrication of highly oriented nanocluster arrays by biomolecular templating. Eur Phys J D 1999;9:45–48.
86. Mertig M, Wahl R, Lehmann M, Simon P, Pompe W. Formation and manipulation of regular metallic nanoparticle arrays on bacterial surface layers: an advanced TEM study. Eur Phys J D 2001;16:317–320.
87. Pompe W, Mertig M, Kirsch R, et al. Formation of metallic nanostructures on biomolecular templates. Zeitschrift für Metallkunde 1999;90:1085–1091.
88. Messner P, Pum D, Sára M, Stetter KO, Sleytr UB. Ultrastructure of the cell envelope of the archaebacteria *Thermoproteus tenax* and *Thermoproteus neutrophilus*. J Bacteriol 1986;166:1046–1054.
89. Hall SR, Shenton W, Engelhardt H, Mann S. Site-specific organization of gold nanoparticles by biomolecular templating. Chem Phys Chem 2001;3:184–186.
90. Sára M, Küpcü S, Sleytr UB. Localization of the carbohydrate residue of the S-layer glycoprotein from *Clostridium thermohydrosulfuricum* L111-69. Arch Microbiol 1989;151:416–420.
91. Györvary ES, O'Riordan A, Quinn A, Redmond G, Pum D, Sleytr UB. Biomimetic nanostructure fabrication: non-lithographic lateral patterning and self-assembly of functional bacterial S-layers at silicon supports. Nano Lett 2003;3:315–319.
92. Kotrba P, Doleckova L, de Lorenzo V, Ruml T. Enhanced bioaccumulation of heavy metal ions by bacterial cells due to surface display of short metal binding peptides. Appl Environ Microbiol 1999;65:1092–1098.
93. Klaus T, Joerger R, Olsson E, Granquist CG. Silver-based crystalline nanoparticles, microbially fabricated. Proc Natl Acad Sci USA 1999;96:13,611–13,614.
94. Rajesh RN, Stringer SJ, Agarwal G, Jones SE, Stone MO. Biomimetic synthesis and patterning of silver nanoparticles. Nat Mat 2002;1:169–172.
95. Rajesh RN, Jones SE, Murray JC, McAuliffe JC, Vaia RA, Stone MO. Peptide templates for nanoparticles synthesis derived from PCR-driven phage display. Adv Funct Mat 2004;14:1–7.
96. Schuster B, Sleytr UB. Single channel recordings of α-hemolysin reconstituted in S-layer supported lipid bilayers. Bioelectrochemistry 2002;55:5–7.

97. Schuster B, Weigert S, Pum D, Sára M, Sleytr UB. New method for generating tetraether lipid membranes on porous supports. Langmuir 2003;19:2392–2397.

98. Gufler PC, Pum D, Sleytr UB, Schuster B. Highly robust lipid membranes on crystalline S-layer supports investigated by electrochemical impedance spectroscopy. Biochim Biophys Acta–Biomem 2004;1661:154–165.

99. Toca-Herrera JL, Krastev R, Bosio V, et al. Recrystallization of bacterial S-layers on flat polyelectrolyte surfaces and hollow polyelectrolytes capsules. Small 2005; 1:339–348.

100. Jahn-Schmid B, Graninger M, Glozik M, et al. Immunoreactivity of allergen (Bet v1) conjugated to crystalline bacterial cell surface layers (S-layers). Immunotechnology 1996;2:103–113.

III
BIONANOELECTRONICS AND NANOCOMPUTING

5

Photoinduced Electron Transport in DNA
Toward Electronic Devices Based on DNA Architecture

Hans-Achim Wagenknecht

Summary

   This chapter briefly summarizes important and basic aspects related to electron transport processes in DNA over long distances. Despite this broad knowledge,
DNA research is still far from a profound and clear understanding of the electronic
properties and electronic interactions in DNA that are crucial for any nanobiotechnological application. In the past, DNA-mediated charge transport has been a subject of
considerable interest with biological relevance in the formation and repair of lesions
and damage in DNA. The most recent developments underscore the significance of
DNA or DNA-like architectures for the development of electronic devices on the
nanoscale. It is clear that there is a great potential for applications of DNA-mediated
charge transport processes in new DNA assays and microarrays for biotechnology, as
well as DNA-inspired devices for nanotechnology.

Key Words: Charge transfer; chromophore; DNA; DNA damage; electron transfer;
fluorescence; hopping; nucleotide; radical; superexchange.

1. DNA AS A FUNCTIONAL Π-SYSTEM
FOR NANOBIOTECHNOLOGY

   Since the first suggestion that the regularly stacked B-form DNA might
serve as a pathway for charge transport was published more than 40 years ago
(1), the potential of DNA as a conducting biopolymer in nanotechnology applications has been a highly controversial scientific dispute. Remarkably, DNA
was considered to be either a molecular wire, semiconductor, or insulator *(2–4)*.
These are the controversial results of both experimental and theoretical work
(2–5). Motivated by the biological relevance of the routes to DNA damage and
also by the controversy, the interest in this subject grew enormously in the
scientific community over the last two decades *(5)*. Research groups of different

From: *NanoBioTechonology: BioInspired Devices and Materials of the Future*
Edited by: Oded Shoseyov and Ilan Levy © Humana Press Inc., Totowa, NJ

chemistry subdisciplines, such as organic chemistry, inorganic chemistry, physical chemistry, and biochemistry, as well as biologists, physicists, and material scientists contributed significantly to this research topic.

Based on these experiments and results, it was possible to obtain a clear picture of charge transport phenomena in DNA. The initially extreme controversy has been resolved by the description of different mechanistic aspects, mainly the superexchange and the hopping mechanism *(6)*. It is now clear that DNA-mediated charge transport processes occur on fast and ultrafast time scales and yield chemical reactions over long distances *(5)*. The oxidative type of DNA-mediated charge transport processes has a significant relevance in the formation of oxidative damage to the DNA, which results in mutagenesis, apoptosis, or cancer *(7–10)*. On the other hand, excess electron transport processes play a growing role in the development of electrochemical DNA chips, e.g., for the detection of single base mutations *(11)*. Moreover, knowledge about excess electron transport in DNA has the potential to be considered for the development for nanotechnological applications, such as new DNA-based electronic devices *(12)*. In this case, the research strays from the natural biological role of DNA by considering this biopolymer a supramolecular architecture that features important properties for nanowires, such as the regular linear structure and the self-complementarity encoded by the DNA base sequence.

2. PHOTOINDUCED OXIDATIVE HOLE VS REDUCTIVE ELECTRON TRANSPORT IN DNA

In order to study charge transport chemistry in DNA in a photoinduced fashion, it is crucial to modify oligonucleotides with suitable chromophores. After the development of the automated DNA phosphoramidite chemistry, the whole spectrum of different methods and protocols for oligonucleotide modifications were applied, developed, and further improved dramatically in order to prepare a structurally well defined functional π-system based on DNA *(13,14)*. Using such well defined DNA donor–acceptor systems, a systematic measurement of the distance dependence and the base sequence dependence of the charge transport processes became accessible.

DNA-mediated charge transport processes can be categorized as either oxidative hole transport or reductive electron transport processes (Fig. 1) *(15,16)*. The description of hole transport is misleading because it includes principally also an electron transport, but in the opposite direction. Hence, both processes are electron migration reactions but with different orbital control. The oxidative hole transport is HOMO-controlled whereas the excess electron transport is LUMO-controlled. This makes clear that this categorization is not just a formalism about the difference in direction of the electron.

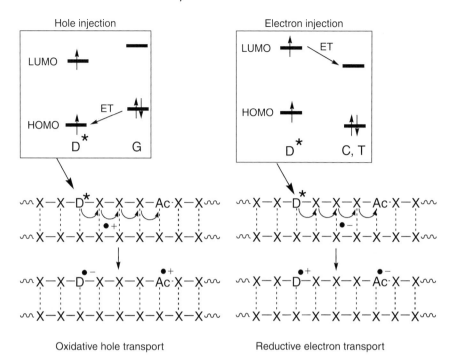

Fig. 1. Comparison of photoinduced oxidative hole transport (HOMO-control) and reductive electron transport (LUMO-control) in DNA. D, donor; A, acceptor; ET, electron transfer.

With respect to oxidative DNA damage as the biological motivation, most research was focused initially on the photochemically induced oxidation of DNA, and furthermore, on the mobility of the created positive radical charge. Such processes can be considered *oxidative hole transport.* On the other hand, the mobility of an excess electron in the DNA base stack can be described as *reductive* or *excess electron transport.* This process occurs if the photoexcited electron of the charge donor is injected into the DNA and then transferred to the final electron acceptor. Several proposals (such as the lack of DNA damage as the result of reductive electron transport and the involvement of all base pairs [C-G and T-A] as intermediate charge carriers) *(15)* strongly support the idea that this type of charge transport has a great potential for application in new nanodevices based on DNA or DNA-inspired architectures. For such applications, a profound understanding of the mechanism and dynamics of excess electron transport processes is crucial. Unfortunately, the mechanistic details of excess electron migration remain somewhat unclear, in contrast to the broad and detailed knowledge about the oxidative charge transport processes.

3. MECHANISM OF LONG-RANGE OXIDATIVE HOLE TRANSPORT IN DNA

For the mechanistic description of DNA-mediated charge transport phenomena over long distances, the hopping model for conducting polymers was applied *(6)*. This interpretation points out that DNA is considered a functional π-biopolymer. Among the four different DNA bases, guanine (G) is most easily oxidized *(17,18)*. Hence, the G radical cation plays the role of the intermediate charge carrier during the hopping process in DNA. The bridge levels of DNA (especially that of G) and the level of the photoexcited donor must be similar in order to inject a hole thermally into the DNA base stack. Subsequently, the positive charge hops from G to G, and can finally be trapped at a suitable charge acceptor. If each single hopping step occurs over the same distance, then the dynamics of hopping display a shallow distance dependence with respect to the number of hopping steps N *(6)*:

$$k_{ET} = P^* N^\eta$$

The value of η lies between 1 and 2 and represents the influence of the medium DNA. Each hopping step itself is a superexchange process *(6)* through the intervening adenine(A)-thymine(T) base pairs, but only if the A-T stretch is not too long (see below). The rate for a single hopping step from G to GG was determined to be $k_{HOP} = 10^6 - 10^8$ s^{-1} *(19)*. Using the site-specific binding of methyltransferase *HhaI* to DNA, a lower limit for hole hopping in DNA k_{HT} >10^6 s^{-1} was measured over 50 Å through the base stack *(20)*. Based on the absence of a significant distance dependence, it was concluded that hole hopping through the DNA is not a rate-limiting step.

Recently, it was supported with experimental evidence that adenines can also play the role of intermediate carriers of the positive charge (Fig. 2) *(21,22)*. Such A-hopping can occur if G is not present within the sequential context, mainly in stretches with at least four subsequent A-T base pairs between the guanines. The oxidation of A by $G^{\bullet+}$ is endothermic. With respect to the low efficiency of this hole hopping step, it was suggested that, once $A^{\bullet+}$ has been generated, the A-hopping proceeds quickly. In fact, the rate of A-hopping has been determined to be $k_{HT} = 10^{10}$ s^{-1} *(23)*. Moreover, it is remarkable that hole transport over eight A-T base pairs is nearly as efficient as the hole transport over only two A-T base pairs *(24)*. In comparison to G-hopping, A-hopping proceeds more quickly, more efficiently, and almost without any distance dependence. Recent calculations support these properties of A-hopping *(25)*. It is known from γ-radiolysis studies that the one-electron oxidation of DNA bases has drastic effects on their acidity. In theory, proton transfer could occur on time scales comparable to charge transport processes and thus have the potential to influence significantly the charge transport

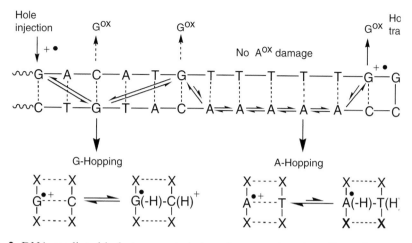

Fig. 2. DNA-mediated hole transport via hopping over domains of guanines or adenines.

efficiency as a result of the separation of spin and charge *(26)*. The question of proton transfer in oxidized $G^{•+}$-C base pairs is crucial for the understanding of hole transport in DNA. The pK_a value of $G^{•+}$ is approx 3.9 *(27)*, and the pK_a value of the complementary DNA base cytosine (C) is very similar (4.5) *(27)*. Hence, there is likely an equally distributed protonation–deprotonation equilibrium in a one-electron oxidized $G^{•+}$-C base pair that is principally reversible but could potentially interrupt hole hopping in DNA. In fact, measurements of the kinetic isotope effect of hole transport in DNA have been performed and provide some evidence for a coupling between hole transport and proton transfer processes *(28)*. The situation is different in the A-T base pair. Oxidized adenine $A^{•+}$ represents a powerful acid with a pK_a of ≤ 1 and T shows an extremely low basicity [pK_a of $T(H)^+$ is -5] *(27)*. Taken together, these facts make it clear that charge and spin keeps located on the A in a $A^{•+}$-T base pair (Fig. 2).

4. MECHANISMS OF REDUCTIVE ELECTRON TRANSPORT IN DNA

As mentioned above, the mechanistic details of excess electron migration in DNA are not yet completely clear. This stands in contrast to the broad knowledge about the oxidative hole hopping, as described in the previous paragraph. The lack of knowledge about the reductive processes has been filled at least partially during the last 3–4 years, but the determination of the electron transport rates in a well defined and suitable donor–acceptor DNA system is still elusive *(16)*. In principle, the mechanisms of oxidative hole transport have been transferred to the problem of excess electron transport

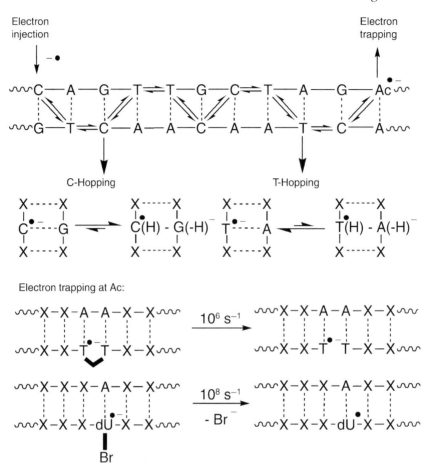

Fig. 3. DNA-mediated reductive electron transport via hopping over thymines and cytosines.

and, accordingly, a hopping mechanism was proposed over long distances (Fig. 3) *(15)*. Even before any experimental work was performed, it was suggested that such an electron hopping involves all base pairs (T-A and C-G) and that the pyrimidine radical anions $C^{\bullet-}$ and $T^{\bullet-}$ function as intermediate electron carriers *(15)*. This proposal is based on thermodynamic arguments represented by the trend for the reducibility of DNA bases: $T \sim C \gg A > G$. This correlation makes clear that the pyrimidine bases C and T are reduced more easily than the purine bases A and G *(18,29)*. It is evident that the situation within the DNA could be significantly different from that of the isolated monomer nucleosides, because calculations have shown that 5′-TTT-3′ and 5′-TCT-3′ should serve as the strongest electron sinks *(30)*. Seidel et al.

measured a complete set of polarographic potentials that are in the range between −2.04 V and −2.76 V *(18)*. In this context, the measured value E(dC/dC$^{\bullet-}$) ≈ E(dT/dT$^{\bullet-}$) ≈−1.1 V provided by Steenken et al. *(31)* is difficult to understand and could reflect the result of a proton-coupled electron transfer. Thus, it is likely that the −1.1 V potential corresponds to E[dC/ dC(H)$^{\bullet}$] and E[dT/ dT(H)$^{\bullet}$]. However, C$^{\bullet-}$ and T$^{\bullet-}$ exhibit a large difference by means of their basicity, which also was described by Steenken *(27)*. Thus, protonation of C$^{\bullet-}$ and T$^{\bullet-}$ by the complementary DNA bases or the surrounding water molecules could possibly interfere with the electron transfer hopping.

5. PHOTOCHEMICAL DNA ASSAYS FOR EXCESS ELECTRON TRANSPORT IN DNA

Before the more recent photochemical DNA assays were published, most knowledge about reductive electron transport in DNA came from γ-pulse radiolysis studies using DNA samples doped with intercalated and randomly spaced electron traps *(32)*. The major disadvantage of this principal experimental setup is that the electron injection and the electron trapping does not occur regioselectively. Nevertheless, it became clear that above 170 K, the electron transport mechanism follows a thermally activated process. These results represented the first experimental support for the proposal of an electron hopping process as discussed in the previous paragraph.

In the most recently developed photochemical DNA assays for electron transport, flavine *(33,34)*, naphthalene diamine *(35–37)*, stilbenediether *(38)*, phenothiazine *(39)*, and pyrene *(40,41)* derivatives have been used as chromophores and photoexcitable electron donors, which were covalently attached to oligonucleotides. It is important to note that they differ in their structure and, more signficantly, in their redox properties.

The major part of these recent photochemical assays focus on the chemical trapping of the excess electron and the corresponding chemical analysis of the resulting DNA strand cleavages. Based on the knowledge about the repair mechanism of DNA photolyases *(42)*, a DNA assay for the investigation of excess electron migration consisting of two DNA modifications was developed by Carell et al. (Fig. 4) *(33,34)*. The first modification was that a flavine (Fl) moiety was incorporated synthetically as an artificial DNA base. It is important to note that the reduced, deprotonated, and photoexcited flavine has a redox potential of −2.8 V capable of reducing all four DNA bases *(43)*. Additionally, a newly designed T-T dimer (T^T) lacking the connecting phosphodiester bridge between the 3′- and the 5′-hydroxy groups of the two adjacent ribose moieties was incorporated. Hence, cycloreversion of the T^T dimer as a result of the one-electron reduction yields a strand break of the oligonucleotide, which can be analyzed and quantified by high-performance

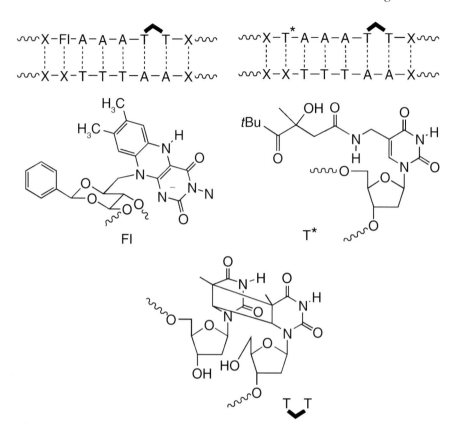

Fig. 4. Photochemical assays for the investigation of reductive electron transport in DNA, part A.

liquid chromatography (HPLC) analysis. It was shown that the amount of T^T dimer cleavage depends rather weakly on the distance to the flavine, indicating a thermally activated electron hopping process exhibiting a shallow distance dependence.

More recently, Giese et al. used a special T derivative (T′) that cleaves in a Norrish-I-type fashion upon photoexcitation and yields the injection of an excess electron into the DNA base stack (Fig. 4) *(29)*. In collaboration with Carell's group, they used the T^T dimer as the chemical probe to measure the efficiency of the reductive electron transport. Very remarkably, they showed that just a single injected electron can cleave and repair more than one T^T dimer.

Another important photochemical DNA assay comes from Rokita's group (Fig. 5) *(35–37)*. They used naphthalene diamine derivatives (Nd), which had been attached to abasic sites in DNA. This photochemical electron donor

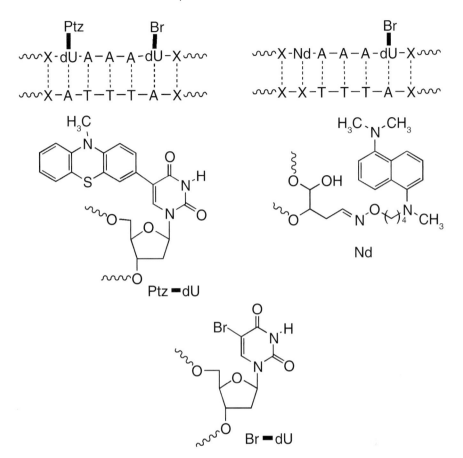

Fig. 5. Photochemical assays for the investigation of reductive electron transport in DNA, part B.

is comparable to the above-mentioned flavine derivative because its reduction potential in the excited state (−2.6 V) also allows the reduction of all four DNA bases. In contrast to Carell's work, 5-bromo-2′-deoxyuridine (BrdU) was used as the chemical electron trap. It is known that BrdU undergoes a chemical modification after its one-electron reduction, which can be analyzed by piperidine-induced strand cleavage *(44,45)*. Hence, the quantification of the strand cleavage yields information about the ET efficiency. It is important to point out that, based on reduction potentials, BrdU is not a significantly better electron acceptor; hence, BrdU represents a kinetic electron trap *(46)*.

In our work, phenothiazine (Ptz) was used as the photochemical electron donor (Fig. 5) *(39)*. The reduction potential of Ptz in the excited state (−2.0 V; *37*) allows the photoreduction of T and C within DNA, but not A and G. Accordingly, we synthesized a Ptz-modified uridine (PtzdU) and incorporated it into oligonucleotides. In the sequences of the corresponding duplexes, the BrdU group as the electron acceptor was placed either two, three, or four base pairs away from the PtzdU group as the photoexcitable electron donor. The intervening base pairs were chosen to be either T-A or C-G. Using this approach, we elucidated the distance and sequence dependence of DNA-mediated ET efficiency. Interestingly, the DNA duplexes with the intervening T-A base pairs show a significantly higher cleavage efficiency compared to the DNA duplexes with the intervening C-G base pairs. In fact, the cleavage efficiency over three A-T base pairs is comparable to that over just one single C-G base pair. From these strand cleavage experiments, it becomes clear that in our assay, T-A base pairs transport electrons more efficiently than C-G base pairs. This implies that $C^{•-}$ is not likely to play a major role as an intermediate electron carrier.

In conclusion, just two different chemical electron traps have been developed and applied—the T^T dimer and BrdU (see Fig. 3). Both chemical probes yield strand cleavage at the site of electron trapping, BrdU only after piperidine treatment at elevated temperature. The main difference between these two electron traps is the time regimes of the radical clocks. Although the exact dynamic behavior has only been examined with the isolated nucleoside monomers, the rates are significantly different: the radical anion of Br-dU loses its bromide at a rate of 7 ns^{-1} (44,45), whereas the radical anion of the T^T dimer splits at a much slower rate of approx 0.6 µs^{-1} (47). This striking difference has important consequences for the elucidation of the distance dependence and DNA base sequence dependence of the excess electron transport efficiency. Hence, it is not surprising that in the assay of Carell et al., the amount of T^T dimer cleavage depends rather weakly on the distance to the electron donor (33,34). On the other hand, using BrdU as the electron trap, a significant dependence of the strand cleavage efficiency on the intervening DNA base sequence has been observed by Rokita's group (35–37) and ours (39). Thus, BrdU seems to be more suitable as a kinetic electron trap because the time resolution is better for exploring the details of a presumably ultrafast electron transport process.

By now, only Lewis, group (38) and ours (40,41,48,49) have focused on the study of the dynamics of DNA-mediated electron transport processes using either stilbene diether-capped DNA hairpins (Sd) or pyrene-modified DNA (Py) duplexes, respectively (Fig. 6). Both chromophores have excited state reduction potentials (Sd: −2.3 V [28], Py: −1.8 V [40]) that are comparable to that of Ptz (see above), selectively initiating electron hopping via C and T. It is remarkable that the measured electron injection rates of the Sd-capped hairpins are larger using T ($>2 \times 10^{12}$ s^{-1}) as the electron acceptor

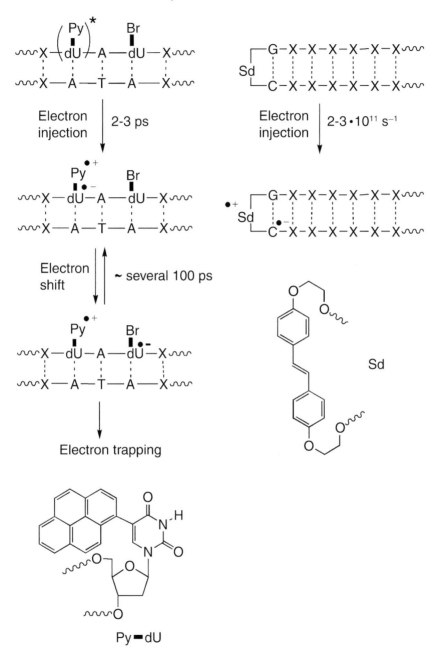

Fig. 6. Photochemical assays for the investigation of reductive electron transport in DNA, part C.

in comparison to C (3.3×10^{11} s^{-1}). This indicates that C represents a better electron trap compared to T by thermodynamic means.

Our group applied femtosecond broadband pump-probe spectroscopy to explore the early time dynamics in DNA modified with 5-pyrenyl-2′-deoxyuridine (PydU; Fig. 6) *(50)*. The measured electron injection rates between 3×10^{11} s^{-1} and 5×10^{11} s^{-1} are independent of the adjacent DNA base sequence. The subsequent electron shift into the base stack is much more sensitive to structural parameters and thereby characterized by a distribution of time constants and different strand cleavage efficiencies. Our results indicate that the electron shift occurs on the time scale of several hundred ps, therefore competing with charge recombination in our duplexes. It is reasonable to assume that subsequent migration steps will be faster because the Coulomb interaction between the excess electron and oxidized Py moiety decreases drastically with distance. Hence, our results provide a lower limit for the rate of reductive ET between single bases in DNA.

In conclusion, these studies suggest an electron hopping mechanism over long distances. In principle, both pyrimidine radical anions, T$^{•-}$ and C$^{•-}$, can play the role of intermediate electron carriers (*see* Fig. 3). The electron hopping via T$^{•-}$ seems to be slightly more favorable because C$^{•-}$ exhibits a much stronger basicity than T$^{•-}$ *(21,39)*. Thus, protonation of C$^{•-}$ by the complementary DNA bases or the surrounding water molecules probably interferes with the electron hopping and decreases the electron transport efficiency and rate, but does not stop electron migration in DNA.

6. ELECTRON TRANSPORT IN DNA CHIP TECHNOLOGY

In the last 10 years, genomic research has required highly parallel analytical approaches such as DNA microarrays and DNA chips. In principle, such systems are segmented, planar arrays of immobilized DNA fragments and are used in a wide field of applications from expression analysis to diagnostic tools *(51–54)*. In the latter case, a reliable detection of point mutations (single-nucleotide polymorphism [SNP]) is crucial for functional genomics *(55,56)*.

Charge transport phenomena show a high sensitivity toward pertubation of the base-stacking, which typically is caused by single base mismatches or DNA damage. Hence, charge transport in DNA represents an important step toward a sensitive electrochemical readout on DNA chips. For this approach, subsets of a critical gene are immobilized as single-stranded oligonucleotides on an electrode, and are modified with a redoxactive probe (Fig. 7). Intact DNA material added to the chip forms correct duplexes. An efficient electron transport between the chip surface and the redoxactive probe can be detected by electrochemical methods, e.g., cyclovoltametry. Base mismatches and DNA lesions significantly interrupt electron transport in DNA, and as a result, the electrochemical signal is lacking.

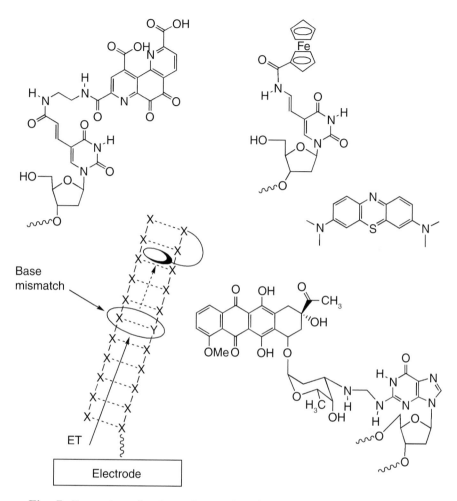

Fig. 7. Examples of redoxactive probes for an electrochemical readout on DNA chips.

One of the most convenient techniques for depositing molecules on surfaces is the use of self-assembled monolayers (SAM) *(57,58)*. In this technique, DNA is attached to an alkyl thiolate linker via the 5′-terminal hydroxy group of the oligonucleotide, which then interacts with the gold electrode to form DNA films as SAM. Daunomycine, pyrrolo-quinoline-quinone, methylene blue, or ferrocene have been applied as redoxactive probes *(11)*. Using this methodology, a broad range of single base mismatches and DNA lesions can be detected without the context of certain base sequences. Hence, electron transport through DNA films offers a new and suitable approach for the development of sensitive DNA sensors and chips. Normally, sensitive gene detection is accomplished by the amplification of the DNA material through

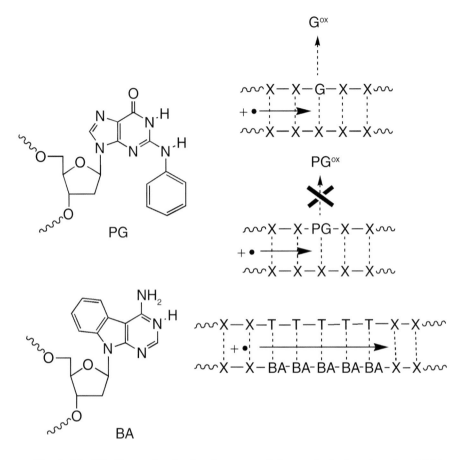

Fig. 8. Modified bases for the development of electronic devices based on DNA-mediated hole transport: PG (top) reduces hole trapping and BA (bottom) increases the hole transport efficiency.

the PCR. Inherent limitations of PCR often prohibit this application. Thus, research in the field of new DNA chips is currently focused on increasing the sensitivity in such a way that the PCR amplification becomes unnecessary.

7. OUTLOOK ON SYNTHETIC NANODEVICES BASED ON DNA-LIKE ARCHITECTURE

The construction of nanoscale electronic devices from conducting polymers or biopolymers remains a challenging task. DNA, beyond its natural biological role, is a supramolecular architecture with important features such as a regular, linear structure, making it a superior material for this research. Moreover, the most remarkable aspect of oligonucleotides is their ability to recognize, with high selectivity, their complementary parts as encoded by the DNA base sequence. DNA-based molecular wires are expected to represent an important nanomaterial that will be applied to new electronic devices.

In principle, both types of charge transport (oxidative and reductive) could be applied to molecular electronics because they both occur on long distances. However, one of the major drawbacks of the oxidative hole transport with respect to electronic devices is the oxidative DNA damage that occurs as a side product of the guanine radical cation (*see* Fig. 2). This hole trapping represents a serious disadvantage. For this reason, Saito's group of developed an *N*-phenylguanosine (PG; Fig. 8) as a substitute for G in duplex DNA. PG represents the first G derivative that maintains hole transport efficiency by avoiding the oxidative degradation *(59)*. Another way to improve the hole transport efficiency is by extending the π-π-stacking interactions. Benzofused DNA bases like BA (Fig. 8) have such an expanded aromatic surface and are able to increase the hole transport efficiency as a result of the enhanced stacking interactions inside the DNA double helix *(60)*.

On the other hand, several proposals strongly support the idea that the excess electron transport has a significantly higher potential for application in electronic devices on the molecular level. First of all, all base pairs (C-G and T-A) are involved as intermediate charge carriers during electron hopping (*see* Fig. 3). In contrast, during hole hopping, domains of G-hopping and A-hopping occur (*see* Fig. 2). Moreover, the proposed lack of DNA damage as the result of reductive electron transport is ideal for the development of DNA-nanowires. It will be interesting to see how this research will develop over the next few years.

REFERENCES

1. Eley DD, Spivey DI. Semiconductivity of organic substances. 9. Nucleic acid dry state. Trans Faraday Soc 1962;58:411–415.
2. Priyadarshy S, Risser SM, Beratan DN. DNA is not a molecular wire: protein-like electron-transfer predicted for an extended π-electron system. J Phys Chem 1996;100:17,678–17,682.
3. Turro NJ, Barton JK. Paradigms, supermolecules, electron transfer and chemistry at a distance. What's the problem? The science or the paradigm? J Biol Inorg Chem 1998;3:201–109.
4. Berlin YA, Burin AL, Ratner MA. DNA as a molecular wire. Superlattices Microstruct 2000;28:241–252.
5a. Schuster GB, ed. (2004) Long-range charge transfer in DNA II. In: Topics in Current Chemistry, Volume 237. Berlin: Springer.
5b. Schuster GB, ed. (2004) Long-range charge transfer in DNA I. In: Topics in Current Chemistry, Volume 236. Berlin: Springer.
6. Jortner J, Bixon M, Langenbacher T, Michel-Beyerle ME. Charge transfer and transport in DNA. Proc Natl Acad Sci USA 1998;95:12,759–12,765.

7. O'Neill P, Frieden EM. Primary free radical processes in DNA. Adv Radiat Biol 1993; 17:53–120.

8. Burrows CJ, Muller JG. Oxidative nucleobase modifications leading to strand scission. Chem Rev 1998;98:1109–1151.

9. Wang D, Kreutzer DA, Essigmann JM, Mutagenicity and repair of oxidative DNA damage: insights from studies using defined lesions. Mutation Res 1998;400:99–115.

10. Kawanashi S, Hiraku Y, Oikawa S. Mechanism of guanine-specific DNA damage by oxidative stress and its role in carcinogenesis and aging. Mutation Res 2001;488:65–76.

11. Drummond TG, Hill MG, Barton JK. Electrochemical DNA sensors. Nat Biotechnol 2003; 21:1192–1199.

12. Porath D, Cuniberti G, Di Felice R. Charge transport in DNA-based devices. Top Curr Chem 2004;237:183–227.

13. Wagenknecht H-A. Synthetic oligonucleotide modifications for the investigation of charge transfer and migration processes in DNA. Curr Org Chem 2004;8:251–266.

14. Grinstaff MW. How do charges travel through DNA?—an update on a current debate. Angew Chem Int Ed 1999;38:3629–3635.

15. Giese B. Long-distance electron transfer through DNA. Annu Rev Biochem 2002;71:51–70.

16. Wagenknecht H-A. Reductive electron transfer and transport of excess electrons in DNA. Angew Chem In. Ed 2003;42:2454–2460.

17. Steenken S, Jovanovic SV. How easily oxidizable is DNA? One-electron reduction potentials of adenosine and guanosine radicals in aqueous solution. J Am Chem Soc 1997;119:617–618.

18. Seidel CAM, Schulz A, Sauer MHM. Nucleobase-specific quenching of fluorescent dyes. 1. Nucleobase one-electron redox potentials and their correlation with static and dynamic quenching efficiencies. J Phys Chem 1996;100:5541–5553.

19. Lewis FD, Liu X, Liu J, Miller SE, Hayes RT, Wasielewski MR. Direct measurement of hole transport dynamics in DNA. Nature 2000;406:51–53.

20. Wagenknecht H-A, Rajski SR, Pascaly M, Stemp EDA, Barton JK. Direct observation of radical intermediates in protein-dependent DNA charge transport. J Am Chem Soc 2001;123:4400–4407.

21. Giese B, Spichty M. Long distance charge transport through DNA: quantification and extension of the hopping model. ChemPhysChem 2000;1:195–198.

22. Giese B, Amaudrut J, Köhler A-K, Spormann M, Wessely S. Direct observation of hole transfer through DNA by hopping between adenine bases and by tunnelling. Nature 2001;412:318–320.

23. Takada T, Kawai K, Cai X, Sugimoto A, Fujitsuka M, Majima T. Charge separation in DNA via consecutive adenine hopping. J Am Chem Soc 2004;126:1125–1129.

24. Giese B, Kendrick T. Charge transfer through DNA triggered by site selective charge injection into adenine. Chem Commun 2002;2016–2017.

25. Li X, Cai Z, Sevilla MD. Energetics of the radical ions of the AT and AU base pairs: A density functional theory (DFT) study. J Phys Chem A 2002;106: 9345–9351.

26. Steenken S. Electron transfer in DNA? Competition by ultra-fast proton-transfer? Biol Chem 1997;378:1293–1297.
27. Steenken S. Electron-transfer-induced acidity/basicity and reactivity changes of purine and pyrimidine bases. Consequences of redox processes for DNA base pairs. Free Rad Res Comm. 1992;16:349–379.
28. Giese B, Wessely S. The significance of proton migration during hole hopping through DNA. Chem Commun 2001;2001:2108–2109.
29. Giese B, Carl B, Carl T, et al. Excess electron transport through DNA: a single electron repairs more than one UV-induced lesion. Angew Chem Int Ed 2004;43:1848–1851.
30. Voityuk AA, Michel-Beyerle M-E, Rösch N. Energetics of excess electron transfer in DNA. Chem Phys Lett 2001;342:231–238.
31. Steenken S, Telo JP, Novais HM, Candeias LP. One-electron-reduction potentials of pyrimidine bases, nucleosides, and nucleotides in aqueous solution. Consequences for DNA redox chemistry. J Am Chem Soc 1992;114: 4701–4709.
32. Cai Z, Sevilla MD. Studies of excess electron and hole transfer in DNA at low temperatures. Top Curr Chem 2004;237:103–128.
33. Behrens C, Burgdorf LT, Schwögler A, Carell T. Weak distance dependence of excess electron transfer in DNA. Angew Chem Int Ed 2002;41:1763–1766.
34. Haas C, Kräling K, Cichon M, Rahe N, Carell T. Excess electron transfer driven DNA does not depend on the transfer direction. Angew Chem Int Ed 2004;43:1842–1844.
35. Ito T, Rokita SE. Excess electron transfer from an internally conjugated aromatic amine to 5-bromo-2′-deoxyuridine in DNA. J Am Chem Soc 2003;125: 11,480–11,481.
36. Ito T, Rokita SE. Criteria for efficient transport of excess electrons in DNA. Angew Chem Int Ed 2004;43:1839–1842.
37. Ito T, Rokita SE. Reductive electron injection into duplex DNA by aromatic amines. J Am Chem Soc 2004;126:15,552–15,559.
38. Lewis FD, Liu X, Miller SE, Hayes RT, Wasielewski MR. Dynamics of electron injection in DNA hairpins. J Am Chem Soc 2002;124:11,280–11,281.
39. Wagner C, Wagenknecht HA. Reductive electron transfer in phenothiazine-modified DNA is dependent on the base sequence. Chem Eur J 2005;11:1871–1876.
40. Amann N, Pandurski E, Fiebig T, Wagenknecht H-A. Electron injection into DNA: synthesis and spectroscopic properties of pyrenyl-modified oligonucleotides. Chem Eur J 2002;8:4877–4883.
41. Kaden P, Mayer-Enthart E, Trifonov A, Fiebig T, Wagenknecht H-A. Real-time spectroscopic and chemical probing of reductive electron transfer in DNA. Angew Chem Int Ed 5;44:1636–1639.
42. Sancar A. Structure and function of DNA photolyase and cryptochrome blue-light photoreceptors. Chem Rev 2003;103:2203–2237.
43. Scannell MP, Fenick DJ, Yeh S-R, Falvey DE. Model studies of DNA photoreapair: reduction potentials of thymine and cytosine cyclobutane dimers measured by fluorescence quenching. J Am Chem Soc 1997;119:1971–1977.
44. Chen T, Cook GP, Koppisch AT, Greenberg MM. Investigation of the origin of the sequence selectivity for the 5-halo-2′-deoxyuridine sensitization of DNA to damage by UV-irradiation. J Am Chem Soc 2000;122:3861–3866.

45. Rivera E, Schuler RH. Intermediates in the reduction of 5-halouracils by e_{aq}^{-1}. J Am Chem Soc 1983;87:3966–3971.

46. Kadysh VP, Kaminskii YL, Rumyantseva LN, Efimova VL, Strandish JP. Khim Geterotsikl Soedin 1992;10:1404–1408.

47. Yeh S-R, Falvey DE. Model studies of DNA photorepair: radical anion cleavage of thymine dimers probed by nanosecond laser spectroscopy. J Am Chem Soc 1997;113:8557–8558.

48. Huber R, Fiebig T, Wagenknecht H-A. Pyrene as a fluorescent probe for DNA base radicals. Chem Commun 2003;1878–1879.

49. Raytchev M, Mayer E, Amann N, Wagenknecht H-A, Fiebig T. Ultrafast proton-coupled electron-transfer dynamics in pyrene-modified pyrimidine nucleosides: model studies towards an understanding of reductive electron transport in DNA. ChemPhysChem 2004;5:706–712.

50. Netzel TL, Zhao M, Nafisi K, Headrick J, Sigman MS, Eaton BE. Photophysics of 2′-deoxyuridine (dU) nucleosides covalently substituted with either 1-pyrenyl or 1-pyrenoyl: observation of pyrene-to-nucleoside charge-transfer emission in 5-(1-pyrenyl)-dU. J Am Chem Soc 1995;117:9119–9128.

51. Niemeyer CM, Blohm D. DNA microarray. Angew Chem Int Ed 1999;38:2865–2869.

52. Blohm DH, Guiseppi-Elie A. New developments in microarray technology. Curr Opin Biotechnol 2001;12:41–47.

53. Pirrung MC. How to make a DNA chip. Angew Chem Int Ed 2002;41:1276–1289.

54. Jung A. DNA chip technology. Anal Bioanal Chem 2002;372:41–42.

55. Strerath M, Marx A. Genotyping—from genomic DNA to genotype in a single tube. Angew Chem Int Ed Engl 2005;44:7842–7849.

56. Nakatani K. Chemistry challenges in SNP typing. ChemBioChem 2004;5:1623–1633.

57. Kumar A, Abott NL, Kim E, Biebuyck HA, Whitesides GM. Patterned self-assembled monolayers and meso-scale phenomen. Acc Chem Res 1995;28:219–226.

58. Porier GE. Characterization of organosulfur molecular monolayers on Au(111) using scanning tunneling microscopy. Chem Rev 1997;97:1117–1127.

59. Okamoto A, Tanaka K, Saito I. Rational design of a DNA wire possessing an extremely high hole transport ability. J Am Chem Soc 2003;125:5066–5071.

60. Nakatani K, Dohno C, Saito I. N2-phenyldeoxyguanosine: modulation of the chemical properties of deoxyguanosine toward one-electron oxidation in DNA. J Am Chem Soc 2002;124:6802–6803.

6
Effective Models for Charge Transport in DNA Nanowires

Rafael Gutierrez and Gianaurelio Cuniberti

Summary

Rapid progress in the field of molecular electronics has led to increasing interest in DNA oligomers as possible components of electronic circuits at the nanoscale. For this, however, an understanding of charge transfer and transport mechanisms in this molecule is required. Experiments show that a large number of factors may influence the electronic properties of DNA. Although full first-principle approaches are the ideal tool for a theoretical characterization of the structural and electronic properties of DNA, the structural complexity of this molecule limits the usefulness of these methods. Consequently, model Hamiltonian approaches, which filter out single factors influencing charge propagation in the double helix, are highly valuable. In this chapter, we review the different DNA models that are thought to capture the influence of some of these factors. We will specifically focus on static and dynamic disorder.

Key Words: Correlated disorder; dissipation; DNA conduction; electron-vibron interaction; static disorder.

1. INTRODUCTION

The increasing demands on the integration densities of electronic devices are considerably limiting conventional semiconductor-based electronics. As a result, new possibilities have been explored in the last decade, leading to the emergence of molecular electronics, which basically relies on the idea of using single molecules or molecular groups as elements of electronic devices. A new conceptual idea advanced by molecular electronics is the switch from a top-down approach, where the devices are extracted from a single large-scale building block, to a bottom-up approach, in which the whole system is composed of small basic building blocks with recognition and self-assembly properties.

From: *NanoBioTechonology: BioInspired Devices and Materials of the Future*
Edited by: Oded Shoseyov and Ilan Levy © Humana Press Inc., Totowa, NJ

A molecule that has recently attracted the attention of both experimentalists and theoreticians is DNA. The observation of electron transfer between intercalated donor and acceptor centers in DNA oligomers in solution over unexpectedly long distances *(1)* led to a revival of interest in the conduction properties of this molecule. Although the idea that DNA might be conducting is rather old *(2)*, there was never any conclusive proof that it could support charge transfers over long distances. This is, however, a critical issue when considering, for example, damage repair during the replication process *(3)*. Apart from the relevance of these and similar experiments for biology and genetics, they also suggested that by appropriately adjusting the experimental conditions, DNA molecules might be able to carry an electrical current. Further, DNA oligomers might be useful as templates in molecular electronic circuits if their self-assembling and self-recognition properties are exploited *(4–6)*. Although many technical and theoretical problems must still be surmounted, it is now possible to carry out transport experiments on single molecules connected to metallic electrodes.

However, despite the many expectations put on DNA as a potential ingredient of molecular electronic circuits, transport experiments on this molecule have revealed some very intriguing and partly contradictory behavior. Thus, it has been found that DNA may be insulating *(7,8)*, semiconducting *(9,10)*, or metallic *(11,12)*. These results demonstrate the high sensitivity of DNA transport to different factors affecting charge motion, such as the quality of the contacts to the metal electrodes, the base-pair sequence, the charge injection into the molecule, or environmental effects (dry vs aqueous environments), among others.

Theoretically, knowledge of the electronic structure of the different building units of a DNA molecule (base pairs, sugar and phosphate groups) is essential for clarifying the most effective transport mechanisms. First-principle approaches are the most suitable tools for this goal. However, the huge complexity of DNA makes *ab initio* calculations still very demanding, so that only comparatively few investigations have been performed *(13–21)*. Further, environmental effects such as the presence of hydration shells and counterions make *ab initio* calculations even more challenging *(14,15,22)*.

In this chapter, we will review a complementary (to first-principle approaches) way to look at DNA, namely, model Hamiltonians. They play a significant role in filtering out possible charge transfer and transport mechanisms as well as in guiding the more involved first-principle investigations. We are not aiming at a thorough review of Hamiltonian-based theories. In fact, because the authors belong to the "physical community," model approaches for charge transfer formulated in the "chemical community" will not be the scope of this chapter. The interested reader can consult refs. *23–28*. We are

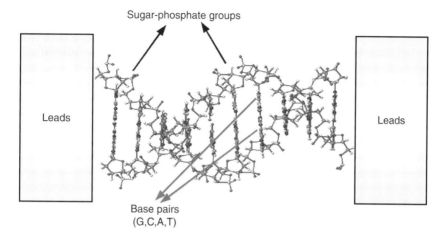

Fig. 1. Schematic representation of a double-stranded DNA oligomer with an arbitrary base-pair sequence and connected to left and right electrodes.

also not considering the infiuence of electron–electron interactions on charge transport, an issue that needs further clarification *(29,30)*. In the next two sections, we discuss models describing the infiuence of static disorder and dynamic effects on charge propagation in DNA. For the sake of presentation, we discuss both factors in different sections. Nevertheless, the reader should be aware that, realistically, the interplay between them is expected to be closer.

2. STATIC DISORDER

DNA oligomers consist of four building blocks (oligonucleotides): adenine (A), thymine (T), cytosine (C), and guanine (G). As is well known, they have specific binding properties, i.e., only A-T and G-C pairs are possible (*see* Fig. 1). Sugar and phosphate groups ensure the mechanical stability of the double helix and protect the base pairs. Because the phosphate groups are negatively charged, the topology of the duplex is only conserved if it is immersed in an aqueous solution containing counterions (Na^+, Mg^+) that neutralize the phosphate groups. Thus, experimenting on "dry" DNA usually means that the humidity has been greatly reduced, but there are still water molecules and counterions attached to the sugar–phosphate mantle.

The specific base-pair sequence is obviously essential for DNA to fulfill its function as a carrier of the genetic code. However, this same fact can be detrimental to charge transport. The apparently random way in which the DNA sequence is composed strongly suggests that a charge propagating along the double helix may basically feel a random potential, leading to backscattering. It is well known that in a one-dimensional (1D) system with uncorrelated disorder, all electronic states are completely localized (Anderson localization). However, correlated disorder with, for example, power-law correlations *(31)* may lead to delocalized states within some special energy windows in the thermodynamic limit, the exact structure of the spectrum being determined by the so-called scaling exponent α. This quantity describes the correlation properties of a random process *(31,32)*, specifically, the length-dependence of the position autocorrelation function: $C(l) \sim l^{-\alpha}$. Thus, $\alpha = 0.5$ corresponds to a pure random walk, whereas other values indicate the presence of long-range correlations and hence, the absence of relevant length scales in the problem (self-similarity).

Some of the main issues to be addressed when investigating the role of disorder in DNA are, in our view, the following. First, is the specific base-pair sequence in DNA completely random (Anderson-like) or do there exist (long-or short-range) correlations? Second, a measure for the degree of confinement of the electronic wave function is given by the localization length ζ *(33)*. Are the resulting localization lengths larger or smaller than the actual length L of the DNA segments studied in transport experiments? For $\zeta \gg L$, the system may appear to be effectively conducting, despite the presence of disorder, although in the thermodynamic limit, all states may remain localized. Clarification of these issues requires close cooperation between experimentalists and theoreticians. In what follows, we review some theoretical studies addressing these problems.

The simplest way to mimic a DNA wire is by assuming that after charge injection, the electron (hole) will basically propagate along one of the strands (the inter-strand coupling being much smaller), so that 1D tight-binding chains can be a good starting point to minimally describe a DNA wire. Roche *(34)* investigated such a model for poly(GC) and λ-phage DNA, with on-site disorder (resulting from the differences in the ionization potentials of the base pairs) and bond disorder $\sim \cos \theta_{n,n+1}$ related to the twisting motion of nearest-neighbor bases along the strand, $\theta_{n,n+1}$ being independent Gaussian-distributed random variables. Poly(GC) displays two electronic bands and thermal fluctuations reduce the transmission peaks and also, slightly, the band widths. The effect of disorder does not appear to be very dramatic. In the case of λ-phage, however, the transmission peaks are considerably diminished in intensity and in number with increasing chain length at zero temperature, because only a few electronic states are not backscattered by the random potential profile of the chain. Interestingly, the average Ljapunov exponent, which is related to the localization length, increases with increasing temperature, indicating that despite thermal fluctuations, many states are still contributing to charge transport.

In an early paper, Roche et al. *(35)* used scaling coefficients (Hurst exponents), which usually indicate the existence of long-range correlations in disordered systems. Their results showed that, e.g., DNA built from Fibonacci sequences has a very small Hurst exponent (indicating strong correlations). Uncorrelated random sequences show strong fragmentation and suppression of transmission with increasing length, whereas in correlated sequences, several states appear to be rather robust against the increasing rate of backscattering. Hence, it may be expected that correlated disorder will be more favorable for long-distance carrier transport in DNA wires.

Another typical example of correlated disorder was presented by Alburquerque et al. *(36)* within a 1D tight-binding model. The authors investigated the quasi-periodic Rudin-Shapiro sequence as well as the human genome Ch22. As expected, the transmission bands became more and more fragmented with increasing number of nucleotides. Although for very long chain lengths, all electronic states did tend to be completely localized, long-range correlations yielded large localization lengths and thus transport might still be supported for special energy points on rather long wires.

Zhu et al. *(37)* formulated an effective tight-binding model including only HOMO and LUMO of poly(GC) together with on-site Coulomb interactions. On-site and off-diagonal disorder, related to fluctuations of the local electrostatic potential *(38)* and to the twisting motion of the base pairs at finite temperatures, respectively, were also included. The main effect of the Coulomb interaction was to first reduce the band gap, so that the system goes over to a metallic state, but finally the gap reappears as a Coulomb-blockade gap. Twisting disorder was apparently less relevant for short wires and low temperatures.

A very detailed study of the localization properties of electronic states in two minimal models of different DNA oligomers [poly(GC), λ-DNA, telomeric DNA] was presented by Klotsa et al. *(39)*: a fishbone model *(40–42)* and a ladder model. Both models fulfill the minimal requirement of showing a band gap in the electronic spectrum, mirroring the existence of a HOMO-LUMO gap in isolated DNA molecules. However, the ladder model allows for the inclusion of interstrand effects as well as of the specific base-complementarity typical of the DNA duplex, an issue that cannot be fully captured by the first model. The authors were mainly interested in environmentally induced disorder. Hence, they assumed that only the backbone sites are affected by it, while the nucleotide core is well screened. Nevertheless, as shown by a decimation procedure *(39)*, disorder in the backbone sites can induce local fluctuations of the on-site energies on the base pairs (gating effect). Uniform disorder (where the on-site energies of the backbones continuously vary over an interval [−W, W], W being the disorder strength) is shown to continuously reduce the localization length, as expected. For binary disorder (on-site energies take only two possible values, $\pm W/2$), as it may arise by the binding of counterions to the backbone sites, the situation is similar up to some critical disorder strength W_c. However, further increase of W leads to unexpected behavior: the localization length on the electronic side bands is suppressed but a new band around the mid-gap with *increasing* localization length shows up. Thus, disorder-induced delocalization of the electronic states is observed in some energy window. This result, obtained within a simple model, may be supported by first-principle calculations *(22)*, which clearly show that the environment can introduce additional states in the molecular band gap.

Most of the foregoing investigations considered on-site disorder only. The influence of off-diagonal short-range correlations was investigated by Zhang and Ulloa *(43)* in λ-DNA. They showed that this kind of disorder can definitely lead to the emergence of conduction channels in finite systems. For some special ratios of the nearest-neighbor hopping amplitudes, there may even exist extended states in the thermodynamic limit. As a consequence, the authors suggested that λ-DNA may show a finite current at low voltages.

Caetano and Schulz *(44)* investigated a double-strand model with uncorrelated disorder along the single strand, but taking into account the binding specificity of the four bases when considering the complementary strand (A-T and G-C). Participation ratios P(E) were computed, which give a measure of the degree of localization of electronic states. P(E) is, for example, almost zero for localized states in the thermodynamic limit. The results suggest that interstrand correlations may give rise to bands of delocalized states, with a participation ratio that does not appreciably decay with increasing length.

3. DYNAMICAL DISORDER

In the previous section, we presented several studies related to the influence of static disorder on the charge-transport properties of different DNA oligomers. Here, we address a second aspect of high relevance, namely the impact of dynamic disorder related to structural fluctuations on charge propagation. Considering the relative flexibility of DNA, one may expect that vibrational modes will have a strong influence on the charge motion via a modification of electronic couplings.

The markedly small decay rates found in electron-transfer experiments *(1)* have led to the proposal that, besides unistep superexchange mechanisms, phonon-assisted hole-hopping might also be of importance *(26)*. The hole can occupy a specific molecular orbital, localized on a given molecular site; it can also, however, extend over several molecular sites and build a polaron, which is basically a lattice deformation accompanying a propagating charge. It results from the energetic interplay of two tendencies: the tendency to delocalize the charge, thus gaining kinetic energy, and the tendency to localize it with a consequent gain in elastic energy. The softness of the DNA molecule and the existence of modes that can appreciably affect the inter-base electronic coupling (like twisting modes or H-stretching bonds) makes this suggestion very attractive *(45,46)*. Conwell and Rakhmanova *(46)* investigated this issue using the Su-Heeger-Schrieffer (SSH) model, which is known to entail rich nonlinear physics and which has been extensively applied to study polaron formation in conducting polymers. The SSH model deals classically with the lattice degrees of freedom while treating the electrons quantum mechanically. The calculations showed that a polaron may be built and be robust within a wide range of model parameters. The influence of random base sequences was apparently not strong enough to destroy it. Thus, polaron drifting may constitute a potential transport mechanism in DNA oligomers.

The potential for the lattice displacements was assumed to be harmonic *(45,46)*. Interstrand modes like H-bond stretching are, however, expected to be strongly anharmonic; H-bond fluctuations can induce local breaking of the double-strand and have thus been investigated in relation to the DNA denaturation problem *(47)*. To investigate this effect, Komineas et al. *(48)* studied a model with strong anharmonic potentials and local coupling of the lattice to the charge density. The strong nonlinearity of the problem led to a *dynamic* opening of bubbles with different sizes that may eventually trap the polaron and thus considerably affect this charge-transport channel.

Zhang et al. *(49,50)* studied a simple model that describes the coupling of torsional excitations (twistons) in DNA to propagating charges and showed that this interaction leads to polaron formation. Twistons modify the interbase electronic coupling, although this effect is apparently weaker than, e.g., in the Holstein model *(51)*, because of the strong nonlinearity of the twistons restoring forces as well as of the twiston–electron coupling. For small restoring forces of the twisting modes and in the nonadiabatic limit ("spring constant" much larger than electronic coupling), the interbase coupling is maximally perturbed and an algebraic band reduction is found that is weaker than the exponential dependence known from the Holstein model. Thus, it may be expected that the polaron will have a higher mobility along the chain.

The observation of two quite different time scales (5 ps and 75 ps) in the decay rates of electron transfer processes in DNA, as measured by femtosecond spectroscopy *(52)*, was the main motivation of Bruinsma et al. *(53)* to investigate the coupling of the electronic system to collective modes of the DNA cage. For this, they considered a tight-binding model of electrons interacting with two modes: a twisting mode, which mainly couples to the interbase π-orbital matrix elements, and a linear displacement coupling to the on-site energies of the radical and acting as a local gating of the latter. In the strong-coupling, high-temperature limit, the hopping matrix elements can be treated perturbatively and build the lowest energy scale. Transport has thus a hopping-like character. In analogy with electron-transfer theories, the authors provide a picture where there are basically two reaction coordinates related to the above-mentioned linear and angular modes. The strong thermal fluctuations associated with the twisting motion are shown to introduce two time scales for electron transfer that can be roughly related to optimal (short) and nonoptimal (long) relative orientation of neighboring base pairs.

In several papers, Hennig et al. *(54,55)* and Yamada *(56)* formulated a model Hamiltonian where only the relative transverse vibrations of bases belonging to the same pair are included. Their calculations showed the formation of stable polarons. Moreover, the authors suggested that poly(GC) should be more effective in supporting polaron-mediated charge transport than poly(AT), because for the latter, the electron-lattice coupling was found to be about one order of magnitude smaller. Although the authors remarked that no appreciable coupling to twisting distortions was found by their semi-empirical quantum chemical calculations, this issue requires further investigation in view of the previously presented results *(49,50,53)*. Disorder did not appear to have a very dramatic influence in this model; the localization length only changed quantitatively as a function of the disorder strength *(56)*.

Asai *(57)* proposed a small polaron model to describe the experimental findings of Yoo et al. *(11)* concerning the temperature dependence of the electrical current and of the linear conductance. Basically, he assumed that in poly(GC), completely incoherent polaron hopping dominates whereas in poly(AT), quasi-coherent hopping, i.e., with total phonon number conservation, is more important. As a result, the temperature dependence of the above quantites in both molecules is considerably different.

In complement to the foregoing research, which mainly addressed individual vibrational modes of the DNA cage, other studies have focused on the influence of environmental effects. Basko and Conwell *(58)* used a semi-classical model to describe the interaction of an injected hole in DNA, which is placed in a polar solvent. Their basic conclusions pointed out that the main contribution was provided by the interaction with water molecules and not with counterions; further, polaron formation was not hindered by the charge–solvent coupling; rather, the interaction increased the binding energy (self-localization) of the polaron by around half an eV, which is much larger than relevant temperature scales. Li and Yan *(59)* as well as Zhang et al. *(60)* investigated the role of dephasing reservoirs in the spirit of the Buettiker-D'Amato-Pastawski model *(61,62)*. Segal et al. showed that a change in the length scaling of the conductance can be induced by the dephasing reservoirs as a result of incoherent phonon-mediated transport, a result known from electron-transfer theories *(63)*. In a similar way, Feng and Xiong *(64)* considered gap-opening a result of the coupling to a set of two-level systems, which simulate low-lying states of the bosonic bath. Gutierrez et al. *(41,42)* discussed electron transport in a "broken-ladder" model in the presence of a strong dissipative environment simulated by a bosonic bath. It was found that the environment can induce virtual polaronic states inside the molecular band gap and thus lead to a change in the low-energy transport properties of the system. In particular, the I-V curves display α non-zero slope at low voltages as a result of phonon-assisted hopping. We note that these latter results are quite similar to those found in *ab initio* calculations, showing that water states can appear between the π-π^* gap *(65)*, thus effectively introducing shallow states similar to those in doped bulk semiconductors. These states may support activated hopping at high temperatures.

We finally mention that the role of nonlinear excitations (solitons, breathers) in the process of denaturation of DNA double strands *(47,66,67)* and in the transmission of "chemical" information between remote DNA segments *(68)* was addressed early on in the literature. Because these approaches are not directly connected with the issue of charge transport in DNA wires between electrodes, we do not go into further detail. They may, however, reveal a novel, interesting mechanism for transport and deserve a more careful investigation.

4. CONCLUSIONS

Although much progress has been made in the past decade in clarifying the relevant transport mechanisms in DNA oligomers, a coherent, unifying picture is still lacking. The experimental difficulties involved in giving reliable transport characteristics of this molecule make the formulation of model Hamiltonians quite challenging. The theoretical research presented in this chapter shows that charge transport in DNA is considerably influenced by both static and dynamical disorder. Long-range correlated disorder can play a role in increasing the localization length beyond the relevant molecular length scales addressed in experiments, thus making DNA effectively appear to be a conductor. This effect may be supported or counteracted by thermal fluctuations arising from internal (vibrations) or external (solvent) modes leading to increased charge localization or to incoherent transport.

The presented models only focus on the equilibrium or low-bias limit of transport. However, real transport experiments probe the molecules at finite voltages and hence, nonequilibrium effects also must be considered. This, of course, makes the mathematical treatment as well as the physical interpretation more involved. Considerable effort has been made recently to deal with this issue *(69–71)*; however, addressing these studies goes beyond the scope of this chapter.

ACKNOWLEDGMENTS

The authors thank R. Bulla, A. Nitzan, R. Römer, and S. Roche for useful suggestions and discussion. This work has been supported by the Volkswagen Foundation and by the European Union under contract IST-2001-38951.

REFERENCES

1. Murphy CJ, Arkin MR, Jenkins Y, et al. Long-range photoinduced electron transfer through a DNA helix. Science 1993;262:1025.
2. Eley DD, Spivey DI. Semiconductivity of organic substances. Part 9: Nucleic acid in the dry state. Trans Faraday Soc 1962;58:411.
3. Friedberg EC. DNA damage and repair. Nature 2003;421:436–440.
4. Dekker C, Ratner M. Electronic properties of DNA. Physics World 2001;August:29.
5. Keren K, Berman RS, Buchstab E, Sivan U, Braun E. DNA-templated carbon nanotube field-effect transistor. Science 2003;302:1380–1382.
6. Mertig M, Kirsch R, Pompe W, Engelhardt H. Fabrication of highly oriented nanocluster arrays by biomolecular templating. Eur Phys J D 1999;9:45–48.
7. Braun E, Eichen Y, Sivan U, Ben-Yoseph G. DNA-templated assembly and electrode attachment of a conducting silver wire. Nature 1998;391:775–778.
8. Storm AJ, Noort JV, Vries SD, Dekker C. Insulating behavior for DNA molecules between nanoelectrodes at the 100 nm length scale. Appl Phys Lett 2001;79:3881–3883.
9. Porath D, Bezryadin A, Vries SD, Dekker C. Direct measurement of electrical transport through DNA molecules. Nature 2000;403:635–638.
10. Cohen H, Nogues C, Naaman R, Porath D. Direct measurement of electrical transport through single DNA molecules of complex sequence. Proc Natl Acad Sci USA 2005;102:11,589–11,593.
11. Yoo K-H, Ha DH, Lee J-O, et al. Electrical conduction through poly(dA)-poly(dT) and poly(dG)-poly(dC) DNA molecules. Phys Rev Lett 2001;87: 198,102–198,105.
12. Xu B, Zhang P, Li X, Tao N. Direct conductance measurement of single DNA molecules in aqueous solution. Nano Lett 2004;4:1105–1108.
13. Felice RD, Calzolari A, Molinari E. Ab initio study of model guanine assemblies: the role of π-π coupling and band transport. Phys Rev B 2002;65: 045104–045113.
14. Barnett RN, Cleveland CL, Joy A, Landman U, Schuster GB. Charge migration in DNA: ion-gated transport. Science 2001;294:567–571.
15. Gervasio FL, Carolini P, Parrinello M. Electronic structure of wet DNA. Phys Rev Lett 2002;89:108,102–108,105.

16. Artacho E, Machado M, Sanchez-Portal D, Ordejon P, Soler JM. Electrons in dry DNA from density functional calculations. Mol Phys 2003;101: 1587–1594.
17. Alexandre SS, Artacho E, Soler JM, Chacham H. Small polarons in dry DNA. Phys Rev Lett 2003;91:108,105–108.
18. Lewis JP, Ordejon P, Sankey OF. Electronic-structure-based molecular-dynamics method for large biological systems: application to the 10 basepair poly(dG)-poly(dC) DNA double helix. Phys Rev B 1997;55:6880–6887.
19. Starikov EB. Role of electron correlations in deoxyribonucleic acid duplexes: Is an extended Hubbard Hamiltonian a good model in this case? Phil Mag Lett 2003;83:699–708.
20. Wang H, Lewis JP, Sankey O. Band-gap tunneling states in DNA. Phys Rev Lett 2004;93:016401–016404.
21. Mehrez H, Anantram MP. Interbase electronic coupling for transport through DNA. Phys Rev B 2005;71:115,405–115,409.
22. Huebsch A, Endres RG, Cox DL, Singh RRP. Optical conductivity of wet DNA. Phys Rev Lett 2005;94:178,102–178,105.
23. Schuster GB. Topics in Current Chemistry, vol. 237. Berlin: Springer, 2004.
24. Nitzan A. Electron transmission through molecules and molecular interfaces. Annu Rev Phys Chem 2001;52:681–750.
25. Nitzan A, Ratner M. Electron transport in molecular wire functions: models and mechanisms. Science 2003;300:1384–1389.
26. Jortner J, Bixon M, Langenbacher T, Michel-Beyerle ME. Charge transfer and transport in DNA. Proc Natl Acad Sci USA 1998;95:12,759–12,765.
27. Jortner J, Bixon M. Long-range and very long-range charge transport in DNA. Chem Phys 2002;281:393–408.
28. Berlin YA, Burin AL, Siebbeles LDA, Ratner MA. Conformationally gated rate processes in biological macromolecules. J Phys Chem A 2001;105:5666–5678.
29. Yi J. Conduction of DNA molecules: a charge-ladder model. Phys Rev B 2003;68:193,103–193,106.
30. Apalkov VM, Chakraborty T. Electron dynamics in a DNA molecule. Phys Rev B 2005;71:033102–033105.
31. Carpena P, Bernaola-Galvan P, Ivanov PC, Stanley HE. Metal-insulator transition in chains with correlated disorder. Nature 2002;418:955–959.
32. Peng C-K, Buldyrev SV, Goldberger AL, Havlin S, Sciortino F, Simons M, Stanley HE. Long-range correlations in nucleotide sequences. Nature 1992;356:168–170.
33. Phillips P. Advanced Solid State Physics. Boulder, CO: Westview Press, 2003.
34. Roche S. Sequence dependent DNA-mediated conduction. Phys Rev Lett 2003;91:108,101–108,104.
35. Roche S, Bicout D, Macia E, Kats E. Long range correlations in DNA: scaling properties and charge transfer efficiency. Phys Rev Lett 2003;91:228,101–228,104.
36. Alburquerque EL, Vasconcelos MS, Lyra ML, de Moura FABF. Nucleotide correlations and electronic transport of DNA sequences. Phys Rev E 2005;71: 21,910–21,916.
37. Zhu Y, Kaun CC, Guo H. Contact, charging, and disorder effects on charge transport through a model DNA molecule. Phys Rev B 2004;69:245,112–245,118.

38. Adessi C, Walch S, Anantram MP. Environment and structure influence on DNA conduction. Phys Rev B 2003;67:081405(R)–081408(R).

39. Klotsa D, Roemer RA, Turner MS. Electronic transport in DNA. Biophys J 2005;89:2187–2198.

40. Cuniberti G, Craco L, Porath D, Dekker C. Backbone-induced semiconducting behavior in short DNA wires. Phys Rev B 2002;65:241,314–241,317.

41. Gutierrez R, Mandal S, Cuniberti G. Quantum transport through a DNA wire in a dissipative environment. Nano Lett 2005;5:1093–1097.

42. Gutierrez R, Mandal S, Cuniberti G. Dissipative effects in the electronic transport through DNA molecular wires. Phys Rev B 2005;71:235,116–235,124.

43. Zhang W, Ulloa SE. Extended states in disordered systems: role of off-diagonal correlations. Phys Rev B 2004;69:153,203–153,207.

44. Caetano RA, Schulz PA. Sequencing-independent delocalization in a DNA-like double chain with base pairing. Phys Rev Lett 2005;95:126,601–126,604.

45. Henderson PT, Jones GHD, Kan Y, Schuster GB. Long-distance charge transport in duplex DNA: the phonon-assisted polaron-like hopping mechanism. Proc Natl Acad Sci USA 1999;96:8353–8358.

46. Conwell EM, Rakhmanova SV. Polarons in DNA. Proc Natl Acad Sci USA 2000;97:4556–4560.

47. Peyrard M, Bishop AR. Statistical mechanics of a nonlinear model for DNA denaturation. Phys Rev Lett 1989;62:2755–2755.

48. Komineas S, Kalosakas G, Bishop AR. Effects of intrinsic base-pair fluctuations on charge transport in DNA. Phys Rev E 2002;65:061905–061908.

49. Zhang W, Govorov AO, Ulloa SE. Polarons with a twist. Phys Rev B 2002;66:060303(R)–060306(R).

50. Zhang W, Ulloa SE. Structural and dynamical disorder and charge transport in DNA. Microelectronics J 2004;35:23–25.

51. Holstein T. Studies of polaron motion. Part I. The molecular-crystal model. Ann Phys NY 1959;8:325–342.

52. Wan C, Fiebig T, Kelley SO, Treadway CR, Barton JK. Femtosecond dynamics of DNA-mediated electron transfer. Proc Natl Acad Sci USA 1999;96:6014–6019.

53. Bruinsma R, Gruener G, D'Orsogna MR, Rudnick J. Fluctuation-facilitated charge migration along DNA. Phys Rev Lett 2000;85:4393–4396.

54. Hennig D. Mobile polaron solutions and nonlinear electron transfer in helical protein models. Phys Rev E 2001;64:041908–041924.

55. Palmero F, Archilla JFR, Hennig D, Romero FR. Charge transport in poly(dG)-poly(dC) and poly(dT)-poly(dA) DNA polymers. New J Phys 2004;6:1–16.

56. Yamada H. Localization of electronic states in chain model based on real DNA sequence. cond-mat/0406040 2004.

57. Asai Y. Theory of electric conductance of DNA molecule. J Phys Chem B 2003;107:4647–4652.

58. Basko DM, Conwell EM. Effect of solvation on hole motion in DNA. Phys Rev Lett 2002;88:098102–098105.

59. Li XQ, Yan Y. Electrical transport through individual DNA molecules. Appl Phys Lett 2001;79:2190–2192.

60. Zhang HY, Li X-Q, Han P, Yu XY, Yan Y-J. A partially incoherent rate theory of long-range charge transfer in deoxyribose nucleic acid. J Chem Phys 2002; 117:4578–4584.
61. Buettiker M. Coherent and sequential tunneling in series barriers. IBM J Res Dev 1988;32:63–75.
62. D'Amato JL, Pastawski HM. Conductance of a disordered linear chain including inelastic scattering events. Phys Rev B 1990;41:7411–7420.
63. Segal D, Nitzan A, Davies WB, Wasielewski MR, Ratner MA. Electron transfer rates in bridged molecular systems. 2. A steady-state analysis of coherent tunneling and thermal transitions. J Phys Chem B 2000;104:3817–3829.
64. Feng J-F, Xiong S-J. Large-bandgap behavior in transport of electrons through individual DNA molecules caused by coupling with a two-level system. Phys Rev E 2002;66:021908–021913.
65. Endres RG, Cox DL, Singh RRP. Colloquium: the quest for high-conductance DNA. Rev Modern Phys 2004;76:195–214.
66. Xiao J-X, Lin J-T, Zhang G-X. The influence of longitudinal vibration on soliton excitation in DNA double helices. J Phys A: Math Gen 1987;20:2425–2432.
67. Yakushevich LV, Savin AV, Manevitch LI. Nonlinear dynamics of topological solitons in DNA. Phys Rev E 2002;66:016614–016627.
68. Hermon Z, Caspi S, Ben-Jacob E. Prediction of charge and dipole solitons in DNA molecules based on the behaviour of phosphate bridges as tunnel elements. Europhys Lett 1998;43:482–487.
69. Chen Y-C, Zwolak M, Ventra MD. Inelastic effects on the transport properties of alkanethiols. Nano Lett 2005;5:621–624.
70. Pecchia A, Carlo AD, Gagliardi A, Sanna S, Frauenheim T, Gutierrez R. Incoherent electron-phonon scattering in octanethiols. Nano Lett 2004;4: 2109–2114.
71. Galperin M, Ratner MA, Nitzan A. Inelastic electron tunneling spectroscopy in molecular junctions: peaks and dips. J Chem Phys 2005;121:11,965–11,979.

7

Optimizing Photoactive Proteins for Optoelectronic Environments by Using Directed Evolution

**Jason R. Hillebrecht, Jeremy F. Koscielecki,
Kevin J. Wise, Mark P. Krebs,
Jeffrey A. Stuart, and Robert R. Birge**

Summary

Genetic engineering has recently emerged as a popular tool for tailoring biological macromolecules to function in nonnative environments. Most protein optimization efforts have advanced in large part as a result of significant advances in the methods and procedures of genetic engineering, most notably, directed evolution. Directed evolution mimics natural selection by combining techniques in genetic modification with differential selection. Most protein engineering research focuses on improving the thermal and chemical properties of enzymatic proteins for pharmaceutical applications. However, the recent emergence of nanobiotechnology has led researchers to broaden the scope of directed evolution. This chapter describes a strategy for tailoring the electronic and photochemical properties of proteins for performance in device applications. Among the many photoactive proteins found in nature, bacteriorhodopsin and its eubacterial counterpart, proteorhodopsin, are two leading candidates for protein-based device applications. The intrinsic stability, branched photochemistry, and photovoltaic properties of bacteriorhodopsin and proteorhodopsin make both proteins excellent candidates for three-dimensional volumetric memories, real-time holographic media, protein-based semiconductor devices, and artificial retinas.

Key Words: Bacteriorhodopsin; directed evolution; photonic devices; hybrid devices; volumetric memories; holographic memories; holography; retinal; photovoltaics.

From: *NanoBioTechonology: BioInspired Devices and Materials of the Future*
Edited by: Oded Shoseyov and Ilan Levy © Humana Press Inc., Totowa, NJ

1. INTRODUCTION

1.1. Rationale for Protein-Based Devices

The increasing demand for speed, miniaturization, and complex architectures has encouraged investigators to search for alternatives to traditional lithography. When fabricating materials no larger than one-thousandth of a micron (nanometer), lithographers frequently use fault-tolerant designs to cope with the high error rates and the thermodynamic interactions that occur between nanoscale structures. The caveat to using fault-tolerant designs is that they rarely match the high level of functional complexity that is observed in biological machinery *(1)*. A key point of the present chapter is that the combination of biotechnology and nanoscale manipulation may provide solutions that have a comparative advantage.

Microorganisms rely on a highly ordered assembly of macromolecules for intercellular communication, motion, metabolism, and information processing. Protein engineers presently use high-throughput screening methods in concert with novel mutagenesis strategies to tailor these biological molecules for performance in protein-based device applications. Current efforts in this field are directed toward the generation of protein-based circuitry, photovoltaic fuel cells, field effect transistors, motion-tracking devices, spatial light modulators, artificial retinas, three-dimensional (3D) removable memories, and holographic associative processors *(2–11)*. These protein-based devices offer a comparative advantage over modern-day semiconductors based on the scale, speed, and efficiency with which these molecules process information.

Over the course of natural evolution, genetic mutations accumulate and are manifested in changes in the structure and function of biological molecules. The accumulation of lethal mutations frequently leads to cell death, whereas the accumulation of beneficial mutations may enable the host organism to overcome a select environmental pressure. If nature can use genetic diversification and selective pressure to create highly efficient, genetically versatile machines, it is expected that they can be modified to function in protein-based devices. Consequently, directed evolution has become a staple of most endeavors to optimize biological macromolecules for performance in nonnative environments.

Directed evolution is a genetic strategy used to optimize the inherent properties of a biological macromolecule via iterative rounds of diversification and differential selection *(12–17)*. Investigators have classically used directed evolution to optimize the thermal stability, chemical stability, and substrate specificity of proteins used in therapeutic and industrial applications *(18–22)*. By using directed evolution to select for atypical phenotypes requires the concurrent development of a novel screening system. Traditionally, hyperthermophilic and psychrophilic microorganisms have been used to express and expose mutant libraries to extreme temperatures *(17,23,24)*. As the stringency of the thermal selection is increased, the surviving genetic variants are isolated and characterized for one of more beneficial mutations. This approach has been used to enhance the melting temperature (T_m) of an enzymatic protein by more than 30°C, relative to the native form of the enzyme *(25)*.

With the advent of bio-nanotechnology, investigators are now looking to optimize photoactive proteins to function in optoelectronic and photovoltaic environments *(22,26–28)*. A selection strategy for optimizing the optoelectronic and thermal properties of photoactive proteins will be discussed in the following sections.

1.2. Bacteriorhodopsin

Among the many macromolecules being investigated for use in bioelectronic devices are bacteriorhodopsin (BR), and its recently discovered cousin proteorhodopsin (PR). BR is a photoactive protein found in the outer membrane of the archaeon *Halobacterium salinarum (29,30)*. Within the lipid bilayer of the membrane, BR monomers are arranged in a 2D hexagonal lattice of trimers *(31)*. The crystalline arrangement of trimers enables BR to trap light in all polarizations and remain functionally active in the most inhospitable environments. Over the past three billion years, BR has adapted to environments with extreme temperatures, pH values, radiation, and salinity.

When environmental oxygen becomes scarce and aerobic respiration demanding, *H. salinarum* expresses BR to generate energy via photosynthesis *(32,33)*. The primary light-absorbing moiety of BR is an all-*trans* retinal molecule that is covalently bound to the protein via a protonated Schiff base linkage (Fig. 1A). Light absorption initiates a photocycle that is coupled to structural changes in the chromophore–protein environment. The primary photochemical reaction involves the conversion of all-*trans* retinal into a high-energy *cis* conformation. Interactions between the high-energy chromophore and the dynamic apoprotein environment are monitored as a series of spectrally discrete intermediates labeled bR, K, L, M, N, and O (Fig. 1B). Each photocycle results in the translocation of a single proton across the membrane of the organism, a process that operates with a quantum efficiency of 0.65 *(34)*.

One of the unique features of the BR photocycle is a branched pathway that is comprised of the P_1, P_2, and the near-permanent Q state (Fig.1B) *(35,36)*. Unless the O state is illuminated with red light, it will thermally decay back to bR. However, if the O state absorbs a photon of red light, it

A

B

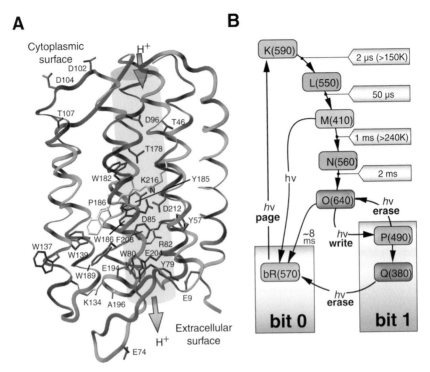

Fig. 1. The structure and photocycle of bacteriorhodopsin. Panel **A** highlights the important amino acids that contribute to the light-driven photocycle. The key intermediates, along with each absorption maximum, of the primary (bR, K, L, M, N, and O) and branched (P and Q) photocycle are shown in panel **B**.

will enter the branched photocycle. It is the branched photochemistry of BR that makes it possible to optically write, read, and erase data from the protein *(10,11)*. The origin of the branched photocycle is unclear. Although it is possible that this photochemistry is an unwanted artifact that has been minimized by evolution, the observation that a significant fraction of random mutants have lower-efficiency branching suggests that nature has not minimized but rather adjusted the efficiency. That is, the branching reaction provides some comparative advantage to the organism. One possibility is that the P and Q states serve as sunscreens to minimize DNA photodamage *(35)*. An alternative is that at high light intensity, it is advantageous to decrease the magnitude of photoactive protein by transferring a portion of the population to a non-proton-pumping state to avoid the generation of too large a pH gradient. Regardless of the biological origin, the presence of a branched photocycle provides an excellent template for 3D data storage as discussed under Subheading 2.

1.3. Rhodopsin

One might expect the visual pigment rhodopsin to be applicable to opto-electronic device applications because of its structural similarities with BR. However, the ejection and subsequent regeneration of the retinal cofactor in rhodopsin poses a major problem that is not easily solved *(34)*. Upon light activation, the chromophore is isomerized from 11-*cis* to all-*trans* and is subsequently ejected from the binding pocket of the protein *(37)*. The retinal cofactor is later transported back to the outer rod segment through a series of enzymatic reactions. The protein remains nonfunctional until the retinal cofactor is regenerated and spontaneously reconstituted into the apoprotein binding pocket, a process that can take up to several minutes. However, there are visual pigments expressed by certain invertebrates that provide binary responses without chromophore expulsion. Further investigation is needed to determine whether the unique optical qualities of these photoactive pigments provide them with a comparative advantage over existing counterparts.

1.4. Proteorhodopsin

PR is a light-transducing protein found in the membranes of an uncultivated clade of γ-proteobacteria known as SAR86 *(38,39)*. The protein was discovered by researchers using environmental sequencing techniques to characterize large-scale samples of DNA from oceanic samples off the coast of Monterey Bay in California *(40,41)*. Analysis of the DNA libraries revealed an open reading frame with a high degree of sequence homology with the bacterio-opsin (*bop*) coding sequence. Flash photolysis studies have confirmed that PR appears to pump protons from the intracellular to the extracellular face of *Escherichia coli* membranes, an observation that suggests that PR serves as a proton pump in vivo *(38,39)*.

Since the initial discovery of PR, investigators have identified several variant forms of the protein. The two primary subgroups of PR are characterized in terms of absorption maxima as green-absorbing (GPR) or blue-absorbing (BPR) variants. Significant discrepancies exist between the photochemical and optical properties of the two primary PR subgroups *(39,42,43)*. The green-absorbing variants possess a photocycle rate (~15 ms) that is reminiscent of BR, whereas the photocycle of the blue-absorbing variants is much less efficient (~150 ms) *(39,42)*. Although additional investigation of PR is needed, photonic device applications may benefit from the unique optical properties inherent to the PR subgroups.

For any of the above-mentioned proteins to function in high-temperature device environments, the thermal stability and inherent lifetime of the protein must be optimized. The following sections will review methods and strategies currently being employed to optimize photoactive proteins for performance in electronic environments.

2. BIOELECTRONIC DEVICE APPLICATIONS

2.1. Protein-Based Memory Architectures

There are three primary architectures under consideration that use proteins as the photoactive element: holographic binary, Fourier-transform associative, and branched-photocycle volumetric *(10,26,28,34,44–46)*. During the 1990s, most of the time and effort was directed toward optimizing the optical and electronic architectures to accommodate the characteristics of the proteins *(10,34,45,46)*. Much of the recent research in this area has been directed toward miniaturization of the memory architecture or optimization of the protein to accommodate lower-power lasers *(26,28)*. The purpose of this section is to introduce the reader to the architectures under study and describe those aspects of the protein that require further optimization. Virtually all of the work on protein memories has been carried out using BR, but our group and many others are currently investigating PR for these applications with encouraging results.

2.2. Holographic Memories

Holographic memories store information within a refractive index gradient that has been created by coherent laser beams that intersect an active volume element within a photoactive medium *(47–50)*. Most holographic optical memories use materials that undergo a change in refractive index in response to light, and the memories work best if the response is linear and results in a diffraction efficiency of 3% or greater *(47–53)*. Retinal proteins like BR and PR are useful photoactive components for holographic memories because these proteins photoconvert to form a blue-shifted species under irradiation. For clarity, we will limit our discussion to BR, but note that PR has very similar properties.

Short-term (real-time) holographic systems use the M state, which as shown in Fig. 2, is actually two states (M_1 and M_2) with identical absorption maxima (~410 nm). For the purposes of discussion, these two states are normally referred to as the M state. For long-term holography, the branched photocycle provides access to the P and Q states. The Q state has a lifetime of many years in the native protein and the long lifetime is preserved in all of the mutants that we have studied. The long lifetime of Q is believed to be associated with the unique properties of this state, which consists of a 9-*cis* retinal chromophore created by hydrolysis of the protonated Schiff-base linkage due to unfavorable steric interactions with the protein residues in the binding site *(35,36)*. Because 9-*cis* retinal can neither enter nor leave the binding site as

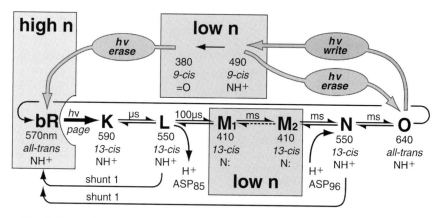

Fig. 2. The primary and branched photocycles of bacteriorhodopsin showing the key stable and metastable states in blocks. The M_1 and M_2 states have variable lifetimes, from 10 to 2000 ms based on the variant, whereas the Q state has a lifetime of many years. Thus, real-time holography uses the M_1 and M_2 states and long-term holographic data storage uses the Q state. The branched-photocycle volumetric optical memory uses bR to represent bit 0 and the P and Q combination to represent bit 1.

a result of steric constraints, the chromophore translocates to an unknown but nearby position within the binding site. Reversion back to bR requires 9-*cis* to all-*trans* isomerization, which has a barrier of approx 190 kJ/mol in the ground state *(10,35)*. The value of the Q state to data storage derives not only from the long lifetime but the efficient photochemical conversion of Q back to bR, which is not only efficient but occurs with high cyclicity under the appropriate conditions *(10,35)*. We note for completeness that the P state is actually two states (P_{525} [or P_1] and P_{445} [or P_2]) that form in sequence but remain in equilibrium under most conditions *(35)*. As is the case for the M state, we will use the term P state to reference both states simultaneously.

The key to creating a useful holographic material is to alter the refractive index in a region of the photoactive component where absorption is at a minimum. The change in refractive index associated with the formation of M and Q is wavelength-dependent and is shown in Fig. 3. Refractive and diffractive properties were calculated based on conversion of a solution of pure bR converted to a 50:50 mixture of the blue-shifted state (either M or Q). Wavelength-dependent diffraction efficiency was calculated using the Kogelnik equations *(54)*, which have been shown to yield reliable results when one bases the calculations on a 50:50 product mixture *(6,34,45,55-57)*. Although the Q state provides a slightly larger diffraction efficiency (8.5% at 665 nm) than that generated using the M state (6.4% at 670 nm), the M state can be formed with much higher efficiency. The distinct advantage of the Q state is lifetime.

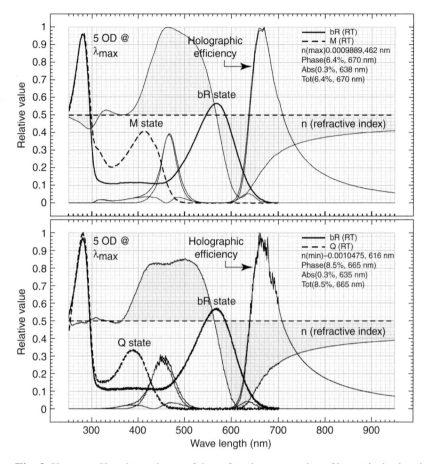

Fig. 3. Kramers-Kronig analyses of the refractive properties of bacteriorhodopsin films with protein concentration adjusted to yield an optical density of 5 at λ_{max} (570 nm). The top figure shows conversion to the M state and the bottom figure shows conversion to the Q state. Refractive and diffractive properties were calculated based on pure bR vs a 50:50 mixture of the blue-shifted state. Note that the Q state provides a slightly larger diffraction efficiency (8.5% at 665 nm) than that generated using the M state (6.4% at 670 nm). The wavelength-dependent refractive index change is also shown.

Space constraints preclude a discussion of holographic memory architectures. The interested reader is directed to the following papers and reviews for a discussion of protein-based holographic memories and other protein-based holographic applications *(6,10,26,57–62)*.

2.3. Branched-Photocycle Memories

Three-dimensional volumetric memories based on BR use the bR state to represent bit 0 and the P and Q states to represent bit 1. The latter states are generated by using the branching reaction shown in Figs. 1 and 2. The write process involves two steps. The first is called paging and involves the use of a tightly focused beam of light to initiate the photocycle in a thin sheet (or page) within a well defined region of the volumetric medium. The second uses an orthogonal write beam of light which has the horizontal and vertical dimensions of the page and onto which the data to be written has been imposed by using a spatial light modulator. The timing of the write beam is adjusted to maximally intercept the O state while rigorously avoiding the K state. The wavelength of the write beam is adjusted to a value which minimizes absorption by bR while maximizing absorption by O, a wavelength in the range of 640 to 680 nm. The write beam then initiates the branching reaction in those regions within the data cuvet where the paging beam and write beam have crossed, and in those regions only. The read process follows a similar sequence but uses a low-power write beam in which the spatial light modulator has turned on all the data bits. The image of the write beam is monitored by a 2D charge-coupled device (CCD) detector that has a resolution identical to the spatial light modulator. In those regions of the page in which no data have been written, the BR photocycle is initiated and O state is formed. The O state preferentially absorbs the light relative to those regions in which P and Q have been formed and when this page is imaged onto the array detector, those voxels with bit 1 are brighter than those with bit 0. The data are erased one page at a time by using a blue laser to the P and Q states and photochemically driving them back to BR. We have significantly oversimplified the details of how the branched-photocycle memory works, and the interested reader should consult ref. *10* for more details on the architecture. What is clear from the above discussion, however, is the importance of having a photochemically efficient branching reaction. The native protein has a low quantum efficiency for the O-to-P photochemical reaction ($\Phi \approx 0.001$) and provides a relatively low maximum concentration of the O state (~3%). The combination requires that a memory designed to use the native protein adopt relatively high-power write lasers (>100 mW). A significant effort in our research laboratory has been directed to improving the efficiency of the branching reaction.

3. OPTIMIZATION STRATEGIES FOR PHOTOACTIVE PROTEINS

3.1. Genetic Diversification Methods

Directed evolution has emerged as a powerful method for tailoring proteins to a variety of industrial, commercial, and therapeutic applications (*18–22*). Although the biochemical characteristics of thousands of enzymatic proteins have been elucidated, only a small percentage of those macromolecules are actually utilized in industrial processes. The reason for this small percentage is that most biological molecules are specialized to function in natural ecosystems rather than synthetic environments. Although researchers have gained momentum in elucidating the structural motifs responsible for enzyme activity, most of the intramolecular interactions contributing to photophysical properties remain unknown. As a result, novel genetic diversification methods and high-throughput screening strategies are needed for the photophysical optimization of photoactive proteins.

The most commonly used method for modifying the genetic coding sequence of a protein is known as site-directed mutagenesis (SDM). This targeted mutagenesis technique is highly efficient at replacing a single amino acid in the primary structure of the protein and provides a comparative advantage over less specific methods that rely on chemicals and ultraviolet (UV) light. However, the enormous number of unique substitutions that are possible in moderate to large proteins make it extremely challenging for investigators to probe the entire genetic landscape using SDM. A protein with 250 amino acids has 4750 single-mutation combinations and 11.2 million double-mutation combinations *(26)*. It would be nearly impossible for any laboratory to investigate several million genetic constructs in a reasonable amount of time using SDM methods. As a result, protein engineers are now turning to global diversification techniques and high-throughput screens to explore and optimize the genetic landscape of any protein of interest. Alternatives to SDM presently include random mutagenesis, semi-random mutagenesis, and a varied collection of in vitro recombination techniques *(63,64)*.

The random mutagenesis and in vitro recombination methods that have been developed in the past decade include the following: random and semi-random mutagenesis, DNA shuffling, the staggered extension process (StEP), and randomly primed recombination, to name a few *(17,18,63–66)*. In contrast to classic mutagenesis techniques, in vitro recombination methods are used to shuffle multiple genes or multiple variants of a single gene, thereby increasing the diversity of the mutant library. Recombination methods offer a comparative advantage over targeted mutagenesis because of the significant degree of sequence space that can be explored. The introduction of genetic diversity into large regions of a gene is especially important when probing for intramolecular associations that contribute to a desired phenotype of interest. Combining these genetic diversification techniques with an effective screening system is the most challenging hurdle now facing protein engineers. The remainder of this chapter will describe a specialized approach that can be applied to the photophysical and thermal optimization of a photoactive protein.

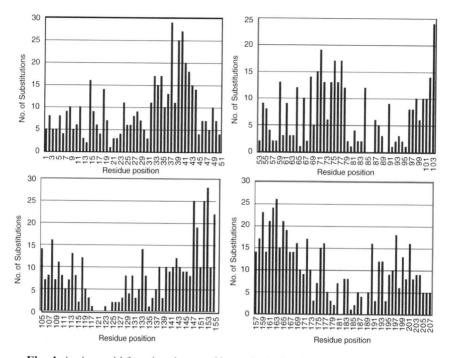

Fig. 4. Amino-acid functional map of bacteriorhodopsin created from a saturation mutagenesis study. Residue positions with no substitutions were deemed sensitive to mutation and did not form a purple membrane when expressed in *Halobacterium salinarum*.

3.2. Photophysical Optimization of a Photoactive Protein

Photoactive proteins typically have individual residues, or an isolated group of residues that contribute to the photochemistry of the molecule. To identify these residues, previous investigations have utilized semi-random (saturation) mutagenesis to create functional maps of the *bop* coding sequence *(26)*. A total of 800 mutants were created from 17 regions (~15 amino acids in length) of the *bop* gene. The variant *bop* genes possessed an average of 2.6 amino-acid changes (these mutants also possessed an additional 1.1 silent amino-acid changes) *(26)*.

The functional maps shown in Fig. 4 illustrate the number of amino-acid substitutions that occurred at each position of the *bop* gene. Interestingly, substitutions were not observed for several residues of known functional relevance to the protein. Many of these residues contribute to the steric and electrostatic environment of the retinal-binding pocket *(67,68)*. Future optimization studies can benefit from these maps by using site-directed saturation mutagenesis to explore the entire mutagenesis landscape at each of the previously identified positions of interest. The mutants that display the most desirable photophysical properties can then be used as a genetic starting point for optimization via directed evolution.

In addition to creating functional maps of the *bop* gene, time-resolved UV-vis spectroscopy was carried out to assess the photokinetic properties of all 800 semi-random mutants. The two primary targets of this photokinetics study include the most blue-shifted (M) and red-shifted (O) intermediates in the BR photocycle. The M state is readily detectable by spectroscopic techniques and has been tailored chemically and genetically for use in holographic devices *(58,59)*. The O state has been genetically modified for performance in binary photonic memory architectures because of its relevance to the branched photocycle of the protein *(35,36)*.

Of the 800 semi-random mutants that were analyzed for improved M- and O-state lifetimes, most possessed photokinetic properties that were comparable to the wild-type protein. Despite the vast number of variants with suboptimal kinetics, the semi-random study yielded a triple mutant (A139G/M145K/L146P) with an M-state lifetime of 1100 ms. It is interesting to note that the only mutant known to have a comparable M-state lifetime occurs at position 96. Mutating the aspartic acid at position 96 (the proton-donating group) to asparagine prolongs the lifetime of the M state to 1050 ms *(69)*. The D96N mutant has been shown to improve the holographic sensitivity of BR by a factor of 100, thus highlighting the value of genetic modification techniques.

Although M-state mutants may provide a comparative advantage for real-time holographic processing, long-term holographic data storage requires a more permanent photo-intermediate state. Figure 3 represents a two-state holographic system that involves the bR resting state of the protein and the near-permanent Q state. A single degree of angular separation is observed between the multiple diffraction peaks shown in Fig. 5. The high resolution and permanent nature of Q-based holographic diffraction media make it an ideal system for storing multiple blocks of data over an extended period of time. Accessing the Q state with high efficiency is the most formidable challenge in developing a Q-based memory system. Current efforts to optimize the branched photochemistry of the protein include the reduction of the all-trans → 9-*cis* barrier via a two-step genetic diversification strategy. This genetic strategy will be used in concert with an in vivo screening system designed to select for BR variants with high Q-state yields.

The primary phase of this two-step optimization strategy focuses on using semi-random mutagenesis to probe for regions of the *bop* gene that contribute to attractive O-state kinetics. By enhancing the lifetime and yield of

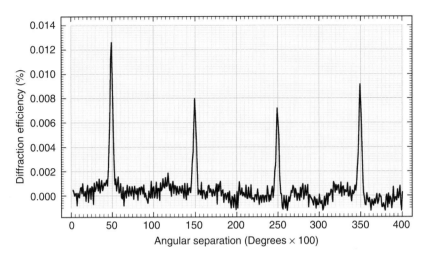

Fig. 5. Q-based diffraction peaks measured in a 1-cm BR poly(acrylamide) cube having an optical density of 1.5. Diffraction peaks were generated from diffraction patterns in the cube having a 1° angular separation.

the O state, the likelihood of accessing the P and Q states via binary photonic activation is improved. The semi-random mutants with the most attractive photokinetic properties will then be used as templates for in vitro recombination and high-throughput screening in the second phase of optimization. The screening system will be designed to reduce the barrier to the branched photocycle by selecting for variants with mutations that favor the 9-*cis* geometry of the chromophore.

The O-state kinetics for the 800 semi-random mutants were analyzed and plotted in Fig. 6. The histograms show that mutations made to residues in the FG-loop region of the protein had the greatest impact on the lifetime of the O state, relative to the native protein. Because certain residues located within the FG-loop region (positions 194 → 208) are known to expedite the release of a proton to the extracellular matrix during the photocycle, introducing genetic diversity into this region has resulted in novel photokinetic properties of the O state. Two of the most commonly studied residues in the FG-loop region are the glutamic acids at positions 194 and 204. Genetic modification of one or both of these residues leads to lengthened O-state lifetimes, relative to the native protein. Not surprisingly, the mutant with the greatest O-state yield was E194A (Q-value = 0.327). The Q-value of a photo-intermediate state is determined by calculating the integral of the time-resolved absorbance trace, normalized to the optical density (at λ max) of the sample. The Q-value of wild-type protein is equal to 0.04. Using E194A as a parental template in

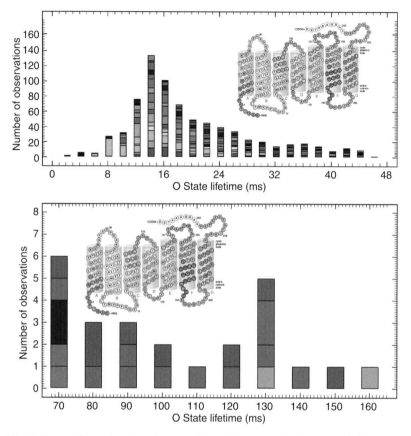

Fig. 6. Histograms showing the correlation between the O-state lifetime and the position of the mutated residue(s). The shaded regions of the bacteriorhodopsin structural model (inset) correspond to the shaded bars of the histogram. Similarly colored regions indicate those residues altering the O-state intermediate. The top histogram shows that most protein variants possess an O-state lifetime similar to the wild type. A number of variants were identified with significantly greater O-state lifetimes, in comparison to the native protein. These lifetimes are shown in the bottom histogram.

future optimization studies may further enhance the yield of O state and improve the access to the branched photocycle.

A sextuple mutant (A196S/I198L/P200T/E204A/T205Q/F208Y) with a 2-s O-state lifetime was also identified in the initial screen of the 800 semi-random mutants. The mutations found in this protein variant illustrate the intricacy of the relationships that exist between neighboring amino acids. Future studies will use site-directed methods to determine which residues are actively contributing to the observed photokinetic properties.

The second phase of the two-step optimization strategy will involve a high-throughput analysis of the *bop* gene by combining in vitro recombination methods with in vivo screening systems. The most promising M- and O-state mutants from phase 1 will be shuffled using in vitro recombination methods. This approach is designed to create novel intramolecular associations that will enhance the branched photochemistry of the protein. The recombinants will then be tested for enhanced Q-state yields by using a novel in vivo screening system. Figure 7 shows that individual colonies of *H. salinarum* can be indirectly screened for modulations in the branched photocycle of BR at 690 nm. The in vivo response data provides information on the amount of O state present and changes in the amount of O state are indirectly associated with the creation of P and Q. As in vivo screening technology matures, a broader range of proteins will be genetically tailored to function in optoelectronic environments, as opposed to their native ecosystems.

3.3. Thermal Optimization of a Photoactive Protein

The performance of a protein in an electronic environment depends on the stability of the macromolecule at elevated temperatures. Optimizing the thermal stability of a biological macromolecule is dependent on a suitable host organism for the expression of mutant proteins. For increasing the thermostability of BR, *Thermus thermophilus* was chosen.

T. thermophilus is an extremely thermophilic bacterium that grows in hot springs and industrial composts *(70)*. This thermophilic organism is highly proficient at incorporating foreign DNA into its genome and is well-suited for temperatures in excess of 80°C *(71)*. Recently, genetic tools were created for the expression of heterologous proteins in *T. thermophilus*. These tools include thermostable antibiotic-resistance selection and high-copy-number expression vectors *(72)*. Such tools make *T. thermophilus* an efficient system for expressing large numbers of mutant constructs at high temperatures.

Protein variants for expression in *T. thermophilus* are constructed using a combination of random mutagenesis and the in vitro recombination methods described above. Isolated protein is then screened in vitro (purified protein) for increased thermostability relative to wild-type protein. Those mutants that retain structural integrity at higher temperatures are isolated, tested for functional stability, and used as starting points for further rounds of mutagenesis and selection. Protein variants that retain functionally stability at the highest thermoselection temperatures are subjected to β-testing in biomolecular electronic devices. The performance of thermostable mutants in device applications determines whether or not additional optimization rounds are required.

The development of photokinetic and thermal screening systems is an important step in optimizing biological macromolecules for performance in hybrid

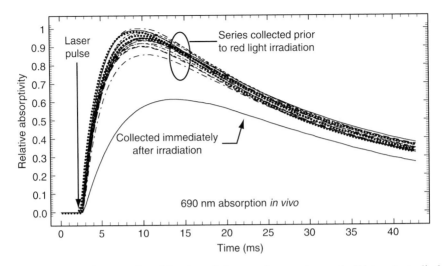

Fig. 7. The O-state photokinetics of the purple membrane (wild-type) studied within a colony of *Halobacterium salinarum*. The O state was measured at 690 nm every 5 min (dashed lines) prior to a 2-h red light (635 nm) irradiation. The solid line represents data collected immediately after red-light irradiation. The decrease in the absorption at 690 nm shows that the changes in the amount of O state are indirectly associated with the creation of Q state.

electronic device applications. The screening systems described above used BR as a template for optimization but can be applied to any photoactive protein.

4. CONCLUSIONS

Biological macromolecules have been optimized to function in natural ecosystems for nearly three billion years. During this time, microorganisms have relied on proteins for motion, intercellular communication, and information processing. Recent advancements in genetic engineering have provided scientists with the tools needed to tailor these molecular machines to nonnative environments. Optimizing the photophysical and thermal properties of photoactive proteins, such as BR, is a significant step in tailoring biological materials for performance in optical and electronic device architectures.

REFERENCES

1. Vsevolodov NN. Biomolecular Electronics. An Introduction Via Photosensitive Proteins. Boston: Birkhauser, 1998.
2. Xu J, Bhattacharya P, Varo G. Monolithically integrated bacteriorhodopsin/semiconductor opto-electronic integrated circuit for a bio-photoreceiver. Biosens Bioelectron 2004;19:885–892.

3. Li Q, Stuart JA, Birge RR, Xu J, Stickrath AB, Bhattacharya P. Photoelectric response of polarization sensitive bacteriorhodopsin films. Biosens Bioelectron 2004;19:869–874.

4. Xu J, Stickrath AB, Bhattacharya P, et al. Direct measurement of the photoelectric response time of bacteriorhodospin via electro-optic sampling. Biophys J 2003;85:1128–1134.

5. Koek WD, Bhattacharya N, Braat JJ, Chan VS, Westerweel J. Holographic simultaneous readout polarization multiplexing based on photoinduced anisotropy in bacteriorhodopsin. Opt Lett 2004;29:101–103.

6. Hampp N. Bacteriorhodopsin: mutating a biomaterial into an optoelectronic material. Appl Microbiol Biotechnol 2000;53:633–639.

7. Chen Z, Birge RR. Protein based artificial retinas. Trends Biotechnol 1993; 11:292–300.

8. Chen Z, Govender D, Gross R, Birge R. Advances in protein-based three-dimensional optical memories. BioSystems 1995;35:145–151.

9. Martin CH, Chen ZP, Birge RR. Towards a bacteriorhodopsin-silicon neuromorphic photosensor. In: Altman RB, Dunker AK, Hunter L, Klein TE, eds. Proc Pacific Symp Biocomputing. Maui: World Scientific, 1997:268–279.

10. Birge RR, Gillespie NB, Izaguirre EW, et al. Biomolecular electronics: protein-based associative processors and volumetric memories. J Phys Chem B 1999;103: 10,746–10,766.

11. Stuart JA, Tallent JR, Tan EHL, Birge RR. Protein-based volumetric memory. Proc IEEE Nonvol Mem Tech (INVMTC) 1996;6:45–51.

12. Arnold F, Moore JC. Optimizing industrial enzymes by directed evolution. Adv Biochem Eng 1997;58:1–14.

13. Rubingh DN. Protein engineering from a bioindustrial point of view. Curr Opin Biotechnol 1997;8:417–422.

14. Kuchner O, Arnold F. Directed evolution of enzyme catalysts. Trends Biotechnol 1997;15:523–530.

15. Miyazaki K, Arnold FH. Exploring nonnatural evolutionary pathways by saturation mutagenesis: rapid improvement of protein function. J Mol Evol 1999; 49:716–720.

16. Olsen M, Iverson B, Georgiou G. High-throughput screening of enzyme libraries. Curr Opin Biotechnol 2000;11:331–337.

17. Arnold F, Wintrode PL, Miyazaki K, Gershenson A. How enzymes adapt: lessons from directed evolution. Trends Biochem. Sci. 2001;26:100–106.

18. Dalby PA. Optimising enzyme function by directed evolution. Curr Opin Struct Biol 2003;13:500–505.

19. Kirk O, Borchert TV, Fuglsang CC. Industrial enzyme applications. Curr Opin Biotechnol 2002;13:345–351.

20. Morawski B, Quan S, Arnold F. Functional expression and stabilization of horseradish peroxidase by directed evolution in *Saccharomyces cerevisiae*. Biotechnol Bioeng 2001;76:99–107.

21. Sterner R, Liebl W. Thermophillic adaptation of proteins. Crit Rev Biochem Mol Biol 2001;36:39–106.

22. Whaley SR, English DS, Hu EL, Barbara PF, Belcher AM. Selection of peptides with semiconductor binding specificity for directed nanocrystal assembly. Nature 2000;405:665–668.
23. Hoseki J, Yano T, Koyama Y, Kuramitsu S, Kagamiyama H. Directed evolution of thermostable kanamycin-resistance gene: a convenient selection marker for *Thermus thermophilus*. J Biochem 1999;126:951–956.
24. Fridjonsson O, Watzlawick H, Mattes R. Thermoadaptation of alpha-galactosidase AgaB1 in *Thermus thermophilus*. J Bacteriol 2002;184:3385–3391.
25. Eijsink VG, Gaseidnes S, Borchert TV, van den Berg B. Directed evolution of enzyme stability. Biomol Eng 2005;22:21–30.
26. Wise KJ, Gillespie NB, Stuart JA, Krebs MP, Birge RR. Optimization of bacteriorhodopsin for bioelectronic devices. Trends Biotechnol 2002; 20:387–394.
27. Seeman NC, Belcher AM. Emulating biology: building nanostructures from the bottom up. Proc Natl Acad Sci USA 2002;99:6451–6455.
28. Hillebrecht JR, Wise KJ, Birge RR. Directed evolution of bacteriorhodopsin for device applications. Meth Enzymol 2004;388:333–347.
29. Oesterhelt D, Stoeckenius W. Rhodopsin-like protein from the purple membrane of *Halobacterium halobium*. Nature (London), New Biol 1971; 233:149–152.
30. Lanyi JK. Bacteriorhodopsin. Internat Rev Cytol 1999;187:161–202.
31. Sato H, Takeda K, Tani K, et al. Specific lipid-protein interactions in a novel honeycomb lattice structure of bacteriorhodopsin. Acta Cryst D Biol Cryst 1999;55:1251–1256.
32. Birge RR. Photophysics of light transduction in rhodopsin and bacteriorhodopsin. Annu Rev Biophys Bioeng 1981;10:315–354.
33. Ebrey TG. Light energy transduction in bacteriorhodopsin. In: Jackson MB, ed. Thermodynamics of Membrane Receptors and Channels. Boca Raton: CRC Press, 1993:353–387.
34. Birge RR. Photophysics and molecular electronic applications of the rhodopsins. Annu Rev Phys Chem 1990;41:683–733.
35. Gillespie NB, Wise KJ, Ren L, et al. Characterization of the branched-photocycle intermediates P and Q of bacteriorhodopsin. J Phys Chem B 2002;106: 13,352–13,361.
36. Popp A, Wolperdinger M, Hampp N, Bräuchle C, Oesterhelt D. Photochemical conversion of the O-intermediate to 9-cis-retinal-containing products in bacteriorhodopsin films. Biophys J 1993;65:1449–1459.
37. Wald G. The molecular basis of visual excitation. Nature 1968;219:800–808.
38. Béjà O, Aravind L, Koonin E, et al. Bacterial rhodopsin: evidence for a new type of phototrophy in the sea. Science 2000;289:1902–1906.
39. Béjà O, Spudich E, Spudich J, Leclerc M, DeLong E. Proteorhodopsin phototrophy in the ocean. Nature 2001;411:786–789.
40. Sabehi G, Beja O, Suzuki MT, Preston CM, DeLong EF. Different SAR86 subgroups harbour divergent proteorhodopsins. Environ Microbiol 2004;6:903–910.
41. Sabehi G, Massana R, Bielawski JP, Rosenberg M, DeLong E, Beja O. Novel proteorhodopsin variants from the Mediterranean and Red seas. Environ Microbiol 2003;5:842–849.

42. Wang WW, Sineshchekov A, Spudich E, Spudich J. Spectroscopic and photochemical characterizatin of a deep ocean proteorhodopsin. J Biol Chem 2003; 278:33,985–33,991.
43. Man D, Wang W, Sabehi G, et al. Diversification and spectral tuning in marine proteorhodopsins. EMBO J 2003;22:1725–1731.
44. Birge RR, Zhang CF, Lawrence AF. Optical random access memory based on bacteriorhodopsin. In: Hong F, ed. Molecular Electronics. New York: Plenum Press, 1989:369–379.
45. Birge RR, Fleitz PA, Gross RB, et al. Spatial light modulators and optical associative memories based on bacteriorhodopsin. Proc. IEEE EMBS 1990;12: 1788–1789.
46. Birge RR. Protein based optical computing and optical memories. IEEE Computer 1992;25:56–67.
47. Knight GR. Page-oriented associative holographic memory. Appl Opt 1974;13: 904–912.
48. Heanue JF, Bashaw MC, Hesselink L. Volume holographic storage and retrieval of digital data. Science 1994;265:749–752.
49. Song QW, Yu FTS. Holographic associative memory system using a thresholding microchannel spatial light modulator. Opt. Eng. 1989;28:533–535.
50. Paek EG, Jung EC. Simplified holographic associative memory using enhanced nonlinear processing with thermoplastic plate. Opt Lett 1991;16:1034–1036.
51. Curtis K, Psaltis D. Recording of multiple holograms in photopolymer films. Appl Opt 1992;31:7425–7428.
52. Gu C, Hong J, McMichael I, Saxena R, Mok F. Cross-talk-limited storage capacity of volume holographic memory. J Opt Soc Am 1992;A9:1978–1983.
53. Qiao Y, Psaltis D. Sampled dynamic holographic memory. Appl Opt 1992; 17:1376–1378.
54. Kogelnik H. Coupled wave theory for thick hologram gratings. The Bell System Technical Journal 1969;48:2909–2947.
55. Birge RR, Izgi KC, Stuart JA, Tallent JR. Wavelength dependence of the photorefractive and photodiffractive properties of holographic thin films based on bacteriorhodopsin. Materials Research Society Proc 1991;218:131–140.
56. Gross RB, Izgi KC, Birge RR. Holographic thin films, spatial light modulators and optical associative memories based on bacteriorhodopsin. Proc SPIE 1992;1662:186–196.
57. Zhang YH, Song QW, Tseronis C, Birge RR. Real-time holographic imaging with a bacteriorhodopsin film. Opt Lett 1995;20:2429–2431.
58. Hampp N, Juchem T. Fringemaker—the first technical system based on bacteriorhodopsin. In: Der A, ed. Nato Science Series I: Life and Behavioural Sciences, Bioelectronic Applications of Photochromic Pigments, vol. 335. Szeged, Hungary: IOS Press, 2000:44–53.
59. Juchem T, Hampp N. Interferometric system for non-destructive testing based on large diameter bacteriorhodopsin films. Optics and Lasers in Engineering 2000;34:87–100.
60. Marcy DL, Vought BW, Birge RR. Bioelectronics and protein-based optical memories and processors. In: Sienko T, Adamatzky A, Rambidi NG, Conrad M, eds. Molecular Computing. Cambridge, MA: MIT Press, 2003:257.

61. Stuart JA, Marcy DL, Birge RR. Photonic and optoelectronic applications of bacteriorhodopsin. In: Der A, ed. Nato Science Series I: Life and Behavioural Sciences, Bioelectronic Applications of Photochromic Pigments, Vol. 335. Szeged, Hungary: IOS Press, 2000:16–29.

62. Birge RR, Parsons B, Song QW, Tallent JR. Protein-based three-dimensional memories and associative processors. In: Ratner MA, Jortner J, eds. Molecular Electronics. Oxford: Blackwell Science Ltd., 1997:439–471.

63. Georgescu R, Bandara G, Sun L. Saturation mutagenesis. Meth. Mol. Biol. 2003;231:75–83.

64. Zhao H, Giver L, Shao Z, Affholter JA, Arnold FH. Molecular evolution by staggered extention process (StEP) *in vitro* recombination. Nature Biotechnol. 1998;16:258–261.

65. Crameri A, Raillart S, Bermudez E, Stemmer WPC. DNA shuffling of a family of genes from diverse species accelerates directed evolution. Nature 1998;391:288–291.

66. Graddis TJ, Remmele RLJ, McGrew JT. Designing proteins that work using recombinant technologies. Curr Pharm Biotechnol 2002;3:285–297.

67. Lanyi JK, Schobert B. Crystallographic structure of the retinal and the protein after deprotonation of the Schiff base: the switch in the bacteriorhodopsin photocycle. J Mol Biol 2002;321:727–737.

68. Luecke H, Schobert B, Richter HT, Cartailler JP, Lanyi JK. Structure of bacteriorhodopsin at 1.55 Å resolution. J Mol Biol 1999;291:899–911.

69. Hampp N, Popp A, Bräuchle C, Oesterhelt D. Diffraction efficiency of bacteriorhodopsin films for holography containing bacteriorhodopsin wild type BRwt and its variants BR_{D85E} and BR_{D96N}. J Phys Chem 1992;96:4679–4685.

70. Beffa T, Blanc M, Lyon PF, et al. Isolation of *Thermus* strains from hot composts. Appl Environ Microbiol 1996;62:1723–1727.

71. Friedrich A, Prust C, Hartsch T, Henne A, Averhoff B. Molecular analysis of the natural transformation machinery and identification of pilus structures in the extremely thermophilic bacterium *Thermus Thermophilus* strain HB27. Appl Environ Microbiol 2002;68:745–755.

72. Moreno R, Zafra O, Cava F, Berenguer J. Development of a gene expression vector for *Thermus thermophilus* based on the promoter of the respiratory nitrate reductase. Plasmid 2003;49:2–8.

8

DNA-Based Nanoelectronics

Rosa Di Felice and Danny Porath

Summary

We discuss the basic inspiration underlying the drive towards using DNA molecules for nanotechnological applications, and focus on their potential use to develop novel nanoelectronic devices. We thus review the current level of understanding of the behavior of DNA polymers as conducting wires, based on experimental and theoretical investigations of the electronic properties, determined by the π-π superposition along the helical stack. First, the importance of immobilizing molecules onto inorganic substrates in view of technological applications is outlined: selected observations by suitable imaging techniques are noted. Then, the emphasis is shifted to investigations of the electronic structure: disappointing evidence for negligible conductivity, from both theory and experiment, on double-stranded DNA molecules, has recently been counterbalanced by clear-cut measurements of high currents under controlled experimental conditions that rely on avoiding nonspecific molecule–substrate interactions and realizing electrode–molecule covalent binding. As a parallel effort, scientists are now tracing the route toward the exploration of tailored DNA derivatives that may exhibit enhanced conductivity. We illustrate a few promising candidates and the first studies on such novel molecular wires.

Key Words: DNA, experiment; nanoelectronics; nanoscience; theory.

1. INTRODUCTION

From a simply operational viewpoint, nanotechnology is a discipline with the objective of fabricating tools, machines, and devices of various kinds on the scale of 10^{-9} m, based on scientific principles that dominate at this scale. Paradigms for nanotechnology can be schematized in two classes: (1) the "top-down" approach consists of obtaining the desired tiny products by sculpting from bulky precursors; (2) the "bottom-up" approach is based on the opposite path, i.e., on assembling the nanoproducts directly

From: *NanoBioTechonology: BioInspired Devices and Materials of the Future*
Edited by: Oded Shoseyov and Ilan Levy © Humana Press Inc., Totowa, NJ

by using nanosized bricks. Nature provides scientists with such elementary nanosized building blocks: atoms and molecules. Thus, while physicists and engineers were working in the last two decades on optimizing the protocols for lithographically imprinting objects at the nanoscale, as well as on improving the tools for manipulating, visualizing, and measuring atoms and molecules (microscopies and spectroscopies), chemists were developing methods to synthesize and assemble molecules with pre-designed functions, reactivity, and controllable recognition. In addition, biologists were providing the chemists with special examples of molecules that in nature behave as precise devices by themselves (motors, switches, converters, assemblers, etc.). Hence, the field of bio-nanotechnology arises as a highly interdisciplinary enterprise that may be defined as the use of biomolecules and bioengineering tools to develop nanotechnology. The target applications span a huge range, from medicine and pharmacology (biocompatible tissue implantation, drug design and delivery) to nanomechanics and nanoelectronics (mechanical, electronic, optical circuits of decreasing size and increasing power), through heterogeneous catalysis and chemical sensing (nanoparticles employed in chemical reactions to reduce pollution by consuming exhaust gases, ion channels). Conceptual proposals *(1,2)* and reviews of the progress achieved so far *(3,4)* can already be found in journals and books. In particular, Merkle highlights the relative roles of self-assembly and directed-assembly *(2)* on the way toward the exploitation of biotechnology tools into nanotechnology.

In this introductory section, we guide the reader to observe examples of (bio)nanodevices occurring in nature, and identify the basic principles that rule the construction and operation of such devices, from which we should draw inspiration for the development of artificial bio-nanodevices (Subheading 1.1); then, we present the main molecular actors on the natural bio-nanotechnological stage, i.e., proteins and DNA (Subheading 1.2), and focus on one of them, DNA, for nanoelectronic applications (Subheading 1.3). Nanoelectronics is only one among a variety of programmable nanotechnological exploitations of biomolecules. Subheading 2 is devoted to explaining the importance of molecular immobilization onto inorganic supports, and to show examples of successful attachment of DNA molecules onto solid surfaces. Subheading 3 then comes to the core issue (electronic properties) related to the use of DNA molecules for the fabrication of devices able to conduct electricity: we report the current level of evidence for the conductivity of native DNA and introduce alternative DNA-like candidates, based on topological and chemical manipulation of the double helix, with potentially enhanced conductivity. Finally, we draw conclusions and speculate on perspectives under Subheading 4.

1.1. Inspiration from Nature as a Factory—Toward Self-Assembled Nanoelectronics

Remarkable *natural* examples of nanodevices exist in various biological contexts: they may be taken as excellent starting points for designing the artificial nanomachines of the future. All of them draw attention to the key mechanism of *self-assembly* to realize complex objects through *supra-molecular chemistry (5)*. We note just a couple of them to illustrate the basic concepts to realize *artificial* nanodevices.

Ribosomes are biological self-assembled plants for the production of functional objects, i.e., proteins *(6,7)*. These fascinating cell components represent the prototype for realizing biomimetic bottom-up strategies for the self-assembly of functional devices using programmed molecular building blocks. Ribosomes behave as protein assemblers, joining the constituents on the basis of *recognition* principles. The natural plant operates in sequential steps that are repeated in a cyclic routine until the desired protein is synthesized, as schematically represented in Fig. 1. First, a ribosome immobilizes a messenger RNA (mRNA) strand by coupling to it through nonchemical interactions: the mRNA strand brings the sequence of the in-fieri peptide chain encoded in a sequence of codons, each of which is a set of three nucleotides and specifies one of the 20 amino acids that constitute natural proteins. Second, a specialized enzyme working outside the ribosome specifically (but not uniquely) associates a given amino acid to a partner transfer RNA (tRNA) molecule by recognizing its terminal codon, and the amino acid is covalently bound at the extremity of the tRNA polymer opposite to the specific codon (*see* Fig. 1). Third, the tRNA with the attached amino acid is captured by the ribosome, and its terminal codon recognizes the complementary codon on the immobilized mRNA strand. Fourth, the tRNA molecule is released and a piece of the peptide chain has been put in place. Fifth, another tRNA with its attached amino acid recognizes the following codon on the mRNA strand, comes close to the previous one and the two neighboring amino acids form peptide bonding, then the tRNA fragment is released and the cycle goes on through steps three to five. In this fascinating spontaneous procedure, covalent interactions are important, but much goes on by virtue of recognition and self-assembly, assisted by the tailored structuring of the participating species (e.g., the formation of the loop that encloses the terminal codon in tRNA; *see* Fig. 1). To mention just a few key recognition steps, crucial ones are the identification of the initial base triplet in mRNA, and the association between complementary mRNA-tRNA codons.

Driven by the observation of these important principles that nature uses to operate its machines, scientists proceeded towards the fabrication of artificial devices based on the same rules.

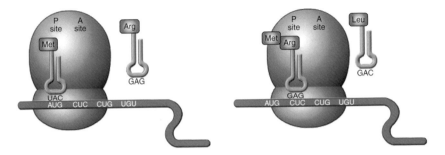

Fig. 1. Two successive steps in the protein production process performed by ribosomes. The bulbed regions represent the large (top) and small (bottom) parts of a ribosome. The horizontal stripe is the messenger RNA (mRNA) whose sequence is translated into an amino acid chain. The amino acids are transported by transfer RNA molecules (tRNA, looped stripes), whose terminal codon can recognize a complementary codon on mRNA. Through cyclic attachment of this kind, the whole sequence on the mRNA molecule is read and translated into the amino acid chain. (From ref. 7, by permission; © 2000 Freeman & Co.)

Rotaxanes and catenanes are natural switches from organic chemistry *(8)*. Figure 2 illustrates schematic models and specific molecular examples of catenanes (left panel) and rotaxanes (right panel). A catenane is essentially constituted of two interlocked rings and can work as a molecular switch *(8)*: in fact, the position of one of the rings can be moved between two stable sites on the companion ring, by either electrochemical (as in the example in Fig. 2) or optical activation. Chemical interaction is responsible for the stability of either site under changing environmental conditions, e.g., under the addition or subtraction of an electron or photon. A rotaxane is constituted by a circular molecular portion around a linear molecule terminated by two caps that prevent the circular part from slipping out. It can similarly act as a switch. Several other examples of natural nanodevices based on organic molecular mimics can be found in recent books and review articles *(8–11)*. Whereas covalent and electrostatic forces control the operation of such devices, self-assembly has been important in achieving chemical synthesis procedures with a high yield.

Rotaxanes and catenanes are particular examples of nanodevices that perform mechanical work *(10)*. However, there is also huge interest and invested effort in constructing other kinds of nanodevices on the basis of the same rules of supramolecular chemistry: by employing elementary building blocks that are able to recognize each other and form complex arrangements and adapting their structure to the environment and to their partners. In particular, self-assembling nanodevices that express an electrical functionality are

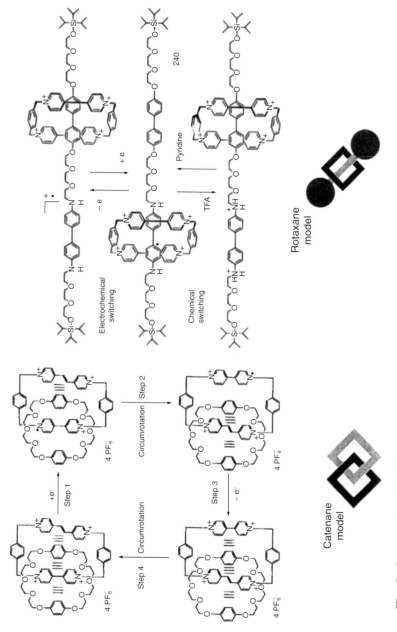

Fig. 2. Catenane-based and rotaxane-based (left and right, respectively) electrochemical switches. The change of position of the mobile ring molecule between two stable sites is activated by reduction-oxidation reactions. (From ref. 8, by permission; © 1999 American Chemical Society.)

145

pursued in view of the development of molecular electronics *(12–14)*. In the following, we focus on investigations aimed at realizing this class of devices.

1.2. Classes of Biomolecules that Perform Specific Tasks: Proteins and DNA

The basic approach toward the realization of self-assembled electrical nanodevices is to employ natural building blocks capable of recognition and structuring, and drive their aggregation in a suitable environment, with the help of artificial "plants" having a role similar to that of ribosomes. Nature again provides us with two classes of ideal candidates, namely proteins and DNA. The former have several biological functions and perform tasks that allow living organisms to operate: among the plethora, there are proteins that behave as mechanical devices (e.g., myosin in muscles *[7,9]*), and others that exchange electrons between interacting partners (e.g., azurin in bacterial respiration *[15,16]*).

In particular, some scientists have been inspired to exploit azurin in artificial electronic nanodevices based on its biological function as an electron shuttle *(17,18)*. Azurin monolayers deposited on a substrate and located between two metal pads have indeed been demonstrated to be capable of performing both diode-like *(17)* and transistor-like *(18)* activity. The extraordinary fact in such devices is that there was no need for expensive fabrication techniques to deposit the active materials through sophisticated lithographies, but the molecules themselves self-assembled spontaneously, by means of docking mechanisms mainly affected by their electrostatic properties *(15,17)*. The mechanism for current flow was interpreted in terms of electron/hole hopping between adjacent proteins *(18)*, by virtue of redox reactions at the copper centers *(19)*. This is a demonstration that the intrinsic redox activity responsible for electron transfer through azurin in the biological environment can be efficiently exploited in an artificial hybrid environment.

The other main class of biomolecules is constituted of nucleic acids. Whereas proteins perform all the active tasks of life and they are indeed natural nanodevices, nucleic acids are containers and translators of information. The genetic information enclosed in DNA is precisely replicated at each cell reproduction, is read by mRNA and translated by tRNA into the protein language (Fig. 1). Native DNA polymers do not perform any functional activity except for coding; therefore they are seen as perfect candidates for biocomputing *(20)*. What are extremely fascinating in DNA are its structuring and recognition, which are crucial properties for the storage and transfer of information and that make nucleic acids an optimal self-assembling material. The basis of such capabilities is the hybridization between complementary strands, which is exact and has a high yield. This feature allows the construction and disruption of very exotic structural motifs *(21,22)* by means of chemical synthesis and bioengineering, in all ranges from one to three dimensions, which can be effectively employed to design and realize nanodevices *(23)*. Indeed, nanomachines based on various conformational transitions have been demonstrated: the transition between the B and Z forms *(24)*, between a double and a triple helix *(25)*, and between a double and a quadruple helix *(26)*.

1.3. What Is the Role of DNA? Can It Be Modified to Include Storage and Transfer of Electrons?

Nature bases its operation on proteins to perform actions of any sort. DNA is the material that encodes and transfers the information to fabricate proteins through a spontaneous process. Therefore, the possible exploitation of DNA in nanotechnology cannot be based on a straightforward translation of an inherent biological action.

On the one hand, various conformational transitions have been proposed to generate motion from DNA and thus realize a mechanical device *(24–26)*; alternatively, a DNA-based mechanical device can even be fueled by DNA by virtue of the elastic response *(27)*. In these examples, DNA is guided to perform a certain action that is not proper for it in nature, based on its recognition and structuring capabilities: in practice, the potentialities of DNA are embedded in its conformation and topology. Other notable applications envisage the two-dimensional (2D) and 3D assembly of complex objects (cubes, octahedra, etc.) made with DNA *(21,22)* onto organized chips to recognize and position other biological materials, with applications in diagnostics and medicine.

On the other hand, scientists are fascinated by the issue of whether or not the DNA tasks of storage/transfer, which are naturally applied to the *property information*, can be artificially directed to the *object electrons*. Consequently, research initiatives were launched worldwide to explore the conductivity of DNA *(12,28–35)*. Alternatively, if measurable currents cannot be sustained by DNA molecules, another interesting strategy is to realize hybrid objects (metal nanoparticles/wires, proteins/antibodies, etc.) in which electrons move and carry current flows, templated by DNA helices at selected locations *(12–14,36,37)*: this route also allows one to embed conducting objects into the hybrid architectures to realize, for example, a carbon nanotube DNA-templated nanotransistor *(14)*. Both of these methods could lead to the development of DNA-based molecular electronics *(38–40)*.

In the rest of this chapter, drawing mainly from our own experience, we discuss selected examples of the experimental and theoretical achievements on the way to understanding the suitability of native DNA and DNA derivatives to sustaining electrical currents in the biopolymer itself, without the need for hybrid components (except for metal ions). The drive toward the exploration of this route is discussed elsewhere *(38)*: although rather speculative, it originated from natural evidence that oxidative damage can be transported through the DNA polymer and influence radiation damage *(41)* and cancer propagation, and continued with the long-standing speculation of a π-way through the helix axis *(42)*.

2. IMMOBILIZATION OF DNA ON SUBSTRATES

The first evident oddity when one starts to work on nanotechnology applications of biomolecules, both for the realization of devices and for the use of suitable instruments to probe their properties, is that the biological objects must interact with the inorganic world. Within our focus on DNA-based nanoelectronic devices, we are mainly concerned with the deposition onto substrates and the contact with metallic leads. Deposition onto substrates is relevant for device applications on the one hand, because, for instance, in field effect transistors (FETs) *(43)*, the conducting material is deposited onto an insulating layer that is then gated, besides being connected between the source and drain electrodes. On the other hand, for various powerful techniques employed by nanotechnologists to image and measure the electrical and mechanical properties of materials, the molecules must be hosted on an insulating or metallic surface, depending on the particular microscope.

In this section, we illustrate the effects of substrate immobilization, with particular attention to whether the DNA molecules maintain their conformation or to what extent they are modified, when they are deposited onto mica for atomic force microscopy (AFM) imaging, and onto gold for scanning tunneling microscopy (STM) imaging. In particular, AFM has become popular in the last two decades for imaging biomolecules, and good reviews describing the principles of operation and particular aspects and results for biological applications of both techniques can be found *(44–52)*. The reader is referred to these reviews, and to references therein, for a comprehensive overview of the variety of applications (from measurement of elastic constants, to unwinding and mechanical response, to morphology) of scanning probe microscopies to nucleic acids. In the following, we focus on the investigation of morphology and electrical response. We point out that both AFM and STM can be applied in spectroscopy mode to measure transport characteristics along and across single molecules (Subheading 3): that is why they are important in connection to DNA-based electronics, to probe the electrical properties of the DNA polymers. In order to not misinterpret the electrical results it is, however, important to understand if and how the measurement setup affects the molecular structure.

2.1. Atomic Force Microscopy Imaging

The clearest results emerging from AFM structural investigation of double-stranded DNA on inorganic surfaces are the following: (1) the images reveal an apparent molecule height smaller *(53,54)* than the native helix diameter in crystallized molecules *(55,56)*; and (2) the surface field forces can induce deformations in terms of persistence length and stability, thus likely affecting the π-stacking and conductivity. We briefly illustrate and comment on this evidence. A similar (but less pronounced) behavior is also expressed by G4-DNA *(57)*, a quadruple helix conformation made of only guanines, which is presented under Subheading 3.3.

Since the early morphological reports *(58,59)*, AFM images of DNA in various forms have revealed molecular heights that are always significantly smaller than the nominal height of approx 2.1 nm. Such discrepancies were initially explained in terms of electrostatic molecule-tip interactions *(60)*. More recently, after substantial instrumental progress and residual evidence of the height-diameter discrepancy in DNA molecules, closer inspection has revealed that indeed the molecules are modified by interaction with the substrate *(53,54)*. In particular, Vesenka and co-workers proposed a model that imputes the flattening of DNA polymers on mica to the various treatments employed to render mica electropositive in order to attach DNA. In the absence of any treatment, the negative phosphate backbone would be repelled by the negatively charged surface. In the presence of positive centers on the surface, the phosphates may, however, interact chemically or electrostatically, disrupting H-bonds on the basis of an energetic balance *(53)*.

By means of spreading-resistance-microscopy (SRM) performed using AFM, Kasumov and collaborators recently showed that the structural distortion experienced by single DNA molecules when they are deposited onto inorganic substrates might have consequences related to the conduction properties *(54)*. In fact, when the molecules were laid on an untreated mica+Pt substrate, the average height measured by AFM was about 1.1 nm and the SRM signal showed negative contrast. On the contrary, when substrate preparation with an organic monolayer was carried out prior to DNA deposition, then the average height measured by AFM was about 2.4 nm and the SRM contrast was positive (*see* Fig. 3). The authors interpreted the SRM change of contrast in terms of a change in molecular conductivity, namely a transition from an insulator to a conductor. The AFM data were also supported by differential conductivity curves at low temperatures. Whereas the change of contrast is not a direct measurement of conductivity and cannot be taken as ultimate proof of the ability to conduct, this work *(54)* undoubtedly proved that direct deposition of DNA onto inorganic supports changes its morphology and its electrostatic response.

The above discussion also suggests that all the electrostatic force microscopy (EFM) and conductive AFM (cAFM) experiments conducted on single DNA molecules deposited on mica *(30,61)* may be largely influenced

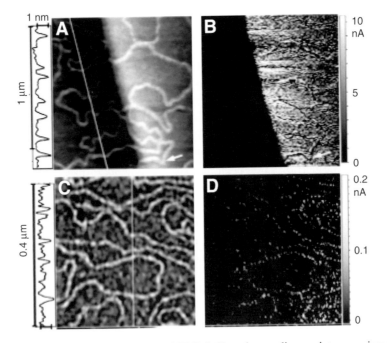

Fig. 3. Atomic force microscopy (AFM) (left) and spreading-resistance-microscopy (SRM) (right) images of single DNA molecules on a mica+Pt substrate. **(A)** AFM image of DNAs on the clean substrate without any treatment prior to DNA deposition. **(B)** SRM image of the same molecules (right bright part of A and B images is Pt). **(C)** AFM picture of DNAs on the substrate treated with an organic penthylamine layer before DNA deposition. **(D)** SRM image of the same molecules, Pt electrode is outside of the image. (From ref. *54*, by permission; © 2004 American Institute of Physics.)

by the effect of substrate-induced deformations, with consequent interruption of the regular helical motif. This conclusion also applies for electrical transport measurements conducted through long DNA molecules (>40 nm) laid on surfaces between electrodes *(12,33)*.

2.2. Scanning Tunneling Microscopy Imaging and Transverse Spectroscopy

Morphological and spectroscopic characterization of DNA molecules by STM is still in its infancy *(62–71)*. We summarize the scant information available to date culled from using this technique on single DNA molecules, emphasizing examples from our work.

Shapir and co-workers recently conducted STM experiments *(62,67)* on uniform-sequence poly(dG)-poly(dC) long DNA molecules synthesized by a novel enzymatic technique *(72)*, as well as on G4-DNA *(57)* and on native DNA. They observed both spontaneous and controlled contrast inversion, and explained this puzzling evidence by a novel mechanism. In the past, contrast inversion in STM images of molecules was attributed to structural deformations induced by the measurement setup. Via interplay between an accurate analysis of the experimental data and theoretical modeling, the authors propose an alternative explanation, namely that the observed contrast is influenced by tunneling through virtual states in the vacuum between the STM tip and the DNA molecule, which are created by the curvature of the field lines induced by irregular charge distribution *(67)*. Figure 4A,B shows selected observations of the discussed phenomenon; Fig. 4C shows a density plot of the height profile, computed for different current settings using the suggested model. In the same work, the authors also demonstrate the dependence of the apparent molecular height on the bias voltage, when the current is kept constant.

The morphological aspects are examined separately by the same authors *(62)*. They imaged poly(dG)-poly(dC) molecules with a nominal length of 4000 bp, immobilized on Au(111): immobilization is a crucial step in imaging DNA, because otherwise, the STM tip can displace the molecules during scanning. As had already been shown in many previous studies, the apparent molecular height depended on the voltage, again proving that STM probes the "electrical" rather than the "geometrical" height, because the technique is sensitive to the density of states of the sample. Figure 5 illustrates high-resolution images, where the longitudinal periodicity is assigned to the pitches along the helices, and the statistical data. This work proves the extreme capabilities of the STM technique for resolving the molecular structure of DNA. Other examples of high-resolution imaging were demonstrated with much higher resolution in works published by the group of Kawai at Osaka, and others.

Although we do not describe here recent works that address the relation between STM spectroscopy of single DNA molecules deposited on metal surfaces and the density of electronic states of such molecules *(68–71)*, we note that also in those experiments, the molecules lie horizontally and therefore nonspecific substrate-molecule interactions may play a role.

3. PROBING THE ELECTRONIC PROPERTIES OF SINGLE DNA MOLECULES

The AFM and STM are also powerful tools for measuring the electrical properties of DNA molecules immobilized onto surfaces. Usually, an imaging analysis by any of these techniques accompanies the spectroscopic investigations, to demonstrate that one is really probing the electrical properties of molecules, rather than of anything else.

However, because the substrate deforms the inherent molecular structure, as discussed above, with consequences to the electronic response, there is a lively effort to develop cAFM and STM experimental setups that avoid longitudinal nonspecific direct molecule–substrate contact *(73,74)*.

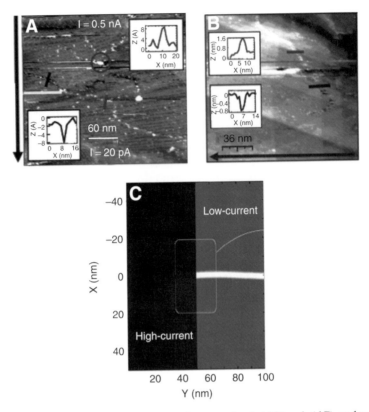

Fig. 4. Scanning tunneling microscopy images of poly(dG)-poly(dC) molecules on Au(111) and computed density plot. **(A)** Image obtained by scanning downward: voltage V_b = 2.8 eV, T = 300 K. The image shows one spontaneous contrast inversion (bright spot marked by the circle) and one controlled contrast inversion induced by changing the current: by decreasing the current from 500 pA to 20 pA, the appearance of the molecule switches from dark to bright. The insets show the relative height profiles in two selected points: the lower (upper) line is at the point marked by a dark (lighter) segment. **(B)** Image obtained by scanning right-to-left: V_b = 2.8 eV, T = 300 K, I_s = 50 pA. The image shows a spontaneous contrast inversion, obtained under no changes of the imaging parameters. The upper and lower insets show the height profiles at the points marked by the upper and lower segments, respectively. **(C)** Computation of the molecular contrast through the proposed model that accounts for tunneling through vacuum states: the molecule appears dark (negative relative height) at high current and bright (positive relative height) at low current. (From ref. *67*, by permission; © 2005 American Chemical Society.)

In addition to cAFM and STM spectroscopies, a few direct electrical transport measurements of short DNA molecules between lithographically defined nanoelectrodes have been published *(29,38)* and are only briefly mentioned here.

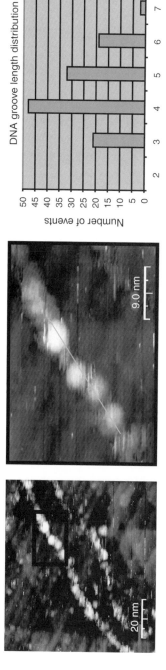

Fig. 5. High-resolution scanning tunneling microscopy images (left and middle) of poly(dG)-poly(dC) molecules immobilized on a gold (*111*) surface by electrostatic trapping, and statistical analysis of the longitudinal periodicity (right). The central image shows a close-up of the left image on the black rectangle. The statistics is done over about 200 sites taken on about 20 molecules. (Adapted from ref. *62*, by permission; © 2006 American Chemical Society.)

3.1. Measurements on DNA Double Helices with Different Experimental Techniques

The experimental approaches mentioned above were applied to explore the electrical response of native DNA molecules (*see* schemes in Fig. 6), along with other methods *(28,75)* that are not addressed here.

As a general observation, we note that recently an overall consensus is developing toward attributing the initial scattered behaviors, insulating to (semi/super)conducting (*see* recent reviews *[38–40]* and references therein), to the variety of experimental conditions. Not only are the measurement technique and molecular phase (single molecule versus networks or bundles *[38]*) important in determining the outcome: how the setup affects the structure of the molecules themselves and modifies the π stack is important as well. How much are the observations due to the target molecule and how much to the modifications which are induced when trying to reveal it?

In relation to these questions, we focus under Subheadings 3.1.1 and 3.1.2 only on the latest reports that are devoted to minimizing the influence of the experimental setup on the molecule, to probe the "intrinsic," rather than "inducted," electron/hole mobility. Although the two selected methods employ different instruments (AFM, STM) and are applied in diverse environments (dry, wet), the principle of measurement is the same, namely a standing-molecule configuration to avoid nonspecific molecule–substrate interactions along the molecular length and to realize covalent molecule-electrode attachment.

3.1.1. Longitudinal Current-Voltage Characteristics by Conductive AFM

cAFM data can be collected by depositing DNA molecules on an insulating substrate (typically mica), covering part of it with a metal electrode, and contacting a metallized tip at various points along a given molecule, thus also investigating the length dependence of the molecule's conductance. cAFM measurements are usually preceded by EFM signal detection *(30,61)*: EFM portrays the electrostatic polarizability. On the one hand, EFM has the great advantage of being a contactless technique, thus avoiding one of the big issues in measuring molecular nano-junctions, namely the role of molecule–electrode contacts relative to the molecular electronic structure. On the other hand, EFM suffers from two significant disadvantages: (1) it is not a direct electrical measurement and always must be accompanied by electrical data; (2) the molecules lie on the surface, thus experiencing possible nonspecific molecule–substrate interactions. cAFM setups, which allow one to reveal current-voltage curves directly, retain the latter disadvantage, while losing the contactless quality (Fig. 6A). Cohen and co-workers *(73)* recently developed cAFM geometries that are simultaneously able to avoid molecule–substrate interactions and control the molecule–electrode binding.

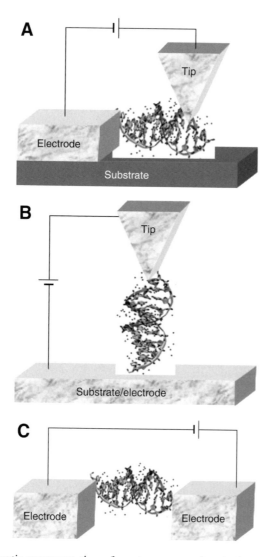

Fig. 6. Schematic representation of most common electrical measurement setups adopted to investigate electrical transport in DNA molecules. (**A**) conductive atomic force microscopy (cAFM) on "laying" molecules: the substrate is insulating, a metal electrode partially covers the target molecule, and voltage is applied between the electrode and the metal-coated tip *(30,61)*. (**B**) cAFM *(73,80)* and scaning tuneling microscopy *(74)* on "standing" molecules on a metal substrate that acts as one electrode: voltage is applied between the metallic or metal-coated scanning tip and the substrate; thiol end-groups are employed to hook the molecule to both measuring electrodes. (**C**) "Bridging" molecules are trapped between two fixed metal electrodes; electrostatic trapping may be used to capture the DNA molecules *(29)*; in principle, thiols can be employed to realize covalent attachment.

These measurement geometries are inspired by a pioneering work on alkanethiol monolayers published a few years ago *(76)*, and are conceptually similar, although with distinct features, to the recent STM-based technique developed by NongJiang Tao at Arizona State University in the United States *(74)*. Other analogous efforts are ongoing worldwide *(77)*.

The two criteria of the methodology reported by Cohen and co-workers *(73)* to control unequivocally the measurement setup are the following: (1) the molecules must be "standing" rather than "laying" on the substrate; and (2) covalent attachment to both measuring electrodes must be achieved. These objectives are realized simultaneously by depositing a thiolated single-strand DNA monolayer on a gold surface, and then arriving with complementary thiolated single-stranded DNA molecules connected to 10-nm gold nanoparticles *(78)*. A metallized AFM tip, covered with Cr and Au successively, then approaches the Au nanoparticles to perform the electrical measurements: contact is established between the Au layer of the tip and the gold nanoparticle. Because the thiol-gold bond is known to be covalent *(79)* (substrate–molecule, molecule–nanoparticle), the covalent attachment at both electrodes is under control, with no arbitrariness. Therefore, measurements done in this way are comparable. In addition, because one end of each double-stranded DNA polymer is attached to the surface and the other to a nanoparticle, the molecules are in a "standing" configuration: not necessarily perpendicular, but definitely not horizontal, such that the helical conformation is not grossly disturbed by the measurement setup. Note that the technique does not guarantee that just one double-stranded DNA molecule is linked to a nanoparticle, but a rough estimate based on the surface area and on the helix dimensions ensures that only a few DNA molecules may be connected in parallel between the two electrodes. By this "standing" cAFM situation, 26-bp double-stranded DNA molecules of nonuniform sequence were measured: currents as high as approx 200 nA at 2V are detected. The I-V curves generally have an S-shape; typical resistances are approx 60 MΩ between −1 and +1 V, as low as 2 MΩ at 2 V. The overall shape of the current-voltage characteristics is highly reproducible in consecutive sets of measurements, although the quantitative details differ. The authors also provide the results of several control experiments done in order to check that the high currents are flowing through the DNA molecules and are not due to any artifact. Figure 7 depicts a scheme of the measurement geometry and a set of I-V curves, along with an AFM image of the nanoparticles on the substrate in the inset.

In a more recent work *(80)*, Cohen and co-workers investigated to a deeper level the role of molecule-electrode contacts within the same cAFM geometry of Fig. 6B, using the 3D mode of the instrument *(81)*. They compared electrical

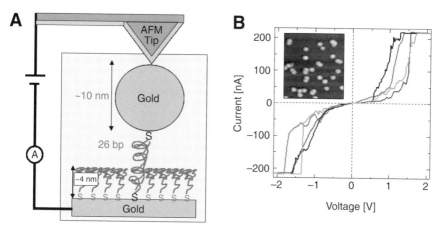

Fig. 7. Twenty-six base-pairs long dsDNA of complex sequence was connected to a metal substrate and a 10 nm metal particle on opposite ends using thiol groups. Schematic representation of the measurement configuration **(A)** and I-V curves, showing high current density **(B)**. The inset (250×250 nm^2) portrays an AFM image of the gold nanoparticles. (Adapted from ref. *73*, by permission; © 2005 National Academy of Sciences USA.)

transport through single-stranded and double-stranded DNA monolayers, with and without upper thiol end-groups. Transport through these systems was also compared to the situation illustrated above, of double-stranded DNA molecules with gold nanoparticles on top and embedded in a single-stranded DNA monolayer. They found that single-stranded DNA monolayers are unable to transport current. For double-stranded DNA monolayers covalently bonded to the substrate but without thiols on the opposite end, charge transport was generally blocked. Double-stranded DNA monolayers with thiols on both ends were, in most cases, able to transport currents as high as those through the double-stranded DNA molecules with the nanoparticles on top. These results are interpreted as evidence that double-stranded DNA molecules have a finite conductivity, but to reveal their conductivity in an actual measurement, it is necessary that covalent contacts be established efficiently to both electrodes, otherwise charge injection is hindered. When the top thiol end-group is missing, one contact cannot be established and consequently, currents are not revealed. When the top thiol end-group is present, contact may be established directly between the metallized tip and the top thiol: this is, however, less controllable than realizing the contact between the top thiol and a gold nanoparticle before double-strand hybridization, followed by establishing direct contact between the tip and the larger nanoparticle during the cAFM measurement (as was done in the first work *[73]; see* previous paragraph). In other words, the tip captures the "large" nanoparticle very easily and the "tiny" S head-group of the thiol less easily, but when the latter capture occurs, the doubly thiolated double-stranded DNA monolayer *(80)* can transport as efficiently as the doubly-thiolated double-stranded DNA molecules decorated by metal nanoparticles and embedded in a single-stranded DNA monolayer *(73)*. The "cost" is, of course, the additional amount of "buffer" that may influence the transport characteristics.

3.1.2. Longitudinal Current-Voltage Characteristics by STM

A similar "standing" molecule setup was earlier realized using the STM.

Xu and co-workers *(74)* developed an STS-based method to measure the I-V curves of molecules in a wet environment: the molecules are captured directly from solution. They studied both uniform poly(GC)-poly(CG) sequences of length varying from 8 to 14 bp, and similar sequences intercalated by AT pairs in the middle with total length changing from 8 to 12 bp and AT length from 0 to 4 bp. The molecules were terminated by $(CH_2)_3$-SH thiol groups at the 3′ ends, to form stable S-Au bonds with the gold substrate. The technique to create molecular junctions proceeds as follows: first, an STM tip is brought into contact with a flat gold surface covered by the DNA solution; then, the tip is retracted under the control of a feedback loop to break the direct tip–electrode contact. During the latter step, the DNA molecules can bridge the tip and the substrate, as shown in the right inset of Fig. 8. For the shortest 8-bp $(GC)_4$ molecules, the results indicated the formation of different junctions with an integer number of molecules: the statistical analysis allowed a determination of the value of the conductance of a single molecule (Fig. 8, left). In addition, the current-voltage curves exhibited a linear behavior and currents of about 100 nA at 0.8 V were detected. The statistical analysis of longer uniform molecules and of nonuniform molecules proved that also in these cases, the molecular junctions realized by retracting the STM tip contain an integer number of molecules: the conductance of a single molecule in the case when two $(GC)_2$ segments are connected by one or two (AT) segments is smaller than in the case of the uniform sequence with only GC pairs. The length dependence of the $(GC)_n$ ($n = 4,5,6,7$) electrical measurements indicated a linear behavior of the conductance, increasing with the inverse in length, a signature of a hopping charge-transfer mechanism. For the $(GC)_2(AT)_m(GC)_2$ polymers, the conductance decreased exponentially with the number of AT pairs, revealing tunneling of the conducting electrons between the two GC portions.

Whereas on the one hand, the two experiments reviewed in this section and in the previous one cannot be interpreted unequivocally, it is remarkable that they undoubtedly show the possibility of measuring significant currents through DNA molecules, when these are not flat on a substrate and are covalently connected to the electrodes.

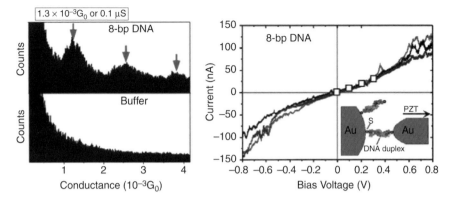

Fig. 8. Left: Conductance histogram over approx 500 measurements, revealing peaks at integer multiples of a value which is identified as the conductance of a single 8 base-pair DNA molecule of uniform poly(GC)-poly(CG) sequence. G_0 is the quantum of conductance. Right: I-V curves of three different junctions formed by the same kind of molecules with the scanning tunnelling microscopy trapping technique illustrated in the inset. (From ref. *74*, by permission; © 2004 American Chemical Society.)

We note some differences between the experimental setups and the results: (1) the molecular sequences were different, as well as the lengths; (2) one experiment was done under ambient conditions *(73,80)*, the other under aqueous conditions *(74)*; (3) the most striking qualitative difference, which still calls for an explanation but is surely influenced by issues (1) and (1), is the presence of a voltage gap in the I-V curves of Fig. 7, and the linear behavior of the current, even around 0 V in Fig. 8.

3.1.3. Longitudinal Current-Voltage Characteristics of Molecules Trapped Between Planar Electrodes

Current-voltage curves of molecules between horizontal electrodes were measured in a number of different experimental setups and for different molecular phases (single molecules, bundles, networks, etc.) *(12,28–35, 38–40,54,61,82)*. Most of these experiments were recently reviewed and we refer the reader to other books and articles *(38–40,82)*. We only note here that also these kinds of measurements, like those discussed under Subheadings 3.1.1. and 3.1.2., revealed that it is possible to elicit significant currents from *short suspended* single DNA molecules *(29)*, as shown in Fig. 9. In contrast, when DNA molecules undergo nonspecific interactions with substrates, the current flow is blocked (e.g., ref. *30*). Besides this clear distinction, residual confusion remains about the detected electrical behavior of DNA in the experimental literature so far, revealing DNA phases from insulating, to conducting, through semiconducting and with a report of superconductivity. This apparent variety should be ascribed to a general lack of accurate control

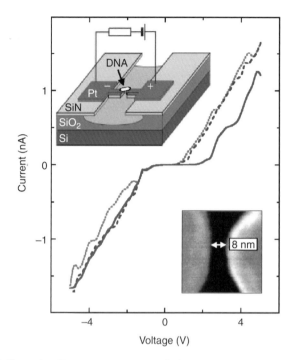

Fig. 9. (A) Current-voltage curves measured at room temperature on a 30-base-pair DNA molecule with uniform poly(dG)-poly(dC) sequence, trapped between two metal nanoelectrodes that are 8 nm apart *(29)*. The upper inset shows a schematic of the sample layout: the molecule is *suspended* between two electrodes by means of an *electrostatic trapping* procedure. The lower inset is a scanning electro microscope image of the two metal electrodes (light area) and the 8 nm gap between them (dark area). Subsequent current-voltage curves (solid, dashed, dotted) show similar behavior and currents of the order of 1 nA around 4 V. More details of this experiment in the context of the recent literature are discussed in recent reviews *(38,40)*. (From ref. *29*, by permission; © 2000 Nature Macmillan Publishers Ltd.)

of the sample preparation, molecule trapping, and probing configuration: indeed, the approaches described in the two preceding sections demonstrate that achieving control represents a significant step forward in performing electrical measurements on DNA molecules.

3.2. State-of-the-Art Evidence for the Conductivity of Native DNA: Summary from Experiments and Theoretical Understanding

It is an arduous task to formulate a critical summary of a widely dispersed set of results obtained under a huge variety of environmental conditions and molecular phases. Originally, several contrasting electrical behaviors from insulating to (semi/super)conducting seemed to emerge. More recently, some clear data has shown that double-stranded DNA molecules can sustain significant currents, whose high values cannot be justified in terms of conventionally evoked charge-transfer mechanisms that mainly rely on an explanation in terms of diffusive hopping. Without presuming to present a comprehensive understanding of all available studies, we just aim to fix some points.

Note that although DNA-based nanoelectronics deals with "dried" molecules on inorganic substrates or between electrodes, the field took inspiration from experiments done in solution to investigate how fast oxidative damage can propagate along a DNA molecule *(83–86)*. The theoretical counterpart to such experiments *(87–89)* is the electron-transfer theory *(90,91)*. Hence, we briefly sketch here the overall understanding from both "electron transfer" and "conductivity" viewpoints.

3.2.1. The Picture that Emerges from "Solution Chemistry" Experiments

The group of Jacqueline Barton at Caltech played an important role in stimulating investigations of the electron-transfer efficiencies through various DNA sequences *(83,84,86)*. Their experiments proceed by injecting a localized hole (oxidized cation through chemical modification) at a specific location, and detecting the rate of transmission of that hole to another specific location. They found that the transfer rates through DNA are much higher than in most proteins, thus initiating the idea that DNA could mediate charge conduction under artificial conditions (e.g., in nanoelectronic devices). In successive works, they also thoroughly analyzed the effect of the environment and the charge transfer through DNA layers deposited on substrates.

In the same way, the group of Bernd Giese in Basel performed similar experiments *(85,87,92)* and showed that the charges are transferred by a coherent superexchange mechanism (direct tunneling) between G and GGG hole traps separated by very few AT base pairs: in such cases, an exponential decay of the hole transfer rate with distance is observed. However, when the number of AT pairs separating the hole traps increases *(93)*, then a transition to a hopping mechanism, with a linear decay behavior of the transfer rate, is inferred by a combination of measured data and theoretical analysis.

We also point out important contributions to the field by the group of Thomas Carell at the LMU-Munich, regarding "electron transfer" rather than "hole transfer." These experiments have been recently reviewed and we refer the reader to the existing literature *(86,94)*.

3.2.2. The Picture that Emerges from "Solid-State Physics" Experiments

These experiments consist of measuring the current-voltage curves and the differential conductance in single molecules or other aggregate phases (bundles, networks). The experiments on single molecules are most suitable for the development of self-assembling nanoelectronics, because control over the structure (and therefore over the desired devices) is possible. Despite the widely scattered data, the following evidence emerges: (1) charges are blocked for single molecules deposited on hard surfaces; and (2) charges can be transported efficiently and high currents can be measured through short DNA segments trapped between electrodes, if nonspecific interactions with the substrate are avoided and the electrode–molecule contacts are optimized, possibly through covalent bonds *(73,74)*. The measured currents exhibit a voltage gap around zero applied bias, and can be as high as hundreds of nA. Such high values cannot be interpreted in terms of a fully incoherent hopping mechanism. It is most likely that the electronic states of the DNA molecule mediate tunneling between the electrodes in a coherent way, although this does not necessarily imply a band-like mechanism as in metals or semiconductors, where a continuum of energy levels exists. Coupling with vibrations probably plays an important role in this "game."

3.2.3. The Picture that Emerges from the Theory and Computation of Charge Transfer Between Localized States

The Marcus-Hush-Jortner theory *(90,91)* expresses the transfer rate between two points in space where a charge (electron or hole) may be localized at the beginning and at the end of the transfer process. Such a quantity is expressed in terms of an electronic coupling term and a nuclear factor. On the basis of the quantum chemical computations of the electronic structure at the Hartree-Fock level *(88,89,95)*, the overall emerging picture is that in an arbitrary complex DNA sequence, the motion of charge is most likely to occur via successive hopping events *(87,96)*. Recently, Bixon and Jortner extrapolated the results of electron-transfer theory to infer the conditions under which currents may flow through a DNA bridge connecting two metal pads, analyzing different schemes of voltage drops *(97)*. Their work clearly establishes some general aspects of the connection between maximal observed currents and structural/electronic features of metal–DNA junctions. The quantitative determination of currents is, however, still inhibited, because the currents depend on several parameters that are unknown at the present level of theory.

3.2.4. The Picture that Emerges from Electronic (Band) Structure Calculations

Clear-cut density functional theory (DFT) electronic structure calculations *(30,98–100)* were performed recently on different periodic DNA sequences and with different basis sets. Despite the revealed differences, due partly to the various approximations and partly to the various investigated molecules, some common results appear. The polymers exhibit a wide HOMO-LUMO gap. The bands are gathered in manifolds of closely spaced energy levels *(38)*: such manifolds unfold into dispersive bands only if the exact helical symmetry is imposed. Coherent band bending due to π-stacking does not occur. Any alteration of a uniform sequence induces charge localization, a structural factor that acts against fast and continuous charge motion. Overall, one can say that a coherent band-like transport mechanism is not predicted on the basis of the polymeric 1D band structure. *See* recent reviews *(38–40,101)* for a more thorough report of other electronic structure calculations.

3.2.5. Summary

The overview in this section is far from being complete. Indeed, we only pointed out those pillar facts that span the various experimental and theoretical approaches, and allow delineating a few clear statements in the complex scenario of charge transfer and transport in DNA. First, although experiments in solution detect fast charge motion through various DNA molecules at a distance, the measured speeds, although high with respect to other biomolecules, are not enough to guarantee efficient transport of electrical currents in devices through arbitrary DNA sequences and lengths. Second, high currents revealed in recent experiments *(73,74)* cannot be explained in terms of the proposed incoherent hopping mechanism *(87)*: at the same time, however, band structure calculations do not sustain a coherent band-like mechanism. Although in recent years, more systematic investigation approaches are shedding light on several controversies concerning charge transfer and transport in DNA, the above-outlined remaining discrepancies call for further in-depth analysis to attain more uniform understanding and control, which are required to enable nanoelectronic applications.

3.3. Toward DNA Derivatives with Enhanced Conductivity

Given the huge scattering of data on the capability of DNA molecules to conduct currents, there remain two primary ways to pursue the route toward DNA-based electronics: either reduce/avoid substrate-induced deformations (standing-molecule or suspended-molecule measurements *[29,73,74,80]*, use of a soft organic buffer *[54]*, etc.), or explore stiffer molecules that retain the structuring and recognition properties of the DNA. Along the latter lines, G4-DNA emerges as a valuable candidate *(57)*, and also bears the promise of better conductivity. Along the former lines, besides continuing to optimize the measurement methods presented under Subheadings 3.1.1 and 3.1.2, it is also interesting to optimize the molecules: metal incorporation *(32,102,103)* (e.g., M-DNA *[32]*) and base alteration (designed to increase the π overlap) *(104,105)* are pursued to enhance the intrinsic conductivity.

3.3.1. Alterations of the Helical Motif: Quadruple Helical G4-DNA

G4-DNA is a quadruple helical structure of homoguanilic or guanine-rich sequences, occurring in a wide variety of natural situations and organisms *(106–108)*. It can play an important role in the telomeric region of chromosomes, and is an excellent prototype case to study supramolecular self-assembly. The *chemico-physical* and *biological* aspects of G4-DNA were recently reviewed *(109)*. But G4-DNA molecules are also currently attracting interest within the *molecular electronics* research community *(38,57)*. This envisaged novel role of G4-DNA is proposed on the basis of some structural features that are expected to influence the electronic characteristics in a device setup. First, the peculiar in-plane arrangement of four guanines forming a tetrad, kept together by a double ring of eight hydrogen bonds (*see* Fig. 10), endows the quadruple helices with a higher stiffness than the DNA double helices: as a consequence, a higher persistence length with respect to B-DNA is indeed reported for G4-DNA molecules deposited on a mica surface *(53,58)*. Second, the high molecular density in a tiny cylindrical section (diameter ~2.5 nm vs ~2.1 nm of B-DNA) also induces a highly packed charge-density distribution and channels for charge motion *(110)*. Third, the experimental evidence that the quadruplex arrangement is stabilized by the presence of metal cations in solution during synthesis *(106–109)* stimulates the idea of a metal-wire accomplishment at the core of the molecule, which in principle could mediate charge motion *(111)*. Fourth, the rotation angle of 30° between adjacent guanines, smaller than in double-stranded guanine-rich DNA sequences, might be an indication of a better π-π superposition (which is expected to improve with decreasing angle *[112]*).

Short G4-wires are well known in a variety of self-assembly schemes (e.g., with a sugar-phosphate backbone, or without such a backbone and with the guanines decorated by alkylic chains) and have been characterized with nanoscopic techniques for awhile now *(53,109)*. Huge efforts have been devoted most recently to the synthesis and characterization of long G4-wires suitable for molecular electronics. The successful demonstration of an efficient synthesis protocol and the accompanying imaging characterization for such materials *(57)* are encouraging signals toward G4-based electronics *(113)* and motivate a continuous interest in the electronic structure of the guanine tetrameric units and the dependence on external factors such as complexation with metal ions and structural deformations *(110,111,114,115)*.

G4-wires can be successfully synthesized using both sugar-phosphate-bonded strands *(57)* and lipophilic guanosine monomers *(107)*, through different chemical protocols. Nonetheless, the resulting self-assembled product presents the same helical fashion: the core of the wires is constituted of stacked planes of four guanines, named G4s, as shown in Fig. 10A. The G-quartets have a square-like symmetry and are separated by 3.4 Å along the stacking direction, with a rotation angle of 30°. Metal ions may be hosted in the central channel resulting from the supramolecular G4 stacking, as revealed, for instance, by the crystal structure of short G4-DNA molecules *(106,108)*.

So far, the G4-wires have been poorly investigated from a nanoelectronics perspective, but such efforts are currently being intensified *(113)*.

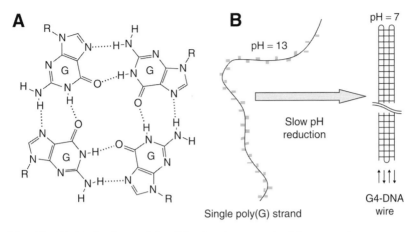

Fig. 10. (A) Each plane of the G4-wires is a tetrad of four guanines connected by eight hydrogen bonds. Besides this basic unit, metals of different chemical nature can be incorporated in the inner hole at various locations in the planes or between the planes. **(B)** Self-assembly process recently demonstrated to obtain long G4-wires from single poly(G) strands *(57)*. (From ref. *57*, by permission; © 2005 Wiley-VCH.)

3.3.1.1. THEORY

Ab initio DFT electronic structure calculations reveal that indeed G4-wires may support enhanced charge motion through the base stack, with respect to double-stranded DNA. Calzolari and co-workers *(101,110,114,115)* showed that the electronic Density of States (DOS) of an infinite periodic G4-wire with K^+ ions in the central cavity resembles that of an "effective" wide band-gap semiconductor* *(116)* with a narrow HOMO manifold. This evidence is accompanied by the formation of partially incoherent electron states delocalized throughout the stack and suitable to host mobile electrons or holes.

---

*Care must be assumed when using for molecule attributes that are defined for solid-state materials. For instance, in condensed matter physics, the term "semiconductor" does not mean only the presence of an energy gap for electronic excitations, but also implies the capability of doping and the presence of coherent bands of Bloch states *(116)*. As such, it should not be adopted for a molecule only on the basis of a HOMO-LUMO gap, unless all the other characteristics are also demonstrated. Such terms are, however, often used in molecular electronics, with the purpose of evoking some principles which physicists are familiar with: it is important that the association remain evocative and not complete. Here and elsewhere, to describe the electronic behavior of G4-wires, the expression "wide band-gap semiconductor" is preceded by "effective" to state, in one word, that although the DOS of a G4-wire appears as that of a semiconductor, the real behavior as a semiconductor is not proven and the term may even prove inappropriate for biomaterials. For instance, in the case of G4-wires, the bands are not formed by coherent Bloch states but are incoherent manifolds of flat, closely spaced energy levels *(see* the original articles for a thorough explanation).

Figure 11 shows the DOS and the iso-density surface plot of the HOMO state manifold.[*] The same authors also find that the main electronic structure parameters that are relevant for conductivity, such as "effective" band gaps and widths, strongly depend on structural conditions (Fig. 11). For instance, axial strain and twist angle are crucial factors to control the π-π overlap *(112,115)*, which in turn is thought to affect the electron-transfer efficiency *(42)*. On the basis of this knowledge, it is in principle possible to design conformational alterations in order to optimally tailor the electronic response to elicit high currents from G4-wires. Of course the demonstration of this possibility rests on feasible chemical synthesis and controlled electrical measurements, still to come.

Because preliminary data indicate that long G4-wires may self-assemble and be stable even in the absence of internal cations, whereas this is not true for short G4-DNA molecules *(108,107,117,118)*, Cavallari and co-workers recently investigated by means of Molecular Dynamics (MD) simulations the relative stability of different G4-wires of finite length as a function of the axial length, the metal species, and stoichiometry *(119)*. They found that metal cations of different species (K^+, Na^+, Li^+) are hosted in the inner cavity at different sites:[†] K^+ (Li^+) ions prefer inter-plane (intra-plane) sites independently of stoichiometry and axial length; Na^+ ions are distributed between inter-plane and intra-plane sites, indirectly suggesting that the migration barrier between different stable locations is lower than for the other species. Furthermore, they find that short G4-wires (four or nine planes) are stable over short times of 3 to 5 ns only in the presence of fully coordinating metal cations, and disrupt if they are under-coordinated, independently of the metal species; the best conserved structures are those with the inner channel full of K^+ ions, followed by those with Na^+ and Li^+ ions, in line with experimental suggestions *(120)*. Most interestingly, the simulated trajectories show that, while the aforementioned short under-coordinated G4-wires undergo a disruption process over short times, long G4-wires (20 planes) maintain the quadruple helical conformation in a stable way even over longer times of approx 20 ns. These results are illustrated in Fig. 12.

The residual open questions from a theoretical viewpoint in the perspective of nanoelectronics applications concern the stability and the electronic performance of long G4-wires in the presence of coordinating cations of

---

[*]Linear combination of 12 wave functions derived from the guanine HOMO. The modulo-12 multiplicity is dictated by the number of guanine molecules in the periodicity unit of a periodic G4-wire, as explained elsewhere *(38,110,114,115)*.

[†]The same behavior was previously reported for extremely short G4-DNA molecules made of four planes.

other metal species, with particular interest in transition metals because of their redox activity, which in principle, may mediate electron transfer through the stack.

3.3.1.2. EXPERIMENT

Short G4-wires have already been deposited onto inorganic substrates and then characterized by AFM a few years ago *(53)*. These early imaging data indicated that the guanine quadruple helix is more resistant than the DNA double helix to substrate-induced deformations. Only very recently, Kotlyar and co-workers invented and implemented a three-step procedure to produce long G4-wires from parent G-strands *(57)*: (1) enzymatic synthesis of long double-stranded poly(G)-poly(C) molecules *(72)*; (2) separation of the poly(G) and poly(C) single strands by pH elevation (pH 13.0) and purification of the obtained strands by size-exclusion high-performance liquid chromatography (HPLC); and (3) poly(G) strand folding into the quadruple helical conformation by slowly lowering the pH to neutrality. Steps 2 and 3 are illustrated in Fig. 10. The aggregation into a quadruple helical form was checked by circular dichroism (CD) *(57)*: the collected CD spectra were successfully compared to those reported earlier for G4-DNA *(121,122)*. In contrast to the G4-wires reported earlier *(106–109,123)*, the strand-folding and wire stability of the long G4-wires synthesized by the new protocol are independent of the presence of cations: this characteristic is probably due to the much greater length and is in agreement with the results of MD simulations discussed above. In addition, they are extremely resistant to heat treatment and utterly insensitive to DNase.*

The novel long G4-wires were thoroughly characterized by acquiring AFM images and making statistics over several images (Fig. 13). The average apparent height of G4-wires (extracted from different molecular cross sections) is about 1.6 nm, about twice that of the parent poly(G)-poly(C) molecules. The height values for the poly(G)-poly(C) are typical of DNA molecules on mica as measured by AFM by Muir and co-workers in the past *(53)* and more recently in Porath's laboratory over thousands of molecules. The apparent height for the G4-wires is typical to measurements on hundreds of G4-wires that were performed in a similar way under similar conditions. The apparent height, lower than the nominal diameter of the molecules, probably results from surface forces and tip pressure applied to the molecules during imaging *(53)*. The increased apparent height of the G4-wires with respect to the poly(G)-poly(C), in spite of the similar diameters of the two molecules, 2.1 nm for B-form DNA and 2.5 nm for G-quartets, as extracted

---

*On the contrary, the DNase enzyme efficiently degrades native dsDNA.

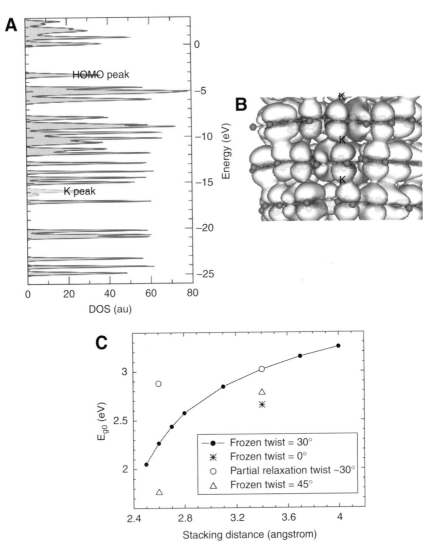

Fig. 11. (A) Density of States (DOS) of an infinite periodic G4-wire with K$^+$ ions in the inner cavity: the shaded area represents the total DOS; the dark gray (pale gray) solid line represents the DOS projected onto atomic orbitals of the guanine bases (K$^+$ ions). The presence of an energy gap between the HOMO peak and the following "effective" bands is clear, which originates the qualitative definition as a wide-bandgap semiconductor. Potassium ions contribute to the DOS with electron states only at very low energies (peak marked at ~ −16 eV), while they do not give states around the energy gap: this means that they do not contribute effectively to electronic excitation upon bias voltages of around few eV, typical in molecular electrical measurements. **(B)** Iso-density surface plot of the wavefunction convolution of 12 electron states associated to the HOMO peak. The inner potassium ions are explicitly labeled. **(C)** Dependence of the main HOMO-LUMO

from X-ray analysis *(109,124)*, indicates higher stiffness and resistance to the surface forces and the pushing AFM tip for the G4-wires. In addition, the persistence length of the G4-wires seems to be approx 100–200 nm, in comparison to approx 50 nm of the poly(G)-poly(C).

Kotlyar and collaborators *(57)* also suggest in their article that the imaged G4-wires are mono-molecular (e.g., one G4-wire is folded from only one parent G-strand[*]) and characterized by a reduced stacking distance with respect to double-stranded DNA of various forms (A-, B-, Z-DNA). Because the theoretical analysis illustrated above indicates that axial compression is a useful tool for tailoring the electronic properties toward enhanced charge motion through the helix *(115)*, the experimental hint of a reduced stacking distance is promising in view of nanoelectronics[*]- G4-wires may be very appealing for nanoelectronic applications if one learns to control the production process to introduce the desired structural changes.

3.3.2. Metal Incorporation: M-DNA Poly(dG)-Poly(dC), M-G4-DNA

In a relatively recent experiment, Rakitin and co-workers claimed that Zn^{2+} ions are able to substitute for hydrogen-bonding protons in Watson-Crick pairs, giving as a result an insulator-metal transition in DNA double helices *(32)*. The "metal-doped" DNA was called M-DNA. While this experiment has not yet been reproduced in any other laboratory, either in the synthesis or in the electrical aspects, it stimulated much discussion and several activities to pursue "metal doping" in DNA molecules.

3.3.2.1. EXPERIMENT

Porath and collaborators recently imaged and measured the electric response of poly(G)-poly(C) molecules obtained via the new synthesis procedure *(72)* and then dosed with silver ions from salts [let us call such

---

Fig. 11. *(Continued)* gap on the stacking distance between adjacent G-quartet planes: the smooth saturation behavior observed for frozen structures is somehow altered when other structural deformations are superimposed to axial strain (structures with only axial strain are marked by dots in the plot; the other symbols mark twist rotations). In any case, the value of the gap can be much reduced by axial compression and augmented twisting (see the triangle at 2.6 Å); the gap reduction is accompanied by enhanced wave function delocalization, not shown here. (Adapted from refs. *110,115*, by permission; © 2004, 2005 American Chemical Society.)

---

[*]Out of the other possibilities with two or four G-strands.

[*]Such indications must be taken with care based on the current status of understanding. They are claimed on the basis of concatenated deductions that are based on hypotheses concerning the synthesis procedure that are not yet conclusively proven.

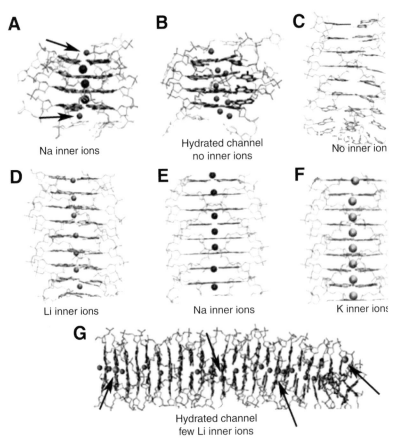

A Na inner ions

B Hydrated channel no inner ions

C No inner ion

D Li inner ions

E Na inner ions

F K inner ions

G Hydrated channel few Li inner ions

Fig. 12. Average structures and snapshots from Molecular Dynamics (MD) simulations of G4-wires of different lengths at 300°K *(118)*. **(A)** A 4-plane G4-wire fully coordinated by Na$^+$ ions: final configuration after a 3-ns MD run, starting from the X-ray crystal structure *(106)*; one Na$^+$ ion leaves the channel but the overall conformation remains intact; the smaller spheres at the lower and upper edges are water molecules (indicated by arrows). **(B)** A 4-plane empty G4-wire: final configuration after a 3-ns MD run, starting from the X-ray crystal structure depleted of metal cations; a strong deformation of the double helix is evident; spheres denote water molecules. **(C)** A 9-plane empty G4-wire: average configuration after a 5-ns MD run; the structure experiences permanent structural deformations mainly at the 3′ end. **(D,E,F)** 9-plane G4-wires fully coordinated by Li$^+$, Na$^+$, K$^+$ ions, respectively: average configurations after 5-ns MD runs; the quadruple helical conformation remains well defined; the different stability sites are evident for the different metal species. **(G)** 20-plane empty G4-wire: final configuration after a 20-ns MD run: few Li$^+$ cations enter the tube from solution but the channel remains definitely under-coordinated; notwithstanding this under-coordination, the quadruple helical arrangement is maintained, at odds with the shorter molecules; most spheres represent water molecules; spheres denoting Li$^+$ ions are marked by arrows. (Adapted from ref. *119*, by permission; © 2006 American Chemical Society.)

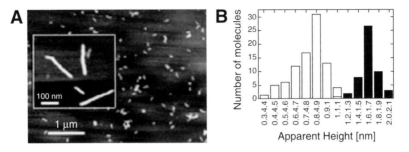

Fig. 13. Atomic force microscopy image (left) of approx 110-nm-long G4-wires that are produced via the new synthesis protocol from approx 550-nm-long parent poly(G)-poly(C) molecules. Inset: enlargement of G4-wires that are made by self-folding of G-strands; their long persistence and uniformity are evident. The right histogram reports the statistical analysis of the apparent height of single, well separated molecules from several images at several areas in each sample; white bars relate to the parent poly(G)-poly(C) double helices (also measured in this study) and dark bars to G4-wires. (Adapted from ref. *57*, by permission; © 2005 Wiley-VCH.)

products M-poly(G)-poly(C)]. The structure of these M-poly(G)-poly(C) polymers is unknown, but optical characterization indicates that the metal ions form complexes with the bases, rather than remaining as counterions in solution. Preliminary results indicate a possible change of the HOMO-LUMO gap for M-poly(G)-poly(C) relative to the parent poly(G)-poly(C).

3.3.2.2. THEORY

Given the absence of convincing structural determination, DFT calculations are hindered. Preliminary calculations were however started on metallized GC pairs with various metals (*see* a sketch in Fig. 14) *(111,125)* and may be continued in the future to predict stacking arrangements. The latter goal is indeed very delicate, and some efforts have already yielded negative outcomes, e.g., the impossibility of obtaining a DFT structure for M-DNA. Refinements in the experiments and in the theory may turn out to be useful. It is also interesting to explore metal complexation with G-quartets. We already said under Subheading 3.2 that transition metal ions added after the synthesis of G4-wires are rather unlikely to penetrate the channel. They can be hosted, however, outside the helix in coordination complexes, and change the electronic structure in a profitable manner.

The preliminary indications available so far on M-DNA-like base pairs pertain to Zn(II)-GC (Fig. 14A). Di Felice and co-workers *(111)* found that these altered GC pairs are very flexible and adjust in a nonplanar conformation. Planarity may be eventually recovered upon stacking with other similarly altered base pairs, but it was not possible to predict reliable helical

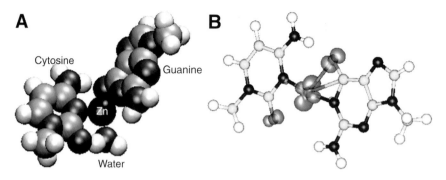

Fig. 14. (A) Structure of the Zn(II)-GC base-pair, according to the metal insertion paradigm indicated in ref. *32*, as obtained from atomic relaxation at the density functional theory (DFT) level *(82)*. White and silver spheres represent H and C atoms, respectively. Dark gray spheres are N and O atoms in the hetero-cycles. The large black sphere at the center stands for the Zn^{2+} ion, which is fourfold coordinated to 2 N atoms and 1 O atom from the bases, plus the O atom of a water molecule. The water molecule has been introduced in the simulated system to respect the tendency of Zn^{2+} ions to fourfold coordination. The strong deviation from planarity is evident. **(B)** Partially occupied orbital with mixed Cu-guanine-cytosine character in a DFT-relaxed Cu(II)-GC base-pair. The geometry is similar to that of Zn(II)-GC pair, but without an extra water molecule.

motifs within the current limits of DFT to compute stacking distances.* The electronic properties of the Zn(II)-GC pair indicate that the frontier orbitals of the nucleobases do not hybridize with the Zn ions: the HOMO and LUMO maintain the same π-guanine and π-cytosine character, respectively, as in the natural GC pair. This outcome suggests that Zn(II) incorporation according to the scheme proposed by Jeremy Lee *(32)* is not likely to substantially alter the electronic properties of the base pairs in the energy range that should be most relevant for charge conduction. However, this does not exclude the fact that completely new phenomena arise between stacked metalated base pairs, not inspected by single metalated base pairs. The same authors also found that, if in the same bonding scheme as in Fig. 14A, namely, with one proton in an H-bond substituted by a metal ion, Zn^{2+} is changed into Cu^{2+}, then the π HOMO of the GC pair hybridize much more significantly with the metal *d* orbitals (Fig. 14B). This is an indication that different metals may be employed to tune the electronic properties of base pairs, if the metal incorporation can be unambiguously demonstrated.

---

*DFT is much more reliable at finding the structure within an individual pair, dictated by ligand- and hydrogen-bonding.

We remark that the information reported in this paragraph is exploratory, and should be sustained by: (1) experimental proof; and (2) theoretical analysis of relative energetics with respect to the unreacted phase, to prove that the explored geometries are meaningful.

3.3.3. Aromatic Expansion of the Bases

Several research laboratories are active worldwide in the design of nucleobase mimics that can form DNA-like double helices along with the natural bases. Different strategies are attempted: (1) simple modification of natural bases by changing just a functional group or a single atom; (2) shape mimics of natural bases by realizing nonpolar isosteres *(126)*; (3) missing DNA bases along the helix stack; (4) substitution of native nucleobases with simple hydrocarbons and heterocycles; (5) insertion of fluorescent species, ranging from inherently fluorescent DNA bases to the addition of hanging flexible tethers connected by linkers; (6) substitution of the natural bases with ligands for metals *(84,103,127)*; and (7) design of novel bases that would pair with other natural or nonnatural bases, even in some cases without forming hydrogen bonds, with the leading rule of enhancing the stacking ability. All such approaches were concisely overviewed *(128)*. One particular case of the latter approach is the pursuing of size-expanded analogs, through the covalent insertion of an aromatic ring co-planar with a natural base, between the pentagonal and the hexagonal rings in purines or aside the heterocycle of pyrimidines (Figs. 15 and 16).

Eric Kool at Stanford University in California is the main promoter, together with his group, of experimental research addressed to the synthesis and characterization of expanded-size base analogs, with emphasis on the exploration of the capability of such analogs to form helical structures and to stack with natural base pairs *(104,105,129–133)*. Other authors followed diverse paths to replace individual base pairs within the overall geometry of natural DNA *(103,134–137)*, but we outline here only Kool's approach.

Kool's group followed a precise path during the last 10 years, composed of consequential steps. The first step was the optimization of the synthesis procedure of single modified bases with the insertion of aromatic rings *(104,129)*: they initially focused on benzoadenine (dxA) and benzothymine (dxT) *(104)*, and only later also worked with guanine and cytosine *(105)*. Then they demonstrated that such base-analogs are able to pair with natural DNA bases by H-bonding, forming base pairs of expanded size *(129,131)*. Finally, they showed that such expanded base pairs stack with each other and with natural base pairs, assembling helical motifs *(104,105,130,132)*.

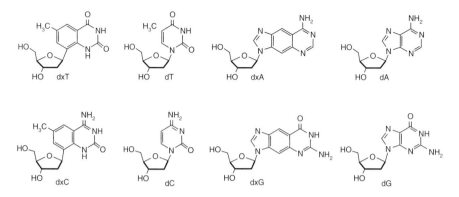

Fig. 15. Structures of the eight nucleosides in expanded-size DNA. Top, from left to right: expanded thymine (dxT), natural thymine (dT), expanded adenine (dxA), natural adenine (dA). Bottom, from left to right: expanded cytosine (dxC), natural cytosine (dC), expanded guanine (dxG), natural guanine (dG). (From ref. *105*, by permission; © 2005 Wiley-VCH.)

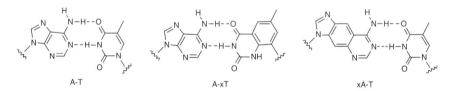

Fig. 16. Proposed structures of hydrogen-bonded expanded-size base pairs formed by expanded adenine with natural thymine (right) and by expanded thymine with natural adenine (middle), compared to the Watson-Crick adenine-thymine pair (left). (From ref. *131*, by permission; © 2004 American Chemical Society.)

Much attention was devoted to understanding the relative stability of the "expanded" DNA double helices with respect to Watson-Crick DNA, as a function of the sequence, and in particular of the alternation of natural and "expanded" base pairs. Thermal denaturation data revealed that dxA-T and dxT-A pairs (Fig. 16) destabilize the double helix when one "expanded" pair is inserted in a 12-bp natural duplex. This observation is attributed to the energy-costly backbone distortions that are needed to accommodate the expanded size within the context of the natural Watson-Crick diameter. However, remarkably different thermal data are obtained for helices composed of only "expanded" pairs. In such cases, no strain is exerted on the backbone because the size is uniformly larger along the axis: consequently, no strain-induced energy loss arises. On the contrary, uniform size-expanded helices containing dxT-A and dxA-T pairs (labeled xDNA) are more stable than natural DNA helices, with stability measured in terms of melting curves and free energy for duplex formation (104). After testing other possible chemical factors for the origin of the added stabilization, the authors concluded that it arises from enhanced stacking either within each strand or across the strands (104).

Liu et al. also performed NMR spectroscopy on xDNA duplexes (130). They found that xDNA duplexes are right-handed and have identical Watson-Crick hydrogen-bonding patterns as those in B-DNA. The duplex diameter is increased by about 3 Å. The number of base pairs per turn of the helix is greater by two units than in natural DNA. The average inter-plane twist is 31°. Most interestingly for potential applications, the inter-plane distance is smaller than in B-DNA, namely approx 3.1 Å instead of approx 3.4 Å. This latter feature is consistent with the interpretation of the higher stability in terms of enhanced stacking. Moreover, it was recently shown that a stacking compression is able to alter the electronic properties of G4-DNA in a way conducive to a more efficient charge motion. Preliminary results indicate that this outcome is also true for xDNA (111,138). Therefore, such xDNA helices may also become appealing in the near future for DNA-based nanoelectronics.

3.3.4. Metal-Ligating Intercalating Planar Chelators

As a final example of possible alterations of DNA helices, we briefly describe an approach that combines the strategies described under Subheadings 3.3.2. and 3.3.3., namely metal incorporation with base modification. Indeed, whereas it is not yet clear if a natural base pair is able to capture a metal ion between the purine and the pyrimidine, some planar aromatic molecules (*see* Fig. 17) are known to act efficiently as metal traps, or chelators (103,139–143). Such alternative pairs lose the extreme capabilities of recognition and structuring of natural DNA. Their potentialities in the context of spontaneous DNA-based nanoelectronics would thus rely on the possibility of inserting them into natural helices, to form a sort of organic heterostructure in which the recognition features would be dictated by the segments of Watson-Crick DNA, and the chelator base pairs would instead contribute tailored electrical/magnetic functionalities given by the succession of metal ions.

A successful attempt to incorporate modified metal-trapping base pairs into a short poly(G)-poly(C) molecule was recently demonstrated by the group of M. Shionoya at the University of Tokyo (103). Tanaka and co-workers used hydroxypyridone bases as chelators, and Cu^{2+} ions bridging two such bases in a pair (Fig. 17A). By electron paramagnetic resonance measurements of short 7-bp DNA molecules with 1 to 5 natural base pairs substituted by synthetic **H**-Cu(II)-**H** pairs, the authors concluded that the central Cu^{2+} ions have an inter-ion distance of 3.7 Å and manifest a ferromagnetic alignment. They

Di Felice and Porath

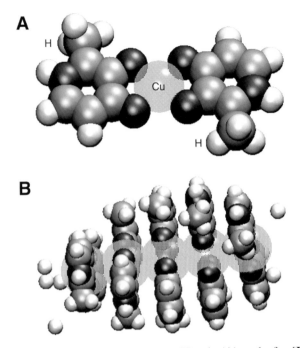

Fig. 17. Relaxed structures of a **H**-Cu(II)-**H** pair **(A)** and of a [**H**-Cu(II)-**H**]$_{5w}$ periodic wire obtained by stacking such mimic base-pairs **(B)**. By **H** we denote the hydroxypyridone base. White and silver spheres represent H and C atoms, respectively. Dark gray spheres are N and O atoms in the hetero-cycles. The large black transparent spheres at the center of each pair stand for the Cu^{2+} ions.

finally suggested that this strategy for arranging metal ions in solution in a controllable manner is promising for the development of metal-based molecular devices such as molecular magnets and wires.

On the stream of this exciting experiment, Zhang and co-workers calculated the electronic structure of the periodic wire shown in Fig. 17B at the spin-polarized DFT level *(127)*. As a result of the computational burden, the exact simulation of the real 7-bp modified DNA molecules investigated in Shionoya's experiment was not feasible. Thus, a simplified model was assumed with all **H**-Cu(II)-**H** synthetic base pairs, and the 5-bp periodicity unit (Fig. 17B) was repeated to form a continuous infinite wire, according to the experimental geometry (3.7 Å stacking distance, 36° axial rotation between adjacent planes). It was found that the frontier orbitals are π-*d* **H**-Cu hybrids, delocalized through the stack essentially at the base sites. Instead, a metallic state (Fig. 18), uniformly distributed through the helix axis at the Cu sites, appears 1.1 eV below the HOMO: this could be excited by voltage application

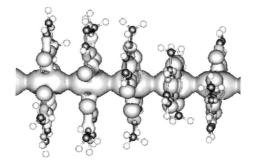

Fig. 18. Isosurface plot of a Cu-delocalized electron state of the $[\mathbf{H}\text{-Cu(II)-}\mathbf{H}]_{5w}$ periodic wire, that lies about 1.1 eV below the highest occupied molecular orbital *(126)*.

at the wire edges in a gated device setup. The ferromagnetic phase was detected as a metastable condition, but it was not possible to resolve energetically if it is more favorable than the anti-ferromagnetic phase. Given the fact that the simulated infinite wire is only a model of the real finite molecules, which neglects, for instance, the backbone and the natural base pairs at the edges, it is not incongruous that the theory does not predict the molecular ferromagnetism. The latter could in fact be caused by asymmetries introduced by the axial edges or even by the interaction with the outer environment.

Most recently, other attempts to intercalate at least one chelator base pair with different bases and with different metals were conducted successfully by Megger's and Carell's groups *(141,142)*.

4. CONCLUSIONS AND PERSPECTIVES

We discussed the basic principles inspiring the use of biological molecules in modern nanotechnology, and then focused on DNA-based nanoelectronics. We showed that, after several initial seemingly inconsistent results on the ability of DNA molecules to conduct electrical currents, recent clear-cut experiments indicate that currents can be transported through short segments if nonspecific longitudinal DNA-substrate interactions are avoided and the DNA–electrode contacts are controlled. The explanation of the transport mechanism calls for further theoretical analysis. In parallel to the investigation of double-stranded Watson-Crick DNA, several ongoing research efforts address promising derivatives that are based on modifications of the bases (metal chelators, aromatic expansion), of the helical conformation (quadruple rather than double helix), and of the H-bond pattern (metal incorporation between two bases in a pair).

We believe that both experimental and theoretical developments, to optimize the measurement configurations on the one hand and to achieve a realistic description of the structure and electronic properties on the other, may eventually contribute to the successful exploitation of these appealing bio-polymers for the development of novel nanoelectronics and nanotechnology applications.

ACKNOWLEDGMENTS

Funding for research was provided by the EC through contracts IST-2001-38951 ("DNA-Based Nanowires"), FP6-029192 ("DNA-Based Nanodevices") and HPRN-CT-2002-00317 ("EXC!TING"); MIUR-IT through FIRB-NOMADE; the INFM parallel computing initiative through the allocation of computing time at CINECA-Bologna; the James Franck foundation; the IPSO organization; the GIF through grant n. I-892-190.10/2005. We are grateful to Arrigo Calzolari, HouYu Zhang, Manuela Cavallari, Anna Garbesi, Elisa Molinari, Hezy Cohen, Errez Shapir, Daniela Ullien and Igor Brodsky for collaboration, assistance and discussions.

REFERENCES

1. Drexler KE. Building molecular machine systems. Nanotechnology 1999;17:5–7.
2. Merkle RC. Biotechnology as a route to nanotechnology. Nanotechnology 1999;17:271–274.
3. Sarikaya M, Tamerler C, Jen AKY, Schulten K, Baynex F. Molecular biomimetics: nanotechnology through biology. Nat Mater 2003;2:577–585.
4. Lowe CR. Nanobiotechnology: the fabrication and applications of chemical and biological nanostructures. Curr Opin Struct Biol 2000;10:428–434.
5. Lehn JM. Perspectives in supramolecular chemistry—from molecular recognition towards molecular information processing and self-organization. Angew Chemie Int Ed 1990;29:1304–1319.
6. Goodsell, DS. Bionanotechnology—Lessons from Nature. Hoboken: Wiley, 2004.
7. Lodish H, Berk A, Zipurski SL, Matsudaira P, Baltimore D, Darnell J. Molecular Cell Biology, 4th ed. New York: Freeman & Co, 2000.
8. Amabilino DB, Stoddart JF. Interlocked and intertwined structures and superstructures. Chem Rev 1995;95:2725–2828.
9. Balzani V, Credi A, Raymo FM, Stoddart JF. Artificial molecular machines. Angew Chemie Int Ed 2000;39:3348–3391.
10. Balzani V, Credi A, Venturi M. Molecular Devices and Machines. A Journey into the Nano World. Weinheim: Wiley-VCH, 2003.
11. Wilson M, Kannangara K, Smith G, Simmons M, Raguse B. Nanotechnology—Basic Science and Emerging Technologies. Boca Raton: Chapman & Hall/CRC, 2002.
12. Braun E, Eichen Y, Sivan U, Ben-Yoseph G. DNA-templated assembly and electrode attachment of a conducting silver wire. Nature 1998;391:775–778.

13. Keren K, Krueger M, Gilad R, Ben-Yoseph G, Sivan U, Braun E. Sequence-specific molecular lithography on single DNA molecules. Science 2002; 297:72–75.
14. Keren K, Berman RS, Buchstab E, Sivan U, Braun E. DNA-templated carbon nanotube field-effect transistor. Science 2003;302:1380–1382.
15. De Rienzo F, Gabdoulline RR, Menziani MC, Wade R. Blue copper proteins: a comparative analysis of their molecular interaction properties. Prot Sci 2000;9:1439–1454.
16. Murphy LM, Dodd FE, Yousafzai FK, Eady RR, Hasnain SS. Electron donation between copper containing nitrite reductases and cupredoxins: the nature of protein-protein interaction in complex formation. J Mol Biol 2002;315:859–871.
17. Rinaldi R, Biasco A, Maruccio G, et al. Solid-state molecular rectifier based on self-organized metalloproteins. Adv Mater 2002;14:1453–1457.
18. Maruccio G, Biasco A, Visconti P, et al. Towards protein field-effect transistors: report and model of a prototype. Adv Mater 2005;17:816–822.
19. Corni S, De Rienzo F, Di Felice R, Molinari E. Role of the electronic properties of Azurin active site in the electron-transfer process. Int J Quantum Chem 2005;102:328–342.
20. Benenson Y, Paz-Elizur T, Adar R, Keinan E, Livneh Z, Shapiro E. Programmable and autonomous computing machine made of biomolecules. Nature 2001;414:430–434.
21. Seeman NC. Nucleic acid nanostructures and topology. Angew Chemie Int Ed 1998;37:3220–3238.
22. Seeman NC. DNA in a material world. Nature 2003;421:427–431.
23. Seeman NC, Belcher AM. Emulating biology: building nanostructures from the bottom up. Proc Natl Acad Sci USA 2002;99:6451–6455.
24. Mao C, Sun W, Shen Z, Seeman NC. A nanomechanical device based on the B-Z transition of DNA. Nature 1999;397:144–146.
25. Brucale M, Zuccheri GP, Samorì B. The dynamic properties of an intramolecular transition from DNA duplex to cytosine-thymine motif triplex. Org Biomol Chem 2005;3:575–577.
26. Alberti P, Mergny JL. DNA duplex-quadruplex exchange as the basis for a nanomolecular machine. Proc Natl Acad Sci USA 2003;100:1569–1573.
27. Yurke B, Turberfield AJ, Mills Jr AP, Simmel FC, Neumann JL. A DNA-fuelled molecular machine made of DNA. Nature 2000;406:605–608.
28. Fink HW, Schönenberger C. Electrical conduction through DNA molecules. Nature 1999;398:407–410.
29. Porath D, Bezryadin A, de Vries S, Dekker C. Direct measurement of electrical transport through DNA molecules. Nature 2000;403:635–638.
30. de Pablo PJ, Moreno-Herrero F, Colchero J, et al. Absence of conductivity in l-DNA. Phys Rev Lett 2000;85:4992–4995.
31. Kasumov AY, Kociak M, Guéron S, et al. Proximity-induced superconductivity in DNA. Science 2001;291:280–282.
32. Rakitin A, Aich P, Papadopoulos C, et al. Metallic conduction through engineered DNA: DNA nanoelectronic building blocks. Phys Rev Lett 2001;86:3670–3673.

33. Storm AJ, van Noort J, de Vries S, Dekker C. Insulating behavior for DNA molecules between nanoelectrodes at the 100 nm length scale. Appl Phys Lett 2001;79:3881–3883.
34. Watanabe H, Manabe C, Shigematsu T, Shimotani K, Shimizu M. Single molecule DNA device measured with triple-probe atomic force microscope. Appl Phys Lett 2001;79:2462–2464.
35. Shigematsu T, Shimotani K, Manabe C, Watanabe H, Shimizu M. Transport properties of carrier-injected DNA. J Chem Phys 2003;118:4245–4252.
36. Berti L, Alesandrini A, Facci P. DNA-templated photoinduced silver deposition. J Am Chem Soc 2005;127:11,216–11,217.
37. Richter J, Mertig M, Pompe W, Mönch I, Schackert HK. Construction of highly conductive nanowires on a DNA template. Appl Phys Lett 2001;78:536–538.
38. Porath D, Cuniberti G, Di Felice R. Charge transport in DNA-based devices. In: Schuster G, ed. Long Range Charge Transfer in DNA II. Topics in Current Chemistry, vol. 237. Berlin: Springer, 2004:183–227.
39. Endres RG, Cox DL, Singh RRP. Colloquium: The quest for high-conductance DNA. Rev Mod Phys 2004;76:195–214.
40. Di Ventra M, Zwolak M. DNA Electronics. In: Singh-Nalwa H, ed. Encyclopaedia of Nanoscience and Nanotechnology, vol. 2. Stevenson Ranch: American Scientific Publishers, 2004:475–493.
41. O'Neill P, Fielden EM. Primary free radical processes in DNA. Adv Radiat Biol 1993;17:53–120.
42. Eley DD, Spivey DI. Semiconductivity of organic substances. Part 9— Nucleic acid in the dry state. Trans Faraday Soc 1962;58:416–428.
43. Young EWA, Mantl S, Griffin PB. Silicon MOSFETs—novel materials and alternative concepts. In: Rainer W, ed. Nanoelectronics and Information Technology. Weinheim: Wiley, 2003:359–386.
44. Hansma HG, Hoh J. Biomolecular imaging with the atomic force microscope. Annu Rev Biophys Biomol Struct 1994;23:115–140.
45. Bustamante C, Keller D. Scanning force microscopy in biology. Phys Today 1995;48:32–38.
46. Ebert P, Szot K, Roelofs A. Scanning probe techniques. In: Rainer W, ed. Nanoelectronics and Information Technology. Weinheim: Wiley, 2003:297–320.
47. Bonnell D. Scanning Probe Microscopy and Spectroscopy: Theory, Techniques, and Applications, 2nd ed. New York: Wiley, 2001.
48. Lindsay SM, Lyubchenko YL, Tao NJ, et al. Scanning tunneling microscopy and atomic force microscopy studies of biomaterials at a liquid-solid interface. J Vac Sci Technol A 1993;11:808–815.
49. Weiss PS. Analytical applications of scanning tunneling microscopy. Trends Anal Chem 1994;13:61–67.
50. Hansma HG. Surface biology of DNA by atomic force microscopy. Annu Rev Phys Chem 2001;52:71–92.
51. Zuccheri G, Samorì B. SFM studies on the structure and dynamics of single DNA molecules. In: Jena B, Hoerber JA, Wilson L, Matsudaira P, eds. Atomic Force Microscopy in Cell Biology. Methods in Cell Biology, vol 68. Amsterdam: Elsevier, 2002:357–395.

52. Alessandrini A, Facci P. AFM: a versatile tool in biophysics. Meas Sci Technol 2005;16:R65–R92.
53. Muir T, Morales E, Root J, et al. The morphology of duplex and quadruplex DNA on mica. J Vac Sci Technol A 1998;16:1172–1177.
54. Kasumov AY, Klinov DV, Roche PE, Guéron S, Bouchiat H. Thickness and low-temperature conductivity of DNA molecules. Appl Phys Lett 2004;84:1007–1009.
55. Saenger W. Principles of Nucleic Acid Structure. Berlin: Springer, 1984.
56. Sinden RR. DNA Structure and Function. San Diego: Academic Press, 1994:12–14:23–25.
57. Kotlyar AB, Borovok N, Molotsky T, Cohen H, Shapir E, Porath D. Long monomolecular guanine-based nanowires. Adv Mater 2005;17: 1901–1905.
58. Vesenka J, Guthold M, Tang CL, Keller D, Delaine E. Substrate preparation for reliable imaging of DNA molecules with the scanning force microscope. Ultramicroscopy 1992;42–44:1243–1249.
59. Bustamante C, Vesenka J, Tang CL, Rees W, Guthold M, Keller R. Circular DNA molecules imaged in air by scanning force microscopy. Biochemistry 1992;31:22–26.
60. Muller DJ, Engel A. The height of biomolecules measured with the atomic force microscope depends on electrostatic interactions. Biophys J 1997;73:1633–1644.
61. Gómez-Navarro C, Moreno-Herrero F, de Pablo PJ, Colchero J, Gómez-Herrero J, Baró AM. Contactless experiments on individual DNA molecules show no evidence for molecular wire behavior. Proc Natl Acad Sci USA 2002;99:8484–8487.
62. Shapir E, Cohen H, Sapir T, Borovok N, Kotlyar AB, Porath D. High-resolution STM imaging of novel poly(dG)-poly(dC) DNA. J Phys Chem B 2006; 110:4430–4433.
63. Kanno T, Tanaka H, Nakamura T, Tabata H, Kawai T. Real space observation of double-helix DNA structure using a low temperature scanning tunneling microscopy. Jpn J Appl Phys 1999;38:L606–L607.
64. Wang H, Tang Z, Li Z, Wang E. Self-assembled monolayer of ssDNA on Au(1 1 1) substrate. Surf Sci Lett 2001;480:L389–L394.
65. Tanaka H, Hamai C, Kanno T, Kawai T. High-resolution scanning tunneling microscopy imaging of DNA molecules on Cu(111) surfaces. Surf Sci Lett 1999;432:L611–L616.
66. Ceres DM, Barton JK. In situ scanning tunneling microscopy of DNA-modified gold surfaces: bias and mismatch dependence. J Am Chem Soc 2003;125: 14,964–14,965.
67. Shapir E, Yi J, Cohen H, Kotlyar AB, Cuniberti G, Porath D. The puzzle of contrast inversion in DNA STM imaging. J Phys Chem B 2005;109: 14,270–14,274.
68. Lindsay SM, Li Y, Pan J, et al. Studies of the electrical properties of large molecular adsorbates. J Vac Sci Technol 1991;9:1096–1101.
69. Iijima M, Kato T, Nakanishi S, et al. STM/STS study of electron density of states at the bases sites in the DNA alternating copolymers. Chem Lett 2005;34: 1084–1085.

70. Xu MS, Endres RG, Tsukamoto S, Kitamura M, Ishida S, Arakawa Y. Conformation and local environment dependent conductance of DNA molecules. Small 2005;1:1–4.
71. Xu MS, Tsukamoto S, Ishida S, et al. Conductance of single thiolated poly(GC)-poly(GC) DNA molecules. Appl Phys Lett 2005;87:083902–1/3.
72. Kotlyar AB, Borovok N, Molotsky T, Fadeev L, Gozin M. In vitro synthesis of uniform poly(dG)–poly(dC) by Klenow exo- fragment of polymerase I. Nucl Acid Res 2005;33:525–535.
73. Cohen H, Nogues C, Naaman R, Porath D. Direct measurement of electrical transport through single DNA molecules of complex sequence. Proc Natl Acad Sci USA 2005;102:11,589–11,593.
74. Xu B, Zhang P, Li X, Tao N. Direct conductance measurement of single DNA molecules in aqueous solution. Nano Lett 2004;4:1105–1108.
75. Omerzu A, Licer M, Mertelj T, Kabanov VV, Mihailovic D. Hole interactions with molecular vibrations on DNA. Phys Rev Lett 2004;93:218101–1/4.
76. Cui XD, Primak A, Zarate X, et al. Reproducible measurement of single-molecule conductivity. Science 2001;294:571–574.
77. Van Zalinge H, Schiffrin DJ, Bates AD, Haiss W, Ulstrup J, Nichols RJ. Measurement of single- and double-stranded DNA oligonucleotides. Chemphyschem 2006;7:94–98.
78. Nogues C, Cohen SR, Daube SS, Naaman R. Electrical properties of short DNA oligomers characterized by conducting atomic force microscopy. Phys Chem Chem Phys 2004;6:4459–4466.
79. Di Felice R, Selloni A. Adsorption modes of cysteine on Au(111): thiolate, amino-thiolate, disulfide. J Chem Phys 2004;120:4906–4914.
80. Cohen H, Nogues C, Ullien D, Daube S, Naaman R, Porath D. Electrical characterization of self-assembled single- and double-stranded DNA monolayers using conductive AFM. Faraday Disc 2006;131:367–376.
81. Gómez-Navarro C, Gil A, Álvarez M, et al. Scanning force microscopy three-dimensional modes applied to the study of the dielectric response of adsorbed DNA molecules. Nanotechnology 2002;13:314–317.
82. Porath D, Lapidot N, Gomez-Herrero J. Charge transport in DNA-based devices. In: Cuniberti G, Fagas G, Richter K, eds. Introducing Molecular Electronics. Lecture Notes in Physics, vol. 680. Heidelberg: Springer, 2005:411–439.
83. Odom DT, Barton JK. Long range oxidative damage in DNA/RNA duplexes. Biochemistry 2001;40:8727–8737.
84. Hall DB, Holmlin RE, Barton JK. Oxidative DNA damage through long range electron transfer. Nature 1996;382:731–734.
85. Meggers E, Michel-Beyerle ME, Giese B. Sequence dependent long range hole transport in DNA. J Am Chem Soc 1998;120:12,950–12,955.
86. Wagenknecht H-A. Charge Transfer in DNA. From Mechanism to Application. Weinheim: Wiley-VCH, 2005.
87. Bixon M, Giese B, Wessely S, Langenbacher T, Michel-Beyerle ME, Jortner J. Long-range charge hopping in DNA. Proc Natl Acad Sci USA 1999;96:11,713–11,716.
88. Voityuk AA, Rösch N, Bixon M, Jortner J. Electronic coupling for charge transfer and transport in DNA. J Phys Chem B 2000;104:9740–9745.

89. Voityuk AA, Jortner J, Bixon M, Rösch. Electronic coupling between Watson-Crick pairs for hole transfer and transport in deoxyribonucleic acid. J Chem Phys 2001;114:5614–5620.
90. Marcus RA, Sutin N. Electron transfers in chemistry and biology. Biochim Biophys Acta 1985;811:265–322.
91. Bixon M, Jortner J. Electron transfer—from isolated molecules to biomolecules. Adv Chem Phys 1999;106:35–202.
92. Giese B. Hole injection and hole transfer through DNA. The hopping mechanism. In: Schuster G, ed. Long Range Charge Transfer in DNA II. Topics in Current Chemistry, vol. 236. Berlin: Springer, 2004:27–44.
93. Giese B, Amaudrut J, Köhler AK, Spormann M, Wessely S. Direct observation of hole transfer through DNA by hopping between adenine bases and by tunnelling. Nature 2001;412:318–320.
94. Behrens C, Cichon MK, Grolle F, Hennecke U, Carell T. Excess electron transfer in defined donor-nucleobase and donor-DNA-acceptor systems. In: Schuster G, ed. Long Range Charge Transfer in DNA II. Topics in Current Chemistry, vol. 236. Berlin: Springer, 2004:187–215.
95. Rösch N, Voityuk AA. Quantum chemical calculation of donor-acceptor coupling for charge transfer in DNA. In: Schuster G, ed. Long Range Charge Transfer in DNA II. Topics in Current Chemistry, vol. 237. Berlin: Springer, 2004:37–72.
96. Jortner J, Bixon M, Langenbacher T, Michel-Beyerle M. Charge transfer and transport in DNA. Proc Natl Acad Sci USA 1998;95:12,759–12,765.
97. Bixon M, Jortner J. Incoherent charge hopping and conduction in DNA and long molecular chains. Chem Phys 2005;319:273–282.
98. Gervasio FL, Carloni P, Parrinello M. Electronic structure of wet DNA. Phys Rev Lett 2002;89:108102.
99. Gervasio FL, Laio A, Parrinello M, Boero M. Charge localization in DNA fibers. Phys Rev Lett 2005;94:158103–1/4.
100. Artacho E, Machado M, Sánchez-Portal D, Ordejón P, Soler JM. Electrons in dry DNA from density functional calculations. Mol Phys 2003;101:1587–1594.
101. Di Felice R, Calzolari A. Electronic structure of DNA derivatives and mimics by density functional theory. In: Starikov E, Lewis JP, Tanaka S, eds. Modern Methods for Theoretical Physical Chemistry of Biopolymers. Amsterdam: Elsevier, 2006:485–507.
102. Fusch EC, Lippert BJ. [Zn3(OH)2(1-MeC-N3)5(1-MeC-O2)3]4+ (1-MeC = 1-Methylcytosine): structural model for DNA crosslinking and DNA rewinding by Zn(II)? J Am Chem Soc 1994;116:7204–7209.
103. Tanaka K, Tengeiji A, Kato T, Toyama N, Shionoya M. A discrete self-assembled metal array in artificial DNA. Science 2003;299:1212–1213.
104. Liu H, Gao J, Lynch SR, Saito YD, Maynard L, Kool E. A four-base paired genetic helix with expanded size. Science 2003;302:868–871.
105. Gao J, Liu H, Kool ET. Assembly of the complete eight-base genetic helix, xDNA, and its interaction with the natural genetic system. Angew Chemie Int Ed 2005;44:3118–3122.
106. Phillips K, Dauter Z, Murchie AIH, Lilley DMJ, Luisi B. The crystal structure of a parallel-stranded guanine tetraplex at 0.95 Å resolution. J Mol Biol 1997;273:171–182.

107. Gottarelli G, Spada GP, Garbesi A. Self-assembled columnar mesophases based on guanine-related molecules. In: Atwood JL, Davies JED, MacNicol D, Vögtle FV, eds. Comprehensive Supramolecular Chemistry, vol 9. Oxford: Pergamon, 1996:483–506.

108. Laughlan G, Murchie AIH, Norman DG, et al. The high-resolution crystal structure of a parallel-stranded guanine tetraplex. Science 1994;265:520–524.

109. Davis GT. G-quartes 40 years later: from 5′-GMP to molecular biology and supramolecular chemistry. Angew Chemie Int Ed 2004;43:668–698.

110. Calzolari A, Di Felice R, Molinari E, Garbesi A. Electron channels in biomolecular nanowires. J Phys Chem B 2004;108:2509–2515; ibid. 13058.

111. Di Felice R, Calzolari A, Zhang H. Towards metalated DNA-based structures. Nanotechnology 2004;15:1256–1263.

112. Di Felice R, Calzolari A, Molinari E, Garbesi A. Ab initio study of model guanine assemblies: the role of π-π coupling and band transport. Phys Rev B 2000; 65:045104-1/10.

113. Cohen H, Shapir T, Borovok N, et al. Polarizability of G4-DNA observed by electrostatic force microscopy measurements. Nano Letters 2007;7:981–986.

114. Calzolari A, Di Felice R, Molinari E, Garbesi A. G-quartet biomolecular nanowires. Appl Phys Lett 2002;80:3331–3333.

115. Di Felice R, Calzolari A, Garbesi A, Alexandre SS, Soler JM. Strain-dependence of the electronic properties in periodic quadruple helical G4-wires. J Phys Chem B 2005;109:22,301–22,307.

116. Kittel C. Introduction to Solid State Physics, 6th ed. New York: Wiley, 1986.

117. Spacková N, Berger I, Sponer J. Nanosecond molecular dynamics simulations of parallel and antiparallel guanine quadruplex DNA molecules. J Am Chem Soc 1999;121:5519–5534.

118. Stefl R, Spacková N, Berger I, Koca J, Sponer J. Molecular dynamics of DNA quadruplex molecules containing inosine, 6-thioguanine and 6-thiopurine. Biophys J 2001;80:455–461.

119. Cavallari M, Calzolari A, Garbesi A, Di Felice R. Stability and migration of metal ions in G4-Wires by molecular dynamics simulations. J Phys Chem B 2006;110:26,337–26,348.

120. Xu Q, Deng H, Braunlin W. Selective localization and rotational immobilization of univalent cations on quadruplex DNA. Biochemistry 1993;32:13,130–13,137.

121. Mergny JL, Phan AT, Lacroix L. Following G-quartet formation by UV-spectroscopy. FEBS Lett 1998;435:74–78.

122. Dapic V, Abdomerovic V, Marrington R, Peberdy J, Rodger A, Trent JO, Bates PJ. Biophysical and biological properties of quadruplex oligodeoxyribonucleotides. Nucl Acids Res 2003;31:2097–2107.

123. Marsh TC, Henderson E. G-wires: self-assembly of a telomeric oligonucleotide, d(GGGGTTGGGG), into large superstructures. Biochemistry 1994;33:10,718–10,724.

124. Forman SL, Fettinger JC, Pieraccini S, Gottarelli G, Davis JT. Toward artificial ion channels: a lipophilic G-quadruplex. J Am Chem Soc 2000;122:4060–4067.

125. Alexandre SS, Soler JM. Private communication.

126. O'Neill BM, Ratto JE, Good KL, Tahmassebi DC, Helquist SA, Morales JC, Kool ET. A highly effective nonpolar isostere of deoxyguanosine: synthesis, structure, stacking, and base pairing. J Org Chem 2002;67:5869–5875.

127. Zhang HY, Calzolari A, Di Felice R. On the magnetic alignment of metal ions in a DNA-mimic double helix, J Phys Chem B 2005;109:15,345–15,348.

128. Kool ET. Replacing the nucleobases in DNA with designer molecules. Acc Chem Res 2002;35:936–943.

129. Liu H, Gao J, Saito YD, Maynard L, Kool ET. Toward a new genetic system with expanded dimensions: size-expanded analogues of deoxyadenosine and thymine. J Am Chem Soc 2004;126:1102–1109.

130. Liu H, Lynch SR, Kool ET. Solution structure of xDNA: a paired genetic helix with increased diameter. J Am Chem Soc 2004;126:6900–6905.

131. Gao J, Liu H, Kool ET. Expanded-size bases in naturally sized DNA: evaluation of steric effects in Watson-Crick pairing. J Am Chem Soc 2004;126: 11,826–11,831.

132. Liu H, Gao J, Kool ET. Helix-forming properties of size-expanded DNA, an alternative four-base genetic form. J Am Chem Soc 2005;127:1396–1402.

133. Lee AHF, Kool ET. A new four-base helix, yDNA, composed of widened benzopyrimidine-purine pairs. J Am Chem Soc 2005;127:3332–3338.

134. Piccirilli JA, Krauch T, Moroney SE, Brenner SA. Enzymatic incorporation of a new base pair into DNA and RNA extends the genetic alphabet. Nature 1990;343:33–37.

135. Rappaport HP. The 6-thioguanine/5-methyl-2pyrimidinone base pair. Nucl Acid Res 1988;16:7253–7267.

136. McMinn DL, Ogawa AK, Wu YQ, Liu JQ, Schultz PG, Romesberg FE. Efforts toward expansion of the genetic alphabet: DNA polymerase recognition of a highly stable, self-pairing hydrophobic base. J Am Chem Soc 1999;126:11,585–11,586.

137. Tae EL, Wu Y, Xia G, Schultz PG, Romesberg FE. Efforts toward expansion of the genetic alphabet: replication of DNA with three base pairs. J Am Chem Soc 2001;123:7439–7440.

138. Migliore A, Varsano D, Corni S, Di Felice R. Electron transfer rates in natural and expanded DNA base pairs. Work in progress.

139. Weizman H, Tor Y. 2,2′-Bypyridine ligandoside: a novel building block for modifying DNA with intra-duplex metal complexes. J Am Chem Soc 2001;123:3375–3376.

140. Meggers E, Holland PL, Tolman WB, Romesberg FE, Schultz PG. A novel copper-mediated DNA base pair. J Am Chem Soc 2000;122: 10, 714–10,715.

141. Zhang L, Meggers E. An extremely stable and orthogonal DNA base pair with a simplified three-carbon backbone. J Am Chem Soc 2005;127: 74–75.

142. Clever GH, Polborn K, Carell T. A highly DNA-duplex-stabilizing metal-salen base pair. Angew Chem Int Ed 2005;44:7204–7208.

143. Switzer C, Sinha S, Kim PH, Heuberger BD. A purine-like nickel(II) base pair for DNA. Angew Chem Int Ed 2005;44:1529–1532.

9

Electrical Manipulation of DNA on Metal Surfaces

Marc Tornow, Kenji Arinaga, and Ulrich Rant

Summary

We review recent work on the active manipulation of DNA on metal substrates by electric fields. This includes the controlled positioning, alignment, or release of DNA on or into dedicated locations and the control of hybridization. In this context, we discuss techniques for immobilizing DNA on metal surfaces and methods of characterizing such hybrid systems. In particular, we focus on electrically induced, conformational changes of monolayers of short oligonucleotides on gold substrates. Such switchable layers allow for molecular dynamics studies at interfaces and have demonstrated large potential in label-free biosensing applications.

Key Words: Biomolecular films; biosensors; conformational changes; DNA-based sensing; molecular dynamics; nano-electromechanical system (NEMS); oligonucleotides; self-assembled monolayers; surface functionalization; switchable layer.

1. INTRODUCTION

Functional biomolecular layers on surfaces have been gaining significant importance as a result of their widespread relevance in surface and physical sciences, molecular biology, and nanobiotechnology. They find diverse applications in the field of biosensors, biomedical diagnostics, functionalized Lab-on-Chip devices, and catalysis of reactions on surfaces.

From a more fundamental point of interest, biomolecular layers have been considered a particular species of organic monolayers that serve as model systems for biophysical studies, molecular dynamics, and fundamental biomedical research involving, for example, the study of regulatory and signaling pathways. Biomolecular layers of most current interest involve protein microarrays as a major tool in proteome research *(1)*, antibody arrays, or even whole-cell arrays, among others.

From: *NanoBioTechonology: BioInspired Devices and Materials of the Future*
Edited by: Oded Shoseyov and Ilan Levy © Humana Press Inc., Totowa, NJ

A particular focus has been put on monolayers of DNA, often in the form of short oligonucleotides that are being tethered to solid substrates. The most widespread applications are fluorescence-based DNA microarrays, commonly termed "DNA chips" in the form of microarrays on activated glass surfaces. These arrays have been extensively used for gene-expression analysis and have found widespread applications in the field of human health such as disease prognosis and drug discovery *(2–6)*. In contrast, biosensors based on DNA *(7)* find more applications in the field of diagnostics or general bioanalytical applications such as screening, e.g., the detection of single-nucleotide polymorphisms (SNPs). They mostly target the judgment of certain disease susceptibilities, or, more generally, make genetic tests feasible. The underlying common architecture of DNA-based sensors is probe strands that are immobilized on a solid surface. Those strands then act as a molecular recognition element for a complementary strand, for proteins or other small molecules. As an important feature of the operation principle, the solid substrate and/or DNA takes part in transducing the recognition event into, for instance, an electrical read-out signal.

In addition to the functionality inherent to the DNA molecules themselves—by virtue of their sequence—any additional active control over their immobilization, steric orientation, hybridization, or molecular recognition properties on the surface is highly advantageous. Such controlled manipulation facilitates applications in diverse fields like the molecular alignment in microelectronics, drug delivery by controlled release, biosensing, or DNA computing. This review will address several of these issues, thereby mostly focusing on the electrical manipulation of layers of short oligonucleotides on gold surfaces, based on the direct interaction of the charged DNA with external electric fields. The largely reported work on DNA immobilized on surfaces other than metal, such as semiconductors, remains beyond the scope of this chapter and will not be described.

Following this short introduction, techniques for immobilizing DNA oligonucleotides on metal surfaces are addressed, concentrating on specific (covalent) chemical bonds to the substrate. Subsequently, methods of characterizing such monolayers are reviewed. The main part of this chapter then focuses on the active manipulation of DNA on metal substrates by electric fields, including, in particular, the reported work on layers of short oligonucleotides. Finally, such manipulation in the context of DNA-based sensing and time-resolved studies of the DNA dynamics are discussed.

2. METAL SURFACE MODIFICATION WITH DNA

2.1. Structure of DNA

The DNA molecule is a linear, oligomer-like chain of nucleotides that are each composed of a ribose sugar, a phosphate group, and an attached nitrogenous heterocyclic base. For details of the structure, the reader is referred to classical text books, e.g., ref. *8*. The genetic information is encoded in the composition sequence of the four bases adenine (A), guanine (G), cytosine (C), and thymine (T). The nucleotide units building up single-stranded (ss)DNA are connected via oxygen atoms, thereby forming phosphodiester linkages. Hybridization of two ssDNA, i.e., double-strand formation, occurs through specific base pairing (A-T) or (G-C) via hydrogen bonds, resulting in a helical structure form of the double-stranded (ds)DNA—the Watson-Crick double helix. In dsDNA, the structure repeats every 3.4 nm along the axis of the helix. Whereas the base pairs are located inside the helix, the phosphate groups outside form the backbone of the DNA. At physiological conditions, the phosphates are ionized, carrying one negative charge per unit. On the scale of a few tens of nanometers, ssDNA appears as a flexible chain owing to a persistence length of approx 1 to 2 nm, depending on the salt concentration in the electrolyte *(9)*. In contrast, dsDNA can be considered a stiff rod (persistence length ~50 nm) in this regime *(10,11)*.

2.2. DNA–Metal Surface Interaction

DNA molecules may bind nonspecifically to metal surfaces by a range of mechanisms, e.g., via the molecule backbone or the bases *(12)*, via electrostatics (image charges) *(13)*, or by van der Waals interaction. Such findings are supported by studies on the formation of ordered organic adlayers on solid surfaces, composed of pyrimidines (e.g., cytosine, thymine, and uracil) and purines (e.g., adenine and guanine) *(14–17)*. In contrast to covalent binding, the nonspecific interaction of DNA bases with surfaces is, in general, weak. It is assumed that monolayers are stabilized by the formation of intermolecular bonds, e.g., hydrogen-bond networks. In the case of thymine, it was found that their interaction or adsorption behavior on gold electrodes can depend on the applied potential *(18)*.

2.3. Immobilization Strategies

A key factor determining the properties of functional DNA layers is the chosen method of immobilization on the supporting metal surface. Attaching molecules by physical adsorption only generally constrains their functionality and accessibility. Various strategies to attach DNA molecules by specific covalent adsorption utilizing a reaction between the metal surface and an anchoring group of the molecules have been developed in recent years. A widely used method employs a thiol (SH)-terminated, molecular linker group to bind to a metal surface via a sulfur–metal bond, well developed in the field of self-assembled monolayers (SAMs) *(19–22)*. In many cases, such SAMs comprising a regular array of molecules form spontaneously on a surface by adsorption from solution, promoted by van der Waals forces and hydrophobic interactions.

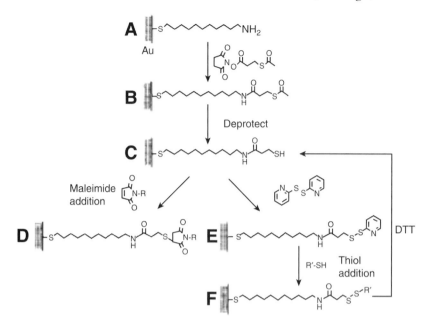

Fig. 1. Schematics of surface functionalization using sulfhydryl-terminated alkan-thiols bound to Au. (Reprinted from ref. *26*, with permission of the American Chemical Society.)

Based on this controllable surface-coating technique, biofunctional inter-faces can be formed in a hybrid approach. By first assembling a monolayer comprised of molecules with functional end groups, in a subsequent step, bio-molecules such as DNA are chemisorbed onto this SAM either directly *(23–26)* or via additional supramolecular interfacial structures *(27)*. Figure 1 shows as an example the modification of a sulfhydryl-terminated alkanethiol SAM.

2.4. Controlling the Surface Density of Immobilized DNA

In general, the chemical adsorption of molecules largely depends on diffu-sion, as has been verified by experiments monitoring the change of the immo-bilized DNA molecule surface density vs time *(28,29)*. Mainly, the measured adsorption kinetics turns out to be in good agreement with the Langmuir adsorption model. Peterson et al. assigned an observed difference in the measured isotherms of ss- vs. dsDNA under identical conditions (Fig. 2) to differences in conformation, flexibility, and electrostatic interaction of both species *(29)*.

Compared to alkyl chains in common SAM systems biomolecules such as DNA and proteins are of relatively large dimensions. Accordingly, in a densely packed monolayer, steric hindrance between biomolecules becomes significant,

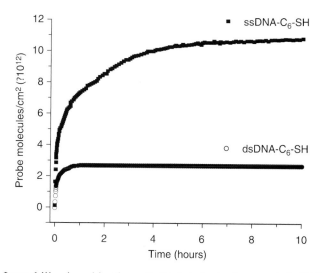

Fig. 2. Immobilization kinetics of thiolated single-stranded DNA (closed squares) and double-stranded DNA (open circles), measured by surface plasmon resonance (SPR) techniqes. (Reprinted from ref. *29*, with permission of Oxford University Press.)

hampering free motion of the biomolecules. On such a surface, it is not easy to enable processes related to reactions between biomolecules (e.g., hybridization), nor is it easy to differentiate between their thermal motion and a controlled manipulation by external forces such as electric fields. In particular, in the field of DNA-based biosensors, it is most important to control the surface density, especially in the low-density regime, with high reproducibility.

As DNA is multiply negatively charged in aqueous solution around neutral pH, the molecules electrostatically repel each other. Such repulsion is reduced by screening effects due to the accumulation of positively charged counterions around the DNA. The screening efficiency depends on the salt concentration of the electrolyte. This way, it is possible to control the closest average distance between DNA molecules assembling on a surface. A corresponding control of the resulting surface density has been demonstrated by Herne and Tarlov *(30)* (*see* Fig. 3), and Peterson et al. *(29)*. Modeling the assembled DNA on the Au surface as a hexagonal monolayer, Arinaga et al. recently obtained good agreement between calculated densities and their measured values, when considering Debye screening and the effective molecular dimensions *(31)*.

2.5. Short Alkanol-Thiol Co-Adsorption Effects

In 1997, Tarlov's group introduced the co-adsorption of short alkanol-thiol molecules, particularly mercaptohexanol (MCH), onto a previously prepared

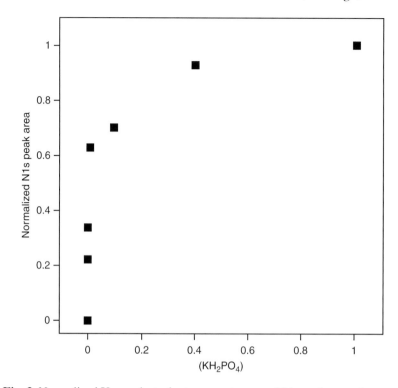

Fig. 3. Normalized X-ray photoelectron spectroscopy N 1s peak areas (corresponding to the DNA surface density) measured on a gold substrate after assembly from
a 1.0 μM thiolated single-stranded DNA solution, as a function of buffer salt concentration. (Reprinted from ref. *30*, with permission of the American Chemical Society.)

layer of DNA oligonucleotides *(30,32)*. It was shown that MCH co-adsorption
can efficiently be employed to control the structure of the DNA layers on the
surface. As depicted in Fig. 4, the process of co-assembly removes and
replaces the loosely bound, physisorbed nucleic acids, and changes the
specifically bound DNA conformation to an upright position, preventing any
further nonspecific interaction of this DNA with the metal surface. Moreover,
remaining areas of previously uncovered Au between DNAs are passivated
electrochemically and physically. Forming the mixed monolayer of MCH and
DNA allows quantifying the surface density of nonlabeled DNA by using an
electrochemical determination method *(33)*, as described in further detail
under Subheading 3. Recently, Rant et al. reported the real-time monitoring of
the release of excess DNA from the surface by co-adsorption of MCH employing the distance-dependent quenching of a fluorescence marker attached to the
oligo strands *(see* Subheading 3.2) *(34)*.

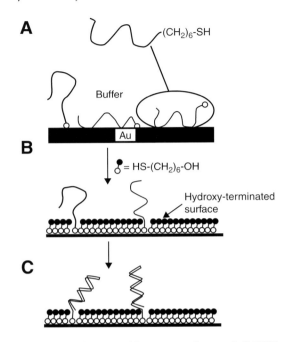

Fig. 4. Backfilling DNA layers with mercaptohexanol (MCH). **A,** HS-single-stranded DNA absorbing in various states on a Au surface; **B,** co-adsorption of MCH removes nonspecifically bound DNA from the surface and, at the same time, detaches specifically (S-Au bond) absorbed DNA, inducing a "stand-up"-type orientation; **C,** after hybridization. (Reprinted from ref. *32,* with permission of the American Chemical Society.)

3. PROBING THE MANIPULATION OF OLIGONUCLEOTIDE LAYERS ON METAL SURFACES

Various techniques have been extensively used to examine the molecular structure of nucleic acid layers bound to metal surfaces. Parameters like molecular surface coverage, surface structure (conformation/orientation), steric interaction, and the DNA-to-metal linker structure directly affect important quantities such as hybridization efficiency; hence, they provide important information about the functionality of the particular DNA layer as a probe in sensing. Among the most important characterization techniques are electrochemical methods involving, e.g., the investigation of the oxidation of DNA bases on a range of different metals such as copper *(35)*. Such oxidation of bases, in particular guanine, was also employed by Thorp et al. for label-free detection of hybridization on indium-tin-oxide (ITO) electrodes *(36)*. A quantification method to deduce the surface coverage of DNA oligonucleotides on Au was developed by Steel et al. *(33)*. Here, the DNA layer is exposed to an electrolyte solution containing a multivalent redox cation, Hexa-amine-ruthenium(III) chloride (RuHex). The negative charge of the phosphate groups is compensated by electrostatically trapping RuHex to its backbone. In a chronocoulometric measurement, the excess charge from the reduction of surface-confined, DNA-bound RuHex is determined. This surface excess charge is proportional to the number of DNA molecules at the surface. Also using redox indicators, various methods have been developed to reduce background signals, e.g., by working with electrically neutral peptide nucleic acid (PNA) instead of DNA as a probe strand *(37)*. The Barton's group used electroactive intercalators for the sensitive detection of single-base mismatches by involving sequence-dependent charge transport through dsDNA that had been tethered to polycrystalline Au electrodes *(38)*. Among other important methods of investigating the surface structure and/or coverage, we mention here radiolabeling *(39)*, neutron reflection *(32)*, X-ray photoelectron spectroscopy (XPS) *(40,41)*, and infrared spectroscopy *(26,42,43)* without discussing them in detail, as that is beyond the scope of this chapter. Instead, we will describe a few more techniques in the following that have been extensively used for monitoring the layer structure response to external manipulation.

3.1. Scanning Probe Techniques

Scanning probe investigations to image DNA on surfaces in general, from a more biological point of interest, have been summarized in a recent review by Hansma *(44)*. Using scanning tunneling microscopy (STM), Rekesh et al. *(45)* were able to visualize short thiolated ss- and dsDNA lying on Au, thereby observing wormlike structures with an unexpectedly short internucleotide spacing for ssDNA. Atomic force microscopy (AFM) measurements were used to investigate the hybridization efficiency of ssDNA on Au *(46)* and for the characterization of dsDNA layers regarding protein interaction *(47)*.

With emphasis on layer conformation changes induced by external electric fields, Kelley et al. used *in situ* electrochemical (EC)-AFM to study layers of 15-mer dsDNA tethered to gold *(48)*. In their work, the electrical potential of the work electrode that supports the DNA monolayer is controlled vs the electrolyte potential, allowing the study of structural changes of the layer as a function of applied electric field. In a similar manner, EC-STM measurements have been used to directly visualize the (voltage-influenced) surface structure of thiol-derivatized DNA oligonucleotide layers on gold surfaces *(49)*. A resolution of the layer structure on single crystal facet Au with single-molecule resolution was reported recently by Wackerbarth et al. *(50)*.

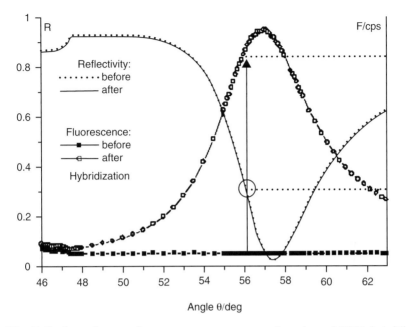

Fig. 5. Surface-plasmon fluorescence spectroscopy detection of DNA hybridization. The angular reflectivity and fluorescence intensity are shown before and after adsorption of fluorescently labeled 15-mer DNA targets onto the probe functionalized surface. As a result of the low added mass (probe strand density maximum $1/40$ nm^2) no change is seen in the reflectivity; however, the excited fluorescence displays a pronounced signal increase. (Reprinted from ref. *53*, with permission of Wiley-VCH.)

3.2. Optical Techniques

The most widespread optical characterization technique of DNA layers on metals is surface plasmon resonance (SPR) *(51)*. Mostly used on thin gold films, SPR detects changes in the refractive index due to changes in the molecular surface coating (e.g., adsorption) with an outstanding sensitivity. Using two-wavelength SPR, fundamental properties of DNA layers such as surface coverage *(28)*, thickness, and dielectric constant can be measured *(52)*. Using surface-plasmon-field-enhanced fluorescence spectroscopy, Knoll's group has achieved very high sensitivities in hybridization detection (*see* Fig. 5). Here, the fluorescence labels of surface-tethered DNA oligos are excited via the evanescent surface plasmon field *(53)*.

Fluorescence labeling and readout provides the basis of today's optical DNA microarray chip technology. These methods are predominantly used to characterize DNA layers on insulating substrates. There, also various sophisticated techniques have been developed that make use of nonradiative energy transfer from one excited dye (donor) to a different fluorophore in close proximity (acceptor), referred to as fluorescence resonance energy transfer (FRET) *(54)*. To characterize DNA layers on metals, a few groups have reported experiments that make use of a closely related phenomenon, namely the fluorescence quenching that occurs when fluorescence dyes reside at small distances z from the metal surface. Here, it is known that the fluorescence quantum yield (which is proportional to the observed fluorescence intensity) exhibits a strong quenching, mainly following a z^3 dependence *(55,56)*. Rant et al. exploited this distance-dependent, nonradiative energy transfer into a metallic surface to investigate the structural properties of oligonuleotide monolayers *(34)*. In their studies, the effect of varying the surface coverage densities of layers of 12- and 24-mer ssDNA on Au was monitored by measuring the fluorescence of a Cy3 dye linked to the top end of the DNAs. This way, information on the average orientation of the strands relative to the surface could be extracted. In particular, an observed accelerated enhancement in fluorescence intensity with increasing DNA coverage could be related to a basic model involving emerging mutual steric interactions of oligonucleotides on the surface.

Making use of the same quenching phenomenon, Perez-Luna et al. *(57)* developed a scheme for biosensing. By immobilizing the analyte of interest (or a structural analog of the analyte) to a metal surface and exposing it to a labeled receptor (e.g., antibody), the fluorescence of the labeled receptor became quenched upon binding because of the close proximity to the metal. Upon exposure to free analyte, the labeled receptor dissociated from the surface and diffused into the bulk of the solution. This increased its separation from the metal, and an increase of fluorescence intensity and/or lifetime of the excited state was observed that indicated the presence of the soluble analyte.

3.3. Quartz Microbalance

Manipulation of DNA on metals can be sensitively monitored by quartz crystal microbalance (QCM) *(58)* through a change in resonance frequency, if associated with a change of mass, density, or elasticity of the molecular surface layer. Owing to the fact that QCM can also be operated in aqueous solution, various biochemical processes like hybridization have been extensively studied *in situ* by this method *(59)*. Several strategies have been developed to further increase the high sensitivity of QCM by, e.g., mass amplification during hybridization *(60)*. Using electrochemical QCM, Wang et al. monitored the electric field-induced release of DNA from a gold surface *(61)*.

4. MANIPULATION ON METAL SURFACES BY ELECTRIC FIELDS

A range of nonelectric methods has been employed to move, position, stretch, and align DNA on surfaces, among which are scanning probe-related techniques *(62–65)*, hydrodynamic flow *(66,67)*, or motor proteins *(68)*. A detailed overview of these fascinating manipulation techniques is beyond the scope of this chapter. In the following, we will review those experimental results dealing with the electrical manipulation of DNA on metals.

4.1. Controlled DNA Immobilization and Hybridization

In the field of DNA-based sensing applications, the control of ssDNA immobilization as well as its accessibility for hybridization play a determining role. We already described some relevant methods for DNA surface functionalization under Subheading 2. However, we have not yet addressed the possibility of influencing such processes by using electric fields. Walti et al. combined the electrochemical desorption of a monomolecular layer (MCH) with the subsequent adsorption of DNA onto the exposed surface *(69)*. Direct electric potential control on DNA was successfully employed to selectively immobilize calf thymus DNA on Au electrodes that had been functionalized with a self-assembled monolayer of 2-amino-ethanethiol in advance. By using cyclic voltammetry (CV), impedance spectroscopy, Auger spectroscopy, and AFM, the expected enhanced immobilization rate for positively charged Au electrodes (decreased for negative charge, respectively) could be confirmed experimentally *(70)*.

The company *Nanogen* developed methods for the accelerated, electric field-directed nucleic acid hybridization on microstructured arrays *(71)*. This group reported on a variety of electronic DNA array devices and techniques that allow electric field-enhanced hybridization to be carried out under special low-conductance conditions *(72)*. Their electronic devices are able to provide controlled electric (electrophoretic) fields that serve as a driving force to move and concentrate nucleic acid molecules (DNA/RNA) to selected microlocation test sites on the device.

By using *in situ* (electrochemical) optical SPR spectroscopy, Heaton et al. *(73)* showed that electrostatic fields can increase or decrease hybridization rates of unlabeled DNA on gold. The electric field control also allowed discrimination between matched and mismatched hybrids by adjustment of the electrode potential: repulsive potentials preferably denatured mismatched hybrids within a short time while leaving the complementary, matching one largely unchanged.

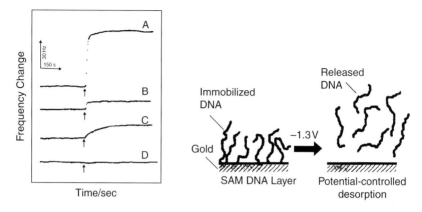

Fig. 6. Voltage step-induced molecule release monitored by resonance frequency change of a quartz microbalance, (A) double-stranded DNA, (B) single-stranded DNA, (C) octadecanethiol, (D) reference (uncoated). Right, schematic of DNA release. (Reprinted from ref. *61*, with permission of the American Chemical Society.)

4.2. Induced DNA Release

The opposing process to immobilization, namely the on-demand release of DNA into solution, has been studied by several groups. With sufficient control, such DNA release is expected to have applications in the field of genetic material delivery to specific locations, e.g., into living cells at appropriate times. As illustrated in Fig. 6, Wang et al. applied considerably large negative voltages (−1.3 V) to gold microelectrodes supporting layers of long (350-bp) thiolated DNA to induce a complete cathodic desorption *(61)*. The authors characterized the desorption behavior by using electrochemical quartz microbalance (ECQM), XPS, and CV blocking experiments.

More recently, Takeishi et al. demonstrated that such bias-induced release of DNA can be controlled by choosing the appropriate chemical linker to the Au surface *(74)*. In their work, the release of short DNA oligonucleotides was confirmed by monitoring the intensity of the fluorescence of cyanine dyes (Cy3) linked to the 5′ end of the DNA (*see* Subheading 3.2 and Fig. 7). The release sets in at bias voltages higher than −0.8 V (i.e., less negative), whereby the threshold voltage was found to depend on the kind of linker, whether -SH, -SS- or −OH (nonspecific), at the DNA 3′-terminal. As studied by the same group, the efficiency of the bias-induced release of thiolated, ss 24-mer oligos also strongly depends on the screening parameters of the electrolyte buffer, i.e., the ionic strength *(75)* (Fig. 8). By varying the concentration of monovalent salt in solution from 3 to 1600 m*M*, it was found that the strength of the electric interaction is predominantly determined by the effective charge of the ssDNA itself. In qualitative agreement with Manning's counterion condensation theory *(76)*, the measured efficiency of desorption stays constant

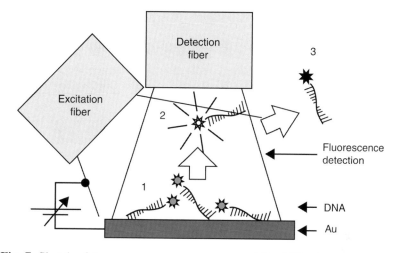

Fig. 7. Sketch of the measurement geometry used for DNA desorption measurements by fluorescent techniques, showing the three main steps of a release process that reveal in the measured fluorescence intensity over time. (1) Dye-labeled oligonucleotides are adsorbed on Au, but fluorescence is quenched as a result of the proximity to the metal surface. (2) Upon application of negative bias to the Au-electrode, the nucleotides are released into solution and float within the volume of fluorescence detection; dye-fluorescence is no longer quenched. (3) Finally, single-stranded DNA have diffused out of the active volume into a large dark reservoir where they cannot be detected. (Reprinted from ref. *75*, with permission of the Biophysical Society.)

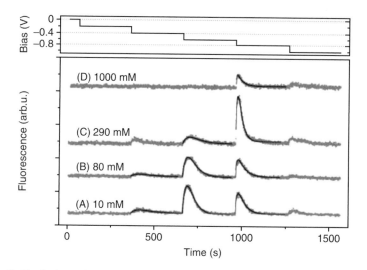

Fig. 8. Evolution of the Cy3-fluorescence intensity with time for different electrolyte salt concentrations (A–D) upon application of a step-like bias sequence to the electrode. Note how the desorption peaks shift to higher on-set voltages and decrease in amplitude when increasing the ionic strength. (Reprinted from ref. *75*, with permission from the Biophysical Society.)

over a wide range of salt concentrations. Only as the Debye screening length is reduced below a value comparable to the axial charge spacing of the DNA the amount of released DNA does drop significantly. The authors assigned this effect to excessive counterion condensation in this regime.

4.3. Stretching and Positioning

The AC dielectrophoretic effect has been widely used to stretch long DNA attached to metal surfaces *(77)*, as described first by Washizu and Kurosawa in 1990 *(78)*. The authors successfully aligned fluorescently labeled lambda phage DNA tethered to aluminum microelectrodes in electric fields of the order of 1 MV/m at frequencies between 40 kHz and 2 MHz. Recently, positioning and stretching of long DNA has been investigated by several groups mainly employing fluorescent techniques for monitoring *(79,80)*, including fluorescence anisotropy techniques *(81)*. Namasivayam et al. investigated the AC dielectrophoretic stretching of thiolated DNA tethered to Au electrodes within a polymer-enhanced buffer medium *(82)*.

Stretching of the DNA macromolecules may serve for a range of applications including size measurements and sorting of lengths >10 kbp, orientation-controlled immobilization on surfaces, investigation of enzyme activity, and DNA sequencing *(83)*. As demonstrated in earlier works, stretched DNA can be cut by a focused laser beam. More recently, Yamamoto et al. reported the defined cutting of DNA by restriction enzymes ("molecular surgery") *(84)*. The authors stretched and attached lambda-DNA between two Al microelectrodes using dielectrophoresis. Latex beads carrying immobilized enzymes were then brought into contact with the DNA using optical tweezer techniques, eventually inducing the enzymatic dissection (Fig. 9).

In DC electric fields, the electrophoretic stretching of tethered lambda-DNA was studied for strands attached to Au posts situated in a silicon microfluidic device after having eliminated electro-osmotic (bulk fluid movement) effects *(85)*. Regarding the potential usage in nanoelectronic circuitry, DNA has been positioned and aligned between lithographically fabricated metal electrodes via dielectrophoresis *(77,86)*. In 2000, Porath et al. successfully positioned 10.4-nm long oligonucleotides between nanogap electrodes to investigate their electrical transport properties *(87)*.

4.4. Orientation Modulation of Short Oligonucleotides

Short, both ss- and dsDNA oligonucleotides tethered to metal surfaces can be effectively manipulated with respect to their average orientation by electrically charging the electrode surface vs electrolyte solution. In 1998, Barton's group reported their investigations of thiol-derivatized monolayers of 15-bp dsDNA on Au using EC-AFM. Depending on the applied electrical

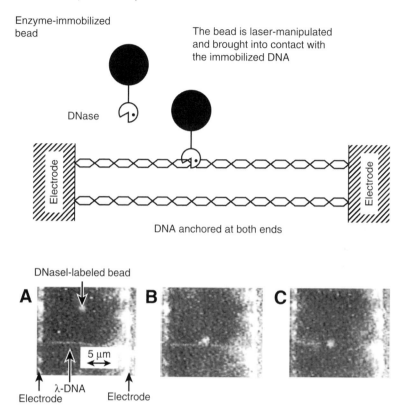

Fig. 9. Top: schematics of enzymatic dissection ("molecular surgery") of stretched DNA using optically manipulated latex beads carrying enzymes. **Bottom: (A)** Fluorescence microscope image of lambda DNA streched between microelectrodes. A single enzyme modified latex bead is seen above the DNA. **(B)** The bead is just touching the DNA. **(C)** The DNA has been cut at the touching position. (Reprinted in part from ref. *84,* with permission of IEEE.)

potentials of the electrode vs a Ag wire electrode, the authors observed a significant change in film thickness (Fig. 10), measured as a step height relative to the uncoated area. The observation corresponded to a change in surface topology where the strands either stand up for negative potentials relative to the potential of zero charge, or lie down on the surface for positive potentials, respectively *(48).*

In a similar manner, the immobilization of thiol-derivatized DNA on a Au (111) single crystal surface has been investigated by EC-STM, focusing on changes in area-scanned surface topology *(49).* In this work, Zhang et al. observed clear orientation changes of ds oligonucleotides when the positive electrode potential was increased, as seen in Fig. 11. However, no changes

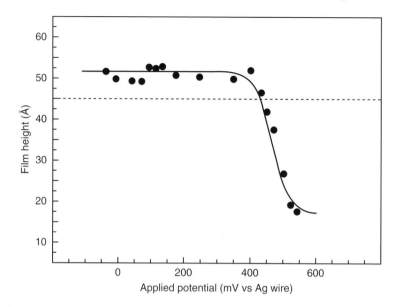

Fig. 10. DNA oligonucleotide monolayer thickness as a function of work electrode potential measured by atomic force microscopy. The dashed line corresponds to the open-circuit value. (Reprinted from ref. *48*, with permission of the American Chemical Society.)

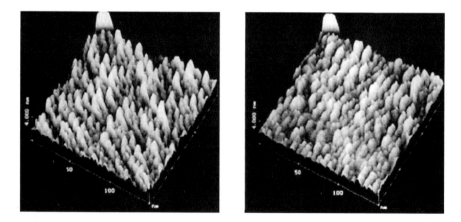

Fig. 11. Electrochemical scanning tunnelling microscopy (150 nm × 150 nm) images of a layer of 15 base-pair ds-DNA on Au in a Tris/HCl/EDTA buffer solution. The potential vs standard calomel electrode (SCE) was 300 mV (left) and 600 mV (right), respectively. (Reprinted in part from ref. *49*, with permission of the American Chemical Society.)

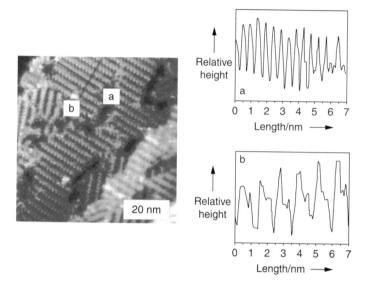

Fig. 12. Left: *in situ* scanning tunnelling microscopy image of a layer of single-stranded poly-adenine (10 bases) measured in 0.01 *M* phosphate buffer. The layer was assembled at a Au electrode potential of –0.61 V vs standard calomel electrode and scanned at 0.21 V. **Right:** relative height profiles (line scans) as indicated in the image on the left. (Reprinted from ref. *50*, with permission of Wiley-VCH.)

appeared in the case of ssDNA, which was attributed to the poor rigidity of the molecules. In contrast, Wackerbarth et al., who studied monolayers of thiolated ss oligo-adenines (10 bases) observed a long-range order domain formation associated with the DNA monolayer at negative electrode voltage *(50) (see* Fig. 12). These authors used an *in situ* STM technique on high-quality single crystalline Au surfaces, making single-molecule resolution feasible.

Recently, Rant et al. *(88)* reported the observation of the orientation control of short oligonucleotide strands tethered to gold surfaces by using the distance-dependent quenching of the fluorescence of an organic dye attached to the top end of the strands *(see* Subheading 3.2.). In their work, the authors demonstrated that for properly chosen electrochemical and surface functionalization conditions, the DNA "switching" mechanism had an outstanding long-term stability, as displayed in Fig. 13. By varying the surface coverage (and thereby the mutual steric interactions between adjacent strands), the system allowed studies ranging from the regime of (an ensemble of) individual strands up to closely packed, polymer brush-like

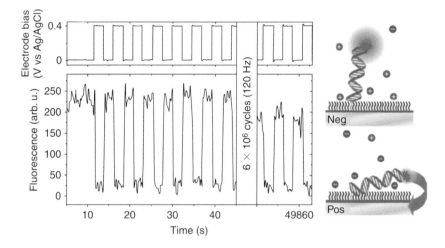

Fig. 13. Electrically induced, persistent switching of a DNA-layer (double stranded, 24-mer) on a Au surface monitored by optical measurements. **Left:** the fluorescence intensity observed from the dye-labeled DNA layer alternates upon periodically reversing the electrode-charge. **Right:** negatively biased electrodes repel the like-wise charged DNA strands, bright fluorescence is emitted from the dye attached to the DNA's top end. Positive surface charge attracts the strands and, as a result of the close proximity to the metal, efficient energy transfer from the excited dye to the Au results in a substantial quenching of fluorescence. Note that the layers maintain their functionality over millions of cycles (>13.8 h), showing outstanding persistence. No indications for desorption of molecules have been found. (Reprinted from ref. *88,* with permission of the American Chemical Society.)

layers (Fig. 14). In general, the complex behavior and interactions of (bio-) polyelectrolytes within the polarized region at liquid/metal interfaces can be addressed directly by employing this technique. Rant et al. *(88)* demonstrated this by changing parameters like salt concentration in the solution and the frequency of the driving signal (*see also* next section). It turned out that the strong electrical field present within the near-surface diffusive double layer is essential for providing the interaction strength required to induce a distinct conformational change. At frequencies too high for the double layer to accumulate, the DNA's orientation is governed by thermal fluctuations and a manipulation of the layer conformation is not feasible. In the case of highly concentrated salt solutions, the electric potential decays too fast into solution (on length scales eventually much shorter than that of the DNA). This leads in a similar way to a situation in which the electrostatic energy of the DNA strands is too small to compete with $k_B T$. A maximum (saturating) fluorescence switching amplitude was observed in the limit of low ionic strength and driving frequency.

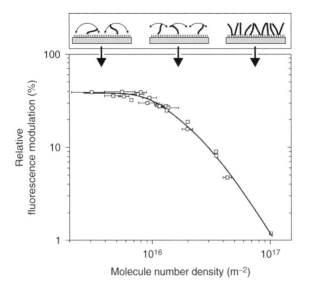

Fig. 14. Influence of the packing density of molecules within layers of 24-mer single-stranded DNA on the observable fluorescence modulation ($\Delta F/F_{avg}$). The top panel illustrates how steric interactions between adjacent molecules constrain the attainable free gyration of individual strands. The solid line is a guide to the eye. (Reprinted from ref. *88*, with permission of the American Chemical Society.)

The reported technique could be extended to systems other than DNA where electrical field-induced molecular layer reorientation has been reported (e.g., ref. *89*).

4.5. Applications of Orientation Modulation in Biosensing

The persistent, electrically driven orientation switching of short DNA oligonucleotide strands on Au surfaces opens new perspectives in bio-nanotechnology applications as a novel detection scheme in biosensing. Any affinity binding of a particular target biomolecule to a receptor linked to the switching DNA strand is expected to directly transduce into an observable change in the switching amplitude and/or dynamics. Here, this modified strand rotation may originate from a change of molecular size or weight, mechanical stiffness, charge, hydrodynamic friction, or a combination of several of these. In their work introducing the long-term stability switching effect, the authors already reported such sensing proof-of-principle investigating the hybridization of switching ssDNA with their complementary strands *(88)*.

The lower part of Fig. 15 displays the measured dynamic response of a switchable ssDNA layer obtained by performing a frequency sweep of the driving AC potential applied to the supporting Au-electrode. The

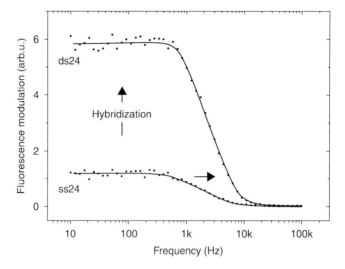

Fig. 15. Response of the fluorescence modulation amplitude as a function of the frequency of the driving electrical AC potentials. Upon hybridization, the double-stranded 24-mer DNA layer (circles) shows substantially enhanced switching compared to the single-stranded conformation (squares). The lines are a guide to the eye. (Reprinted in part from ref. *88*, with permission of the American Chemical Society.)

signal decrease observed at high frequencies ($> \approx 10\text{kHz}$) can be assigned to the fact that there the formation of the electrical double layer ceases to follow the electrical excitation, as already described in the previous section. Having monitored the frequency dependence of the ssDNA layer, the same sample was exposed to its complementary strand *in situ*, resulting in a remarkable change of the observed switching behavior: the modulation amplitude exhibits a substantial increase upon hybridization, which can be assigned to the distinct mechanical flexibilities of both molecular systems.

As anticipated, the sensing mechanism can be generalized to other target molecules, such as proteins. In a proof-of-principle experiment, this has already been demonstrated using the biotin-streptavidin model system. There, switching 48-mer dsDNA was labeled at its top end with both the Cy3 fluorophore and a biotin linker molecule. A distinct change in fluorescence modulation amplitude upon binding of streptavidin at concentrations down to below 100 p*M* has been observed (*90*).

4.6. Time-Resolved Molecular Dynamics Studies

Apart from its apparent potential in biosensing applications, the electrically induced orientation switching can provide fundamental insight into the most

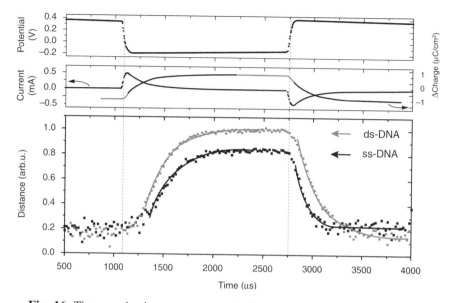

Fig. 16. Time-resolved measurements of the mechanical switching of a 48-mer oligonucleotide layer in electrolyte solution induced by an electrical AC driving potential applied to the Au substrate. The two topmost graphs show the electrochemical response of the electrode, whereas the distance of the DNA's top end to the surface is depicted in the bottom graph. The potential of the Au electrode is measured vs a Ag/AgCl reference electrode; the current measured to the Pt reference electrode has been integrated to yield the charge which accumulates at the electrode interface upon applying a bias step. Relative information of the distance of the DNA's top end to the surface (lowest graph) is inferred from the measured fluorescence intensity $(d \sim F^{1/3})$. Solid lines are single exponential fits to the data. After measurements were performed on single-stranded DNA (black squares), the layer was hybridized with strands of complementary sequence (double-stranded, grey circles). (Reprinted from ref. *91*, with permission of the Biophysical Society.)

intriguing molecular dynamics of linear polyelectrolytes within the polarized region of a charged metal–electrolyte interface. In recent experiments, Rant et al. succeeded in measuring the response to alternating repulsive and attractive electric surface fields by time-resolved measurements employing a box-car integrator technique *(91)*. Representative data comparing ss and ds 48-mer DNA are shown in Fig. 16. As anticipated from the switching mechanism described in the previous sections, the molecules undergo conformational changes driven by the electric fields that build up at the interface by the accumulation of excess ions from the solution. The authors found the field-induced kinetics of attraction and repulsion of stiff double strands to show similar time constants of the order of 230 μs.

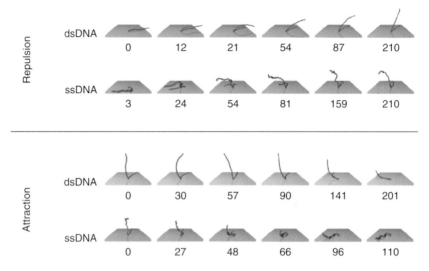

Fig. 17. Time sequence showing snap-shots of the strands kinetics obtained from hydrodynamic simulations, by Y. W. Kim and R. R. Netz. The numbers denote timing information in units of 10^3 simulation steps. (Reprinted from ref. *91*, with permission of the Biophysical Society.)

Surprisingly, however, the more flexible single strands lie down significantly faster (115 µs) than they stand up (240 µs). Hydrodynamic simulations performed by the group of R.R. Netz found good quantitative agreement regarding the relative timescales of the up-and-down motions for ss and dsDNA. The findings could be rationalized, revealing two different kinetic mechanisms: upon applying attractive potentials, the dsDNA strands, which can be considered stiff polymers, undergo rotation around the anchoring pivot point; flexible polymers, on the other hand, are pulled to the surface segment by segment, which makes this process the most efficient and, hence, the fastest. In contrast, the repulsion process is largely determined by diffusion of molecular segments outside the short-ranged surface interaction region. Figure 17 illustrates this distinct behavior by showing a time sequence of snap-shots of the strand kinetics.

5. CONCLUSION AND OUTLOOK

The active manipulation of DNA on metal surfaces by electric fields has been established as a versatile methodology in recent years, covering a wide range from basic science to application-oriented research and device development. The most prominent fundamental studies include investigations of the basic mechanical, geometric, and charge-transfer properties of DNA. Largely driven by prospects in applications such as DNA-based sensors or

DNA arrays, much control has been acquired regarding the positioning, alignment, or release of DNA on or into dedicated locations and the efficiency of hybridization on surfaces. In particular, layers of oligonucleotides that can be reversibly switched with great long-term stability have gained much interest recently. Such switchable layers have demonstrated large potential to conduct fundamental studies on the molecular dynamics of polyelectrolytes at interfaces. At the same time, this system provides the basis for an entirely new detection scheme in label-free biosensing.

Regarding applications, there is an increasing demand to realize new label-free sensing systems for the highly specific detection of biomolecules in (bio-)chemical analysis, medical research, and diagnostics. Eventually, such systems are being envisaged to integrate the preparation and separation of analyte molecules, monitoring, specific detection, and quantification on a single device, preferably with all-electronic in- and output signal processing (Lab-on-Chip concept). We believe that in particular, biomolecular switching layers comprised of DNA oligonucleotides may become a new active component of such integrated nanobiotechnological devices in the future.

ACKNOWLEDGMENTS

We gratefully acknowledge Gerhard Abstreiter, Shozo Fujita, and Naoki Yokoyama for their continuous support of this work and many useful discussions. We thank Roland R. Netz and Y. Woon Kim for a most fruitful collaboration. This work was financially supported by the Fujitsu Laboratories of Europe and in part by the DFG via SFB563. M.T. gratefully acknowledges funding by the BMBF under grant 03N8713 (Junior Research Group "Nanotechnology").

REFERENCES

1. Kambhampati D. Protein Microarray Technology, 1st ed. Weinheim: Wiley-VCH, 2004.
2. Hoheisel JD. Oligomer-chip technology. Trends Biotechnol 1997;15(11):465–469.
3. Niemeyer CM, Blohm D. DNA microarrays. Angew Chem Int Ed 1999;38(19):2865–2869.
4. Xiang CC, Chen YD. cDNA microarray technology and its applications. Biotechnol Adv 2000;18(1):35–46.
5. Schena M, Heller RA, Theriault TP, Konrad K, Lachenmeier E, Davis RW. Microarrays: biotechnology's discovery platform for functional genomics. Trends Biotechnol 1998;16(7):301–306.
6. Heller MJ. DNA Microarray technology: devices, systems, and applications. Annu Rev Biomed Eng 2002;4:129–153.

7. Tarlov MJ, Steel AB. DNA-based sensors. In: Rusling JF, ed. Biomolecular Films. New York: Marcel Dekker Inc., 2003;545–608.

8. Voet D, Voet JG, Pratt CW. Fundamentals of Biochemistry: Life at the Molecular Level, 2nd ed. New York: Wiley, 2006.

9. Tinland B, Pluen A, Sturm J, Weill G. Persistence length of single-stranded DNA. Macromolecules 1997;30:5763–5765.

10. Taylor WH, Hagerman PJ. Application of the method of phage T4 DNA ligase-catalyzed ring-closure to the study of DNA structure. II. NaCl-dependence of DNA flexibility and helical repeat. J Mol Biol 1990;212(2):363–376.

11. Smith SB, Finzi L, Bustamante C. Direct mechanical measurements of the elasticity of single DNA molecules by using magnetic beads. Science 1992; 258(13 Nov):1122.

12. Kimura-Suda H, Petrovykh DY, Tarlov MJ, Whitman LJ. Base-dependent competitive adsorption of single-stranded DNA on gold. J Am Chem Soc 2003;125(30):9014–9015.

13. Netz RR, Andelman D. Neutral and charged polymers at interfaces. Phys Rep 2003;380:1–95, and references therein.

14. Allen MJ, Balooch M, Subbiah S, Tench RJ, Siekhaus W, Balhorn R. Scanning tunneling microscope images of adenine and thymine at atomic resolution. Scanning Microscopy 1991;5(3):625–630.

15. Boland T, Ratner BD. Two-dimensional assembly of purines and pyrimidines on Au(111). Langmuir 1994;10:3845.

16. Tao NJ, DeRose JA, Lindsay S. Self-assembly of molecular superstructures studied by in situ scanning tunneling microscopy: DNA bases on gold (111). J Phys Chem 1993;97(4):910.

17. Hölzle MH, Wandlowski T, Kolb DM. Structural transitions in uracil adlayers on gold single crystal electrodes. Surface Sci 1995;335:281.

18. Roelfs B, Bunge E, Schröter C, Solomun T, Meyer H, Nichols RJ, Baumgärtel H. Adsorption of thymine on gold single-crystal electrodes. J Phys Chem B 1997;101(5):754–765.

19. Whitesides GM, Bain CD. Modeling organic surfaces with self-assembled monolayers. Angew Chem Int Ed 1989;28(4):506–512.

20. Ulman A. Thin Films: Self-Assembled Monolayers of Thiols. London: Academic Press Inc., 1998.

21. Love JC, Estroff LA, Kriebel JK, Nuzzo RG, Whitesides GM. Self-assembled monolayers of thiolates on metals as a form of nanotechnology. Chem Rev 2005;105(4):1103–1169.

22. Ulman A. Formation and structure of self-assembled monolayers. Chem Rev 1996;96:1533–1554.

23. Peelen D, Smith LM. Immobilization of amine-modified oligonucleotides on aldehyde-terminated alkanethiol monolayers on gold. Langmuir 2005;21(1): 266–271.

24. Johnson PA, Levicky R. Polymercaptosiloxane anchor films for robust immobilization of biomolecules to gold supports. Langmuir 2003;19(24):10,288–10,294.

25. Xu XH, Bard AJ. Immobilization and hybridization of DNA on an aluminum(III) alkanebisphosphonate thin-film with electrogenerated chemiluminescent detection. J Am Chem Soc 1995;117(9):2627–2631.
26. Smith EA, Wanat MJ, Cheng YF, Barreira SVP, Frutos AG, Corn RM. Formation, spectroscopic characterization, and application of sulfhydryl-terminated alkanethiol monolayers for the chemical attachment of DNA onto gold surfaces. Langmuir 2001;17(8):2502–2507.
27. Knoll W, Yu F, Neumann T, Schiller S, Naumann R. Supramolecular functional interfacial architectures for biosensor applications. Phys Chem Chem Phys 2003;5(23):5169–5175.
28. Georgiadis R, Peterlinz KP, Peterson AW. Quantitative measurements and modeling of kinetics in nucleic acid monolayer films using SPR spectroscopy. J Am Chem Soc 2000;122:3166–3173.
29. Peterson AW, Heaton RJ, Georgiadis RM. The effect of surface probe density on DNA hybridization. Nucl Acids Res 2001;29(24):5163–5168.
30. Herne TM, Tarlov MJ. Characterization of DNA probes immobilized on gold surfaces. J Am Chem Soc 1997;119:8916–8920.
31. Arinaga K, Rant U, Tornow M, Fujita S, Abstreiter G, Yokoyama N, to be published.
32. Levicky R, Herne TM, Tarlov MJ, Satija SK. Using self-assembly to control the structure of DNA monolayers on gold: a neutron reflectivity study. J Am Chem Soc 1998;120:9787–9792.
33. Steel AB, Herne TM, Tarlov MJ. Electrochemical quantitation of DNA immobilized on gold. Anal Chem 1998;70:4670–4677.
34. Rant U, Arinaga K, Fujita S, Yokoyama N, Abstreiter G, Tornow M. Structural properties of oligonucleotide monolayers on gold surfaces probed by fluorescence investigations. Langmuir 2004;20(23):10,086–10,092.
35. Singhal P, Kuhr WG. Ultrasensitive voltammetric detection of underivatized oligonucleotides and DNA. Anal Chem 1997;69(23):4828–4832.
36. Napier ME, Loomis CR, Sistare MF, Kim J, Eckhardt AE, Thorp HH. Probing biomolecule recognition with electron transfer: electrochemical sensors for DNA hybridization. Bioconjugate Chem 1997;8(6):906–913.
37. Palecek E, Fojta M, Tomschik M, Wang J. Electrochemical biosensors for DNA hybridization and DNA damage. Biosensors & Bioelectronics 1998; 13(6):621–628.
38. Kelley S, Boon E, Barton J. Jackson N, Hill M. Single-base mismatch detection based on charge transduction through DNA. Nucl Acids Res 1999;27(24): 4830–4837.
39. Steel AB, Levicky RL, Herne TM, Tarlov MJ. Immobilization of nucleic acids at solid surfaces: effect of oligonucleotide length on layer assembly. Biophys J 2000;79(2):975–981.
40. Petrovykh DY, Kimura-Suda H, Tarlov MJ, Whitman LJ. Quantitative characterization of DNA films by X-ray photoelectron spectroscopy. Langmuir 2004;20(2):429–440.

41. Leavitt AJ, Wenzler LA, Williams JM, Beebe TPJ. Angle-dependent X-ray photoelectron spectroscopy and atomic force microscopy of sulfur-modified DNA on Au(111). J Phys Chem 1994;98(35):8742–8746.
42. Petrovykh DY, Kimura-Suda H, Whitman LJ, Tarlov MJ. Quantitative analysis and characterization of DNA immobilized on gold. J Am Chem Soc 2003;125: 5219–5226.
43. Boncheva M, Scheibler L, Lincoln P, Vogel H, Akerman B. Design of oligonucleotide arrays at interfaces. Langmuir 1999;15(13):4317–4320.
44. Hansma HG. Surface biology of DNA by atomic force microscopy. Annu Rev Phys Chem 2001;52:71–92.
45. Rekesh D, Lyubchenko Y, Shlyakhtenko L, Lindsay S. Scanning tunneling microscopy of mercapto-hexyl-oligonucleotides attached to gold. Biophys J 1996;71(2):1079–1086.
46. Huang E, Satjapipat M, Han SB, Zhou FM. Surface structure and coverage of an oligonucleotide probe tethered onto a gold substrate and its hybridization efficiency for a polynucleotide target. Langmuir 2001;17(4):1215–1224.
47. O'Brien JC, Stickney JT, Porter MD. Preparation and characterization of self-assembled double-stranded DNA (dsDNA) microarrays for protein : dsDNA screening using atomic force microscopy. Langmuir 2000;16(24):9559–9567.
48. Kelley SO, Barton JK, Jackson NM, et al. Orienting DNA helices on gold using applied electric fields. Langmuir 1998;14(24):6781–6784.
49. Zhang Z-L, Pang DW, Zhang R-Y. Investigation of DNA orientation on gold by EC-STM. Bioconjugate Chem 2002;13:104–109.
50. Wackerbarth H, Grubb M, Zhang J, Hansen AG, Ulstrup J. Dynamics of ordered-domain formation of DNA fragments on Au(111) with molecular resolution. Angew Chem Int Ed 2004;43:198–203.
51. Knoll W. Optical Characterization of organic thin films and interfaces with evanescent waves. MRS Bull 1991;16(7):29–39.
52. Peterlinz KA, Georgiadis R. Two-color approach for determination of thickness and dielectric constant of thin films using surface plasmon resonance spectroscopy. Optics Communications 1996;130(4–6):260–266.
53. Neumann T, Johansson ML, Kambhampati D, Knoll W. Surface-plasmon fluorescence spectroscopy. Advanced Functional Materials 2002;12(9):575–586.
54. Didenko V. DNA probes using fluorescence resonance energy transfer (FRET): designs and applications. BioTechniques 2001;31:1106.
55. Chance RR, Prock A, Silbey R. Molecular fluorescence and energy transfer near interfaces. Adv Chem Phys 1978;37:1–65.
56. Barnes WL. Fluorescence near interfaces: the role of photonic mode density. J Modern Optics 1998;45(4):661–699.
57. Perez-Luna VH, Yang SP, Rabinovich EM, et al. Fluorescence biosensing strategy based on energy transfer between fluorescently labeled receptors and a metallic surface. Biosensors & Bioelectronics 2002;17(1–2):71–78.
58. Buttry DA, Ward MD. Measurement of interfacial processes at electrode surfaces with the electrochemical quartz crystal microbalance. Chem Rev 1992;92(6):1355–1379.

59. Janshoff A, Galla HJ, Steinem C. Piezoelectric mass-sensing devices as biosensors—an alternative to optical biosensors? Angew Chem Int Ed 2000; 39(22):4004–4032.

60. Patolsky F, Lichtenstein A, Willner I. Electronic transduction of DNA sensing processes on surfaces: amplification of DNA detection and analysis of single-base mismatches by tagged liposomes. J Am Chem Soc 2001;123(22):5194–5205.

61. Wang J, Rivas G, Jiang M, Zhang X. Electrochemically induced release of DNA from gold ultramicroelectrodes. Langmuir 1999;15:6541.

62. Rief M, Clausen-Schaumann H, Gaub HE. Sequence-dependent mechanics of single DNA molecules. Nat Struct Biol 1999;6(4):346–349.

63. Harris SA. The physics of DNA stretching. Contemp Phys 2004;45(1):11–30.

64. Seitz, M. Force spectroscopy. In: Niemeyer CM, Mirkn CA, eds. Nanobiotechnology. Weinheim: Wiley-VCH, 2004:404–428.

65. Bustamante C, Bryant Z, Smith SB. Ten years of tension: single-molecule DNA mechanics. Nature 2003;421:423–427.

66. Doyle PS, Ladoux B, Viovy JL. Dynamics of a tethered polymer in shear flow. Phys Rev Lett 2000;84(20):4769–4772.

67. Braun E, Eichen Y, Sivan U, Ben-Yoseph G. DNA-templated assembly and electrode attachment of a conducting silver wire. Nature 1998;391(Feb 19): 775–778.

68. Diez S, Reuther C, Dinu C, et al. Stretching and transporting DNA molecules using motor proteins. Nano Lett 2003;3(9):1251–1254.

69. Walti C, Wirtz R, Germishuizen WA, et al. Direct selective functionalization of nanometer-separated gold electrodes with DNA oligonucleotides. Langmuir 2003;19(4):981–984.

70. Ge C, Liao J, Yu W, Gu N. Electric potential control of DNA immobilization on gold electrode. Biosensors & Bioelectronics 2003;18:53–58.

71. Edman C, Raymond D, Wu D, et al. Electric field directed nucleic acid hybridization on microchips. Nucl Acids Res 1997;25(24):4907–4914.

72. Gurtner C, Tu E, Jamshidi N, et al. Microelectronic array devices and techniques for electric field enhanced DNA hybridization in low-conductance buffers. Electrophoresis 2002;23(10):1543–1550.

73. Heaton RJ, Peterson AW, Georgiadis RM. Electrostatic surface plasmon resonance: direct electric field-induced hybridization and denaturation in monolayer nucleic acid films and label-free discrimination of base mismatches. Proc Nat Acad Sci USA 2001;98(7):3701–3704.

74. Takeishi S, Rant U, Fujiwara T, et al. Observation of electrostatically released DNA from gold electrodes with controlled threshold voltages. J Chem Phys 2004;120(12):5501–5504.

75. Rant U, Arinaga K, Fujiwara T, et al. Excessive counterion condensation on immobilized ssDNA in solutions of high ionic strength. Biophys J 2003;85: 3858–3864.

76. Manning GS. The molecular theory of polyelectrolyte solutions with applications to the electrostatic properties of polynucleotides. Q Rev Biophys 1978;2: 179–246.

77. Zheng LF, Brody JP, Burke PJ. Electronic manipulation of DNA, proteins, and nanoparticles for potential circuit assembly. Biosensors & Bioelectronics 2004;20(3):606–619.
78. Washizu M, Kurosawa O. Electrostatic manipulation of DNA in microfabricated structures. IEEE Trans Ind Appl 1990;26(6):1165–1172.
79. Dewarrat F, Calame M, Schönenberger C. Orientation and positioning of DNA molecules with an electric field technique. Single Mol 2002;3(4): 189–193.
80. Germishuizen WA, Walti C, Wirtz R, et al. Selective dielectrophoretic manipulation of surface-immobilized DNA molecules. Nanotechnology 2003;14(8): 896–902.
81. Suzuki S, Yamanashi T, Tazawa S, Kurosawa O, Washizu M. Quantitative analysis of DNA orientation in stationary AC electric fields using fluorescence anisotropy. IEEE Trans Ind Appl 1998;34(1):75–83.
82. Namasivayam V, Larson RG, Burke DT, Burns MA. Electrostretching DNA molecules using polymer-enhanced media within microfabricated devices. Anal Chem 2002;74:3378–3385.
83. Washizu M, Kurosawa O, Arai I, Suzuki S, Shimamoto N. Applications of electrostatic stretch-and-positioning of DNA. IEEE Trans Ind Appl 1995; 31(3):447.
84. Yamamoto T, Kurosawa O, Kabata H, Shimamoto N, Washizu M. Molecular surgery of DNA based on electrostatic micromanipulation. IEEE Trans Ind Appl 2000;36(4):1010–1017.
85. Ferree S, Blanch HW. Electrokinetic stretching of tethered DNA. Biophys J 2003;85(4):2539–2546.
86. Holzel R, Gajovic-Eichelmann N, Bier FF. Oriented and vectorial immobilization of linear M13 dsDNA between interdigitated electrodes—towards single molecule DNA nanostructures. Biosensors & Bioelectronics 2003;18(5–6): 555–564.
87. Porath D, Bezryadin A, de Vries S, Dekker C. Direct measurement of electrical transport through DNA molecules. Nature 2000;403(6770):635–638.
88. Rant U, Arinaga K, Fujita S, Yokoyama N, Abstreiter G, Tornow M. Dynamic electrical switching of DNA layers on a metal surface. Nano Lett 2004;4(12): 2441–2445.
89. Lahann J, Mitragotri S, Tran TN, et al. A reversibly switching surface. Science 2003;299(5605):371–374.
90. Rant U, Arinaga K, Tornow M, Fujita S, Yokoyama N, Abstreiter G, to be published.
91. Rant U, Arinaga K, Tornow M, et al. Dissimilar dynamic behavior of electrically manipulated single and double stranded DNA tethered to a gold surface. Biophys J 2006; 90:3666–3671.

10
Nanocomputing

Jennifer Sager, Joseph Farfel, and Darko Stefanovic

Summary

Nanocomputing encompasses any submicron devices and technologies applied to any computational or related tasks. A brief survey is given, and emphasis is placed on biomolecular devices that use nucleic acids as their substrate. Computational self-assembly of DNA and DNA-based enzymatic computing are surveyed in greater detail. The foremost implementation challenge for computation, namely, DNA word design, is also surveyed.

Key Words: DNA computing; DNA self-assembly; DNA word design; enzymatic computing; nanocomputing; universal computation.

1. INTRODUCTION

Computing as we know it is based on the von Neumann stored program concept and its ubiquitous implementation in the form of electronic instruction processors. For the past three decades, processors have been fabricated using semiconductor integrated circuits, the dominant material being silicon, and the dominant technology complementary metal–oxide–semiconductor (CMOS). Relentless miniaturization has been decreasing feature size and increasing both the operating frequency and the number of elements per chip, giving rise to so-called Moore's law. Indeed, vast amounts of raw computational power are now available in every personal computer sold, at a very modest cost. By improving the processes and materials and using new geometries, the semiconductor industry expects to be able to continue this trend for at least another decade, according to its common Roadmap document *(1)*. Whereas a 90-nm node is characteristic of current processes (implying that the semiconductor industry is already operating in the nanotechnology domain), it is expected that 18 nm will be reached by 2018. Beyond that lie fundamental limits of the technology, principally the problem of heat dissipation *(2,3)*

From: *NanoBioTechonology: BioInspired Devices and Materials of the Future*
Edited by: Oded Shoseyov and Ilan Levy © Humana Press Inc., Totowa, NJ

inherent in devices in which an electronic charge is used for state representation. Alternatives to CMOS fabrication *(4–8)* are being sought at the level of devices, such as single-electron transistors *(9,10)*, carbon nanotubes *(11)*, silicon nanowires *(12–14)*, molecular switches *(15–17)*, nanomagnets *(18)*, quantum dots *(19)*, chemically assembled electronics *(20–29)*, chemical logic gates with optical outputs *(30–34)*, and three-dimensional (3D) semiconductor integration *(35)* (predicted much earlier *[36]*). Alternative architectures are also being explored, such as amorphous computing *(37)*, spatial computing *(26,38)*, blob computing *(39,40)*, cell matrix computing *(41)*, chaos computing *(42)*, and the entire field of quantum computing.

Thus, while we need not fear a scarcity of computing cycles, the prospect of the eventual demise of Moore's law has given impetus to a great variety of research into new computational substrates. A separate chapter in this volume treats nanoelectronics, that is, work that aims to, more or lesss seamlessly, extend the viability of microelectronic technologies beyond the lifetime of CMOS processes. Here, we focus on research over the past decade that has been less concerned with continuity, and that attempts to achieve computational effects through the application of biochemical principles in new and unexpected ways. Our main focus is on various computing paradigms using DNA. We examine in which sense they perform computation and interpret them in terms of conventional mathematical notions of computation. We also examine their commonalities, in particular the question of DNA word design.

DNA computation in its original formulation *(43–49)* seeks to employ the massive parallelism inherent in the small scale of molecules to speed up decision problems. The essential property of nucleic acids, specific hybridization (formation of the double helix) *(50–53)*, is either exploited to encode solutions as long strings of nucleotides, generate large numbers of random strings, and check them in a small number of steps that are often manual, such as PCR (though more reliable detection is now available *[54]*), or to construct solutions directly through oligonucleotide self-assembly. A number of nondeterministic polynomial-time (NP)-complete decision problems have been rendered in this fashion *(55–58)*, and encodings for general computation *(59–63)* and combinatorial games *(64)* have also been proposed. A limitation of the approach is the need for large amounts of nucleic acid *(65)*; with amounts currently feasible (and the low speed of operations), it has been difficult to outperform electronic computers. Another limitation has been in imperfect specificity of nucleic acid hybridization. The research in this area *(66–73)* has ranged from the physico-chemical constraints on usable nucleotide strings (e.g., melting points; secondary structure) to tools for systematic string generation *(74)*; this is reviewed under Subheading 8.3.

Further variations on the theme of DNA computation have included using proteins instead of nucleic acids for a larger alphabet *(75)*, hairpin computation *(76)*, and sophisticated forms of self-assembly *(77)*, to avoid manual operations, and for cellular computation in which cells (real or simulated) are viewed as elementary computational elements, with some form of communication among multiple cells *(51,77–93)*.

Although early on, it was believed that DNA computing might be a competitor to electronics in solving hard computational problems, the focus has now shifted to the use of DNA to compute in environments where it is uniquely capable of operating, such as in smart drug delivery to individual cells *(94,95)*.

Our review of biochemically based computing, necessarily limited in scope, is organized according to the manner in which the principle of specific hybridization is exploited. Under Subheading 2, we consider how large 2D and 3D structures are built in a programmable fashion through molecular self-assembly. Under Subheading 3, we treat approaches in which short strands representing logic signals specifically bind to activate particular enzymatic reactions in a reaction network. Finally, the pervasive subproblem of the design of good DNA sequences for computation is treated under Subheading 4.

2. COMPUTING USING STRUCTURAL SELF-ASSEMBLY OF DNA

One of the most interesting and useful paradigms in biomolecular computation is molecular self-assembly. *Self-assembly* is the spontaneous formation of ordered structure out of structural building blocks that encode within themselves information about both what they are and how they can fit together. Useful computation can occur if the rules that govern how certain types of blocks may attach to other types of blocks are intelligently selected.

In the molecular case, the building blocks that self-assemble are normally DNA molecules. DNA is perfect for self-assembly because pieces of DNA may be linked together in very programmable and predictable ways. In fact, we can construct many different building-block structures with DNA, and we can program how these blocks attach to each other to achieve infinitely variable superstructures—indeed, DNA self-assembly has even been proven to be capable of universal computation.

2.1. Building Blocks

The most familiar form of DNA is the double-stranded molecule. These molecules consist of two backbones, which wrap around each other in a double-helix (50) and are connected by Watson-Crick complementary bonds

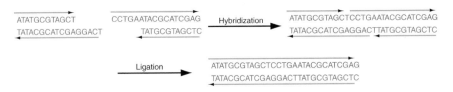

Fig. 1. Hybridization of the single-stranded sticky ends extending from double-stranded DNA molecules. After the base pair bonding occurs in hybridization, the backbones of the two dsDNA molecules may be joined by ligation.

between the amino acids A, C, T, and G (adenine, cytosine, thymine, and guanine). Watson-Crick complementary bonding refers to the fact that these four amino acids form two pairs of acids that bind very strongly to each other—A binds to T, and C binds to G.

Double-stranded DNA, or dsDNA, may be used as a building block for self-assembly. In order for pieces of dsDNA to self-assemble, however, they must have outreaches of single-stranded (ss)DNA at their ends. We call these extending segments *sticky ends*, because a segment of ssDNA will bind (stick) to another segment of ssDNA that contains a sequence of amino acids that is Watson-Crick complementary to its own sequence. If multiple pieces of dsDNA have sticky ends on both sides, they can link together to form a long chain. The initial bonding of the amino acids of one piece of ssDNA to another is called *hybridization*. After hybridization, the pieces may complete their attachment through a process called *ligation*, where the DNA backbone is extended and connected. See Fig. 1 for an illustration of these reactions between pieces of dsDNA with extending sticky ends *(96)*.

Pieces of dsDNA are linear, and therefore are inadequate building blocks for the construction of any 2D or 3D structures. This has led researchers to use other types of DNA molecules, beyond the standard double-helix, for producing complex structures. The first type of molecules are called junction molecules.

Junction molecules are formed when two strands of dsDNA undergo reciprocal exchange (recombination), whereby they fuse together at what is called a branched junction, or Holliday junction (*see* Fig. 2) *(97)*. In reciprocal exchange, the strands of DNA fuse by exchanging connections at a particular site. This may happen between dsDNA molecules of the same or opposite polarity, and although either polarity combination yields the same structure after one crossover, different structures are achieved if more exchanges occur (molecules are of the same polarity if they are arranged such that the two strands that undergo reciprocal exchange have the same orientation of their 3′ and 5′ ends). A junction molecule may be constructed with an arbitrary number of arms, and there is no known limit to this number *(97)*. Figure 2

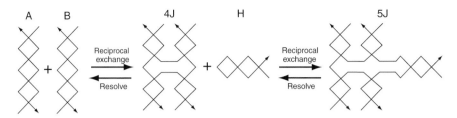

Fig. 2. The opposite-polarity dsDNA molecules A and B undergo reciprocal exchange to form the four-arm branched junction 4J. The four-arm junction 4J then undergoes reciprocal exchange with the hairpin molecule H to form the five-arm branched junction molecule 5J.

shows a five-arm junction made from a four-arm junction and a hairpin DNA molecule. We may link junction molecules together to form more complicated structures if we extend a bit of ssDNA off each arm of a junction molecule, creating sticky ends on the arms. Molecules with topologies resembling the edges of a cube and a truncated octahedron have been demonstrated *(97)*. However, structures made out of singly branched junctions are relatively flexible, and so it is impossible to characterize the actual 3D structure of these molecules. To create predictable complex structures from DNA molecules, more rigidity is needed than that provided by branched junctions. Another class of molecules called DNA crossovers offers this rigidity.

A *DNA crossover molecule* is a structure consisting of two dsDNA molecules, where each dsDNA molecule has a single strand that crosses over to the other molecule (*see* Fig. 3) *(96)*. This is just reciprocal exchange between the two molecules happening at multiple sites. The two most significant types of crossover molecules are *double-crossovers*, or *DX* molecules, and *triple-crossovers*, or *TX* molecules. DX molecules are made up of two pieces of dsDNA, with two crossover locations *(93)*. TX molecules are made up of three pieces of dsDNA with four crossovers *(62)*. We may extend sticky ends off DX and TX molecules to link them together, and call the linkable molecules *tiles*, in the manner of Wang tiles, which are discussed in the next section. These DX and TX tiles are sufficiently rigid to create very complex, stable, and beautiful 2D and 3D nanostructures via self-assembly, and, with intelligent selection of how different pieces may attach, this assembly may also be used to perform computation.

2.2. Computation

Erik Winfree was the first to discover that planar self-assembly of DNA molecules can perform universal computation *(48)*. This discovery was made based on the insight that DX molecules may be regarded simply as *Wang tiles*. Wang tiling is a mathematical model where square unit tiles are

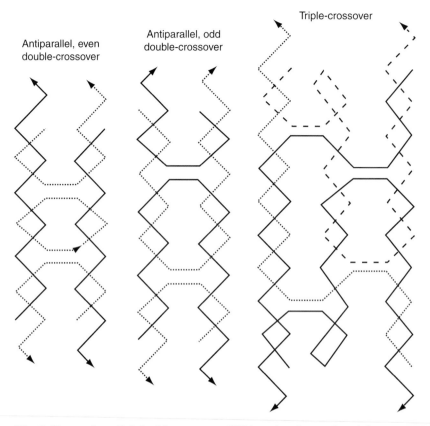

Fig. 3. Two antiparallel double-crossover DNA molecules, and a triple-crossover molecule. The even and odd labels on the double-crossovermolecules refer to the number of helical half-turns between the two crossovers (two in the left molecule, three in the middle molecule). The double-crossover molecules are formed when two crossovers occur between two double-stranded DNA molecules, whereas the triple-crossover molecule is formed when four crossovers occur between three double-stranded DNA molecules. The different line styles represent different contiguous single DNA strands in the new molecule. The four extended strands on each molecule are "sticky ends" that can be used to connect DNA tiles together.

labeled with specific symbols on each edge. Each tile is only allowed to associate with tiles that have matching symbols. We can construct DNA molecules that are analogous to Wang tiles (call these DNA tiles) by creating a molecule with a rigid, stable body and open, sticky ends for attachment to other tiles. The DX and TX molecules are both ideal for this. The sticky ends of DNA tiles may be labeled with certain sequences of amino acids, which are analogous to the symbols labeling the sides of Wang tiles. This labeling allows the sticky ends to bind only to tile ends that have a complementary sequence of base pairs; this corresponds to the rule that restricts Wang tiles to only associate with tiles that have matching symbols. It has been shown that Wang tiles, when designed with a certain set of symbols, are capable of universal computation, and because DNA molecules can represent Wang tiles, it was shown that universal computation could also be accomplished by self-assembling DNA tiles *(93,98)*.

The biggest advantage of computing with self-assembly, compared to other molecular computing paradigms, is that it avoids the many tedious laboratory steps required with other computation methods. The reason for this is that if DNA tiles are designed to correctly specify the desired steps in a computational problem, the only structures to form from these tiles will be the desired, valid solutions of the problem. Because only valid solutions are encoded in the resulting structures, one needs only to design and form the tiles from DNA strands, allow the tiles to self-assemble, and then read the output. Of course, reading the output usually involves at least two main steps, such as ligation of reporter strands embedded in the tiles, and subsequent separation and PCR. However, the number of total steps in performing computation with self-assembly remains very low.

The first example of computing performed by DNA self-assembly was a four-bit cumulative exclusive disjunction (XOR) *(62)*. The function XOR takes two binary input bits and returns a zero if the inputs are equal and a one if they are not equal. The cumulative XOR takes Boolean input bits x_1, ..., x_n, and computes the Boolean outputs y_1, ..., y_n, where $y_1 = x_1$, and for $i > 1$, $y_i = y_{i-1} \text{XOR} x_i$. The effect of this is that y_i is equal to the even or odd parity of the first i values of x. The cumulative XOR calculation was performed via the self-assembly of TX molecules. Eight types of TX molecule were needed: two corner tiles, two input tiles, and four output tiles. The types were different only in that they had different labels (sequences of amino acids) on their sticky ends, and, in some cases, different numbers of sticky ends. The corner tiles were used to to connect a layer of input tiles to a layer of output tiles. The two input tiles represented $x_i = 0$ and $x_i = 1$. The four output tiles were needed because there are two ways to get each of the two possible outputs of a bitwise XOR. So, one output tile represents the state where we have output bit $y_i = 0$ and input bits $x_i = 0$ and $y_{i-1} = 0$, while another tile represents the state where we have output bit $y_i = 0$ and input bits $x_i = 1$ and $y_{i-1} = 1$. Similarly, the other two output tiles represent the two states where $y_i = 1$. The actual computation of the XOR operation is accomplished by harnessing the way the output tiles connect to the input tiles. Each output tile (y_i) will only attach to a unique combination of one input tile (x_i) and one output tile (y_{i-1}), and will leave one sticky end open that represents its own value (y_i) so that another output tile may attach to it. For example, the output tile signifying $y_i = 1$, $x_i = 0$, and $y_{i-1} = 1$ has the value 1, and will only connect to an input tile with value 0 and an output tile with value 1. With this system, only the output tiles that represent the correct solution to the problem will be able to attach to the input tiles.

Another example of computation using self-assembled DNA tiles is the binary counter created by Rothemund and Winfree *(63)*. The counter uses seven different types of tiles: two types of tiles representing 1, two types representing 0, and three types for the creation of a border (corner, bottom, and side tiles). The counter works by first setting up a tile border with the border tiles—it is convenient to think of the "side" border tiles as being on the right, as then the counter will read numbers from left to right. The border structure forms before the rest of the counter because of the properties of border tiles: two border tiles bind together with a double bond, while all other tiles bind to each other and to border tiles with a single bond. Doubly bound tiles have a very low tendency to detach from each other, whereas singly bound tiles detach relatively easily. Because any tile except a border tile must bind to two additional tiles in order to have two bonds, but a border tile and another border tile of the correct type will form a double bond with each other, a stable border forms before other stable formations, composed of nonborder tiles, are created. The bottom and side border tiles are designed such that the only tile that may bind in the border's corner (to both a side and a bottom border tile) is a specific type of 1 tile. Only one of the 0 tiles may bind to both this 1 tile and the bottom of the border, and this type of 0 tile may also bind to itself and the bottom of the border, and thus may fill out the left side of the first number in the counter with leading zeros. Now, the only type of tile which may bind both above the 1 in the corner and to the right side of the border is the other type of 0 tile, and the only tile which may bind to the left of it is a 1 tile—we get the number 10, or two in binary. The tile-binding rules are such that this can continue similarly up the structure, building numbers that always increment by one. Figure 4 shows a more intuitive picture of this device's operation.

DNA self-assembly has also been used to solve the Boolean formula satisfiability (SAT) problem. This has been done with both string (linear) assembly of DX or TX tiles and with graph self-assembly of duplex and branched junction molecules *(99)*. In the string-assembly solution, the DNA tiles have a width (the number of helixes that are fused together) equal to the number of clauses in the SAT problem. Each variable involved in the problem has two tiles, one representing its being true, and one representing its being false. A variable's "true" tile has a hairpin structure in each clause where the variable appears, and no hairpin in clauses where its complement appears (where the variable is false). The same applies for a variable's "false" tile.

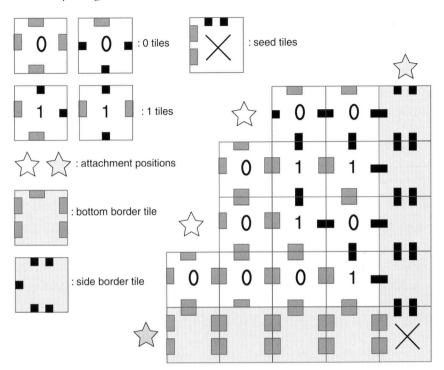

Fig. 4. A binary counter in the process of self-assembly. The seed tile starts off the assembly. The right side and bottom border tiles connect to each other with double bonds, whereas all the other tiles connect with single bonds. A tile needs two single bonds (or one double bond) to form a stable attachment to the structure; the marked attachment positions show where a tile can form a stable attachment.

When all the tiles are mixed together (including a "start" and an "end" tile), they all join together to form only valid solutions of the SAT problem.

2.3. Complex Nanostructures

In addition to performing computation, DNA tiles can self-assemble to create very complex 2D and 3D geometrical structures. 2D periodic lattices have been constructed of both DX and TX DNA tiles *(51)*. Both types of lattice have been observed through atomic force microscopy, to see that the desired geometric structure is actually being self-assembled. To assist in visualizing the structure, a lattice made of TX molecules may be designed in such a way that rows of molecules contain loops of DNA that protrude perpendicularly to the plane of the lattice. These rows can be placed at regular distances that can be designated with high accuracy (in the lab, stripes were seen at 27.2 nm when they were expected at 28.6 nm) *(51)*. The stripes can be seen even more clearly when metallic (normally gold) balls are affixed to the tiles making up the stripes *(96)*.

Recently, researchers have proposed methods of making complex nanoscale 3D fractals. Specifically, a method has been proposed by which the Sierpinski cube fractal could be produced using DNA self-assembly *(100)*. The recursive algorithm for generating a Sierpinski cube fractal is as follows: take a cube, divide it evenly into 27 smaller cubes, and remove the most interior cube as well as the middle cubes on the large cube's six faces. Research has shown that theoretically, the cube may be produced by using Mao triangles based on DX molecules. However, the cube has not yet been produced in the lab.

2.4. Errors and Error Correction

Atomic force microscopy has allowed us to view self-assembled DNA structures and investigate whether or not they are forming properly. There is indeed great success, but this has also allowed us to see that there are problems with reliably building large, error-free structures. The self-assembled binary counter, for example, is error-prone in its current incarnation, only counting to 7 or 8 accurately *(101)*. In fact, there is a 1 to 10% error rate for each tile binding in all 2D structures constructed without any error correction or error avoidance techniques *(102)*. This can lead to disastrous results in many computations; such error rates come with the new territory of biological computation, and are not a problem that traditional computer scientists are at all accustomed to dealing with.

There are three main kinds of assembly error *(103)* (*see* Fig. 5 for visual examples of each). The first kind of error is a *mismatch error*, where sometimes tiles become locked in the assembly in the wrong place. A tile can attach to a corner in the assembly's fringe by binding to one tile at the corner, but mismatching with the other. Normally, a tile in such a state would fall off the assembly, because two bonds (or one strong bond, as is the case with border tiles) are necessary for a tile to be locked in a stable position. However, if other tiles attach around it before it falls off, it may be bound to enough tiles to be locked in a stable, but incorrect position in the assembly. It is easy to see that just one tile locked in an incorrect position will throw the binary counter completely off course, as the assembly of each row of digits in the counter is dependent upon the previous row. Although some other self-assembled patterns may be less sensitive, the fact remains that even one erroneously placed tile can greatly impact the structure of an assembly.

The second kind of error is a *facet error*. This happens when a tile attaches to a facet (a portion of the boundary apart from the built interior structure) rather than to a desired attachment site at a corner in the structure's frontier, and more tiles bind it into place. Even though no mismatches occur, an incorrect structure can be formed this way.

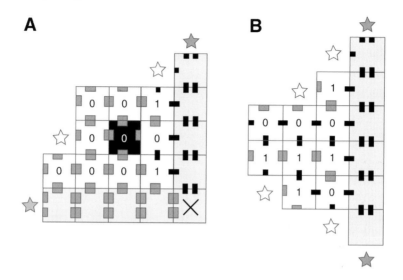

Fig. 5. Errors present in two binary counters in the process of self-assembly. In A, the tile highlighted in black is *mismatched*, but has been locked into place by other tiles binding correctly around it. Hence, in this case, our counter counts to one, then zero, then one again; obviously, a similar error can be arbitrarily serious, destroying the counter's count. In B, we see two types of errors. The boundary tiles have formed without growing off of a seed tile (corner tile); this is a *nucleation error*. Also, although there are no mismatches, the rule tiles have begun counting at eight (or more, depending on whether more ones or zeros bind to the frontier) and are continuing forward and backward. This is because they started assembling on a facet (edge) rather than in the corner as in A. This constitutes a *facet error*.

The third kind of error is a *spurious nucleation error*. This occurs when the assembly begins growing from a tile other than the special "seed" tile (normally the corner of the lattice). For example, a portion of the interior can spontaneously come together without any boundary tiles at all. More commonly, however, a stretch of boundary tiles will bind together without being bound to the seed tile. A section of linked boundary tiles floating around without a seed tile to set up the assembly structure is a perfect recipe for facet errors, because the seed tile, which links two boundary lines together, is necessary to create the first desired binding site for the main body of the lattice, in the corner where the boundaries meet. Any binding of tiles to a boundary line not linked to a seed tile constitutes facet error. Avoiding spurious nucleation when running a self-assembly algorithm is analogous to providing correct inputs to the beginning of a computer program: in other words, growing from the seed tile makes the algorithm begin with the desired input.

All of these errors occur because of the reversible, kinetic way in which DNA molecules in solution react and bind together. Although two bonds (or one double bond) are indeed required to hold a tile in a stable spot in an assembly, in reality, there are many times when a tile will attach to the assembly with only one bond, and hang on for a little while, sometimes allowing itself to be locked into place with further bonds. Likewise, it is also possible that even the strong double bonds may be reversed, and break apart, at times. A kinetic Tile Assembly Model (kTAM) has been created (by Winfree and others) to simulate reversible tile interactions. The kTAM approximates perfect, abstract self-assembly with strength threshold τ (given as a property of the tile program) when $G_m = \tau G_s - \varepsilon$, with G_m being the monomer tile concentration and G_s being the sticky-end bond strength; ε is the error rate. The model defines the forward rate of crystal growth (association) of particular tiles as $r_f = k_f e^{-G_m}$, where k_f is a reaction constant. The backward rate of growth (dissociation) of a tile which makes bonds with total strength b is $r_{r,b} = k_f e^{-bG_s}$. The free energy of a nucleus of tiles is defined as $\Delta G = (bG_s - nG_m)kT$, where b is the total bond strength, n is the number of tiles, k is Boltzmann's constant, and T is temperature. These measures help determine under what conditions assembly steps are energetically favorable (and thus have higher probability of occurring at any given point in time).

Perhaps obviously, we can account for most errors just by slowing down the rate at which structures assemble. Research has shown, however, that mismatch errors occur at a rate that is at least proportional to the square root of the speed of assembly *(77)*. Thus, in order to reduce the rate of error by some reasonable amount, we must slow the rate of assembly down tremendously, by greatly decreasing the temperature and/or the monomer concentration. Better solutions are being investigated, then, for lowering error rates.

The most promising methods involve using *proofreading tiles (104–106)*. These methods can greatly help in controlling both mismatch and facet errors. Proofreading tiles are extra tiles added to a tile set that are used to store information redundantly, so it is harder to lock errors in place in a forming structure. This type of error correction forces errors to be co-localized, so that many more erroneous tile bindings must occur before one wrong tile is locked in place. This greatly increases the probability that an individual wrong tile will fall off the assembly before growth continues around it, thus substantially reducing the error rate in building the assembly. Each tile is replaced by a block of tiles, where the bond between each pair of tiles in the block is unique *(105)*. When using a simple 2×2 array of proofreading tiles, the tile set for a given problem is four times larger in size, but the error rate is 10^4 times lower *(104)*. Originally, the internal binding between proofreading tiles was very simple, but Chen and Goel have improved upon this to produce the "snake" proofreading method. A snake tile set forces the assembly process to double, or "snake" back onto itself when binding each proofreading block, making it less likely that an entire block will be bound incorrectly to the growing

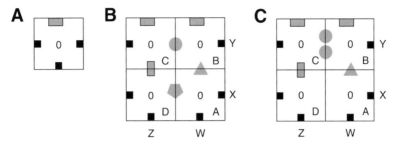

Fig. 6. A tile, A, and 2 × 2 proofreading tile sets representing it, B and C. B is the simple redundant representation. The assembly is growing from right to left and top to bottom (as in the binary counter example). Note that all four tiles are connected to each other with unique types of single-bonds. C is the improved, "snake" proofreading tile set representation, so named because its formation snakes back upon itself. We can see that the snake tileset greatly decreases the chance of a facet nucleation error (when a tile binds to some facet instead of at a corner, and is then locked into place by another tile). Recall that a tile must be attached with two bonds (or one double-bond) to be a stable part of the structure. If tile *A* in B formed a single-bond with *Z*, for example, it could be locked in place by tile *D* binding (in a stable, two-bond manner) to its left, and so the error propagates to the left after only one single-bond facet nucleation (*A* binding to *Z*). However, with the snake tileset, there can be no bond between *A* and *D*. In order for the block in C to grow, *A* must bind to *B*, which binds (with a stable, double-bond) to *C*, which then binds to *D*. Thus, the set in C would require two undesired single-bonds in very close proximity (namely, *A* to *Z* and *B* to *A*) before only double-bonds are required to lock the error in place (*C* to *B*, etc.). The probability of this happening is very small.

structure *(104)*. With either type of proofreading tile set, the mismatch and facet error rates can be made arbitrarily small by using larger and larger tile sets (although this produces larger and more redundant self-assembled lattices, of course). Figure 6 shows an example of each type of proofreading tile set.

The "zig-zag" boundary tile set helps prevent spurious nucleation errors by forcing border tiles to bind to seed tiles before binding to each other *(103)*. This border tile set makes it more energetically favorable for border tiles to bind correctly, so a complete border structure (with seed in place) is set up before the rest of the structure begins growing. The zig-zag border construction method can be combined with the proofreading tile sets mentioned earlier to yield a self-assembled creation that is robust to all three types of error.

3. ENZYMATIC DNA COMPUTING

In this section, we focus on the approach to biochemical computing—either digital or analog, depending on the interpretation—in which signals are represented by concentrations of designated molecular species. Although such systems can be devised with protein enzymes, here we look at smaller DNA enzyme molecules. Deoxyribozymes are enzymes made of DNA that catalyze DNA reactions such as the cleavage of a DNA strand into two or the ligation of two strands into one. Cleaving enzymes (known as phosphodiesterases) can be modified to include allosteric regulation sites to which specific control molecules can bind and so affect the catalytic activity. There is a type of regulation site to which a control molecule must bind before the enzyme can complex with (i.e., bind to) the substrate, thus the control molecule promotes catalytic activity. Another type of regulation site allows the control molecule to alter the conformation of the enzyme's catalytic core, such that even if the substrate has bound to the enzyme, no cleavage occurs; thus, this control molecule suppresses or inhibits catalytic activity. This allosterically regulated enzyme can be interpreted as a logic gate, the control molecules as inputs to the gate, and the cleavage products as the outputs. This basic logic gate corresponds to a conjunction, such as e.g., $a \wedge b \wedge \neg c$, here assuming two promotory sites and one inhibitory site, and using a and b as signals encoded by the promotor input molecules and c as a signal encoded by the inhibitor input molecule. Deoxyribozyme logic gates are constructed via a modular design *(107)* that combines molecular beacon stem-loops with hammerhead-type deoxyribozymes (Fig. 7).

A gate is active when its catalytic core is intact (not modified by an inhibitory input) and its substrate recognition region is free (owing to the promotive inputs), allowing the substrate to bind and be cleaved. Correct functioning of individual gates can be experimentally verified through fluorescent readouts *(108)*.

Note that the gates use oligonucleotides as both inputs and outputs, so cascading gates is possible without any external interfaces (such as e.g., photoelectronics). The inputs are compatible with sensor molecules *(109)* that could detect cellular disease markers. Final outputs can be tied to the release of small molecules. Two gates are coupled *in series* if the product of an "upstream" gate specifically activates a "downstream" gate. All products and inputs (i.e., external signals) must be sufficiently different to minimize the error rates of imperfect oligonucleotide matching, and they must not bond to one another; we examine this problem in the next section. A series connection of two gates, the upstream being a ligase and the downstream being a phosphodiesterase, has been experimentally validated *(110)*.

Multiple elementary gates have been constructed, so there is a large number of equivalent ways that any given Boolean function can be realized—equivalent in terms of digital function, but not in speed or cost of realization. For instance, a single four-input gate may be preferable to a cascade with three two-input gates. Clearly construction of deoxyribozyme logic circuits bears a resemblance to traditional low-level logic design, but, perhaps because the technology has not matured, with many more options to explore.

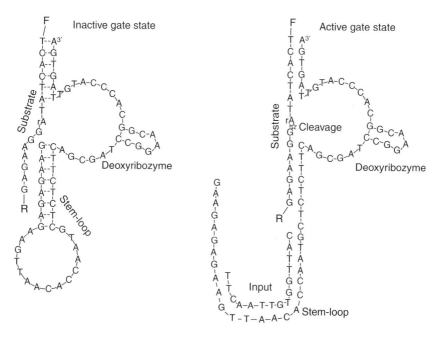

Fig. 7. A YES gate, in which an "input" oligonucleotide activates a deoxyribozyme by opening an inhibitory stem.

3.1. Simple Enzymatic Circuits

Deoxyribozyme logic gates have been used to build computational devices. A half-adder was achieved by combining three two-input gates in solution *(111)*. A half-adder computes the sum of two binary digits (bits); there may be a carry. It can be implemented using an XOR gate for the sum bit and an AND gate for the carry bit. The XOR gate, in turn, is implemented using two AND-NOT gates (gates of the form $x \wedge \neg y$). The two substrates used are fluorogenically marked, one with red tetramethylrhodamine (T), and the other with green fluorescein (F), and the activity of the device can be followed by tracking the fluorescence at two distinct wavelengths. The results, in the presence of Zn^{2+} ions, are shown in Fig. 8. When both inputs are present, only the green fluorescein channel (carry bit) shows a rise in fluorescence. When only input i_1 is present or only input i_2 is present, only the red tetramethylrhodamine channel (sum bit) rises. With no inputs, neither channel rises. Thus, the two bits of output can be reliably detected and are correctly computed.

3.2. Enzymatic Game Automata

Using deoxyribozyme logic gates, an automaton for the game of tic-tac-toe has been constructed *(112)*. To understand how this was achieved, we first

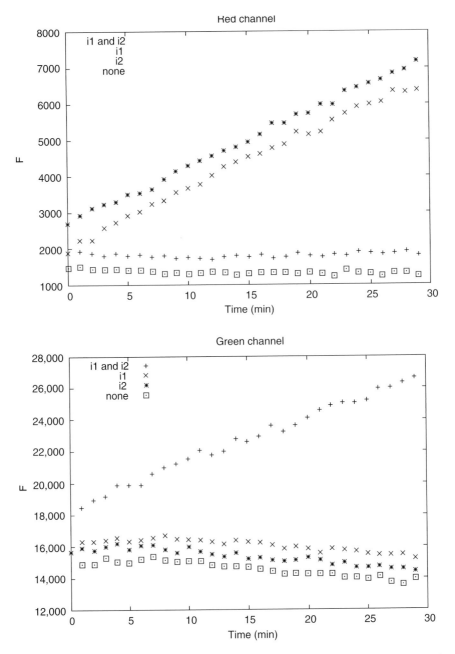

Fig. 8. Observed fluorescence change in a half-adder deoxyribozyme logic circuit: the red tetramethylrhodamine channel is shown on the left; the green fluorescein channel is shown on the right.

briefly examine the structure of that game. A *sequential game* is a game in which players take turns making decisions known as *moves*. A *game of perfect information* is a sequential game in which all the players are informed before every move of the complete state of the game. A *strategy* for a player in a game of perfect information is a plan that dictates what moves that player will make in every possible game state. A *strategy tree* is a (directed, acyclic) graph representation of a strategy. The nodes of the graph represent reachable game states. The edges of the graph represent the opponent's moves. The target node of the edge contains the strategy's response to the move encoded on the edge. A leaf represents a final game state, and can, usually, be labeled either win, lose, or draw. Thus, a path from the root of a strategy tree to one of its leaves represents a game.

In a tree, there is only one path from the root of the tree to each node. This path defines a set of moves made by the players in the game. A player's *move set* at any node is the set of moves made by that player up to that point in a game. For example, a strategy's move set at any node is the set of moves dictated by the strategy along the path from the root to that node. A strategy is said to be *feasible* if, for every pair of nodes in the decision tree for which the opponent's move sets are equal, one of the following two conditions holds: (1) the vertices encode the same decision (i.e., they dictate the same move), or (2) the strategy's move sets are equal. A feasible strategy can be successfully converted into Boolean logic implemented using monotone logic gates, such as the deoxyribozyme logic gates.

In the tic-tac-toe automaton, the following simplifying assumptions are made to reduce the number and complexity of needed molecular species. The automaton moves first and its first move is into the center (square 5, Fig. 9). To exploit symmetry, the first move of the human, which must be either a side move or a corner move, is restricted to either square 1 (corner) or square 4 (side).

The game tree in Fig. 10 represents the chosen strategy for the automaton. For example, if the human opponent moves into square 1 following the automaton's opening move into square 5, the automaton responds by moving into square 4. If the human then moves into square 6, the automaton responds by moving into square 3. If the human then moves into square 7, the automaton responds by moving into square 2. Finally, if the human then moves into square 8, the automaton responds by moving into square 9, and the game ends in a draw.

This strategy is feasible; therefore, following a conversion procedure, it is possible to reach a set of Boolean formulae that realize it, given in Table 1. (For a detailed analysis of feasibility conditions for the mapping of games of strategy to Boolean formulae, *see* ref. *113*.) The arrangement of deoxyribozyme logic gates corresponding to the above formulae is given in Fig. 11. This is the

$$\begin{array}{c|c|c}
1 & 2 & 3 \\
\hline
4 & 5 & 6 \\
\hline
7 & 8 & 9
\end{array}$$

Fig. 9. The tic-tac-toe game board.

initial state of the nine wells of a well-plate in which the automaton is realized in the laboratory.

The play begins when Mg^{2+} ions are added to all nine wells, activating only the deoxyribozyme in well 5, i.e., prompting the automaton to play its first move into the center. After that, the game branches according to the opponent's inputs. A representative game is shown in Fig. 12. As the human opponent adds input to indicate his moves, the automaton responds with its own move, activating precisely one well, which is shown enlarged. The newly activated gate is shown in light green. The bar chart shows the measured change in fluorescence in all the wells. Wells that are logically inactive (contain no active gates) have black bars, and wells that are logically active have green bars (the newly active well is light green).

3.3. Open Systems and Recurrent Circuits

The first oscillatory chemical reaction was discovered by Belousov in the 1950s, but for awhile, remained little known *(114)*. Once this Belousov-Zhabotinsky reaction became better known and its mechanisms were understood *(115–117)*, it inspired treatments of chemical computation devices, made out of hypothetical large systems of coupled chemical reactions with many stable states *(118–126)*; moreover, information-theoretic connections were made with Maxwell's daemon *(127)*, and, chaotic behavior having been observed, with unpredictability *(128–130)*. Chemical reactions, owing to diffusion, have a spatial component in addition to the temporal. Therefore, the oscillatory Belousov-Zhabotinsky reaction gives rise to waves *(131)*; this was used to implement computation on a prefabricated spatial pattern by wave superposition *(132–134)*. Recently, an oligonucleotide periodic system has been shown *(135)* (*see also* ref. *136*).

It has been suggested that computational devices based on chemical kinetics are Turing-equivalent *(137)*, but one must consider the inherently finite number of reactions and molecular species possible *(138)*, and the difficulty

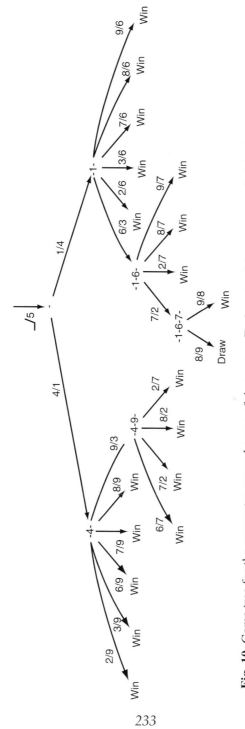

Fig. 10. Game tree for the symmetry-pruned game of tic-tac-toe. Each state of the automaton is labelled according to the inputs seen on the path to it. Each edge is labelled a/b, where b is the output that is activated on input a.

Table 1
Boolean Formulae Resulting from the Tic-Tac-Toe Game Tree Express the Dependence of Outputs o on the Inputs i

$o_1 = i_4$

$o_2 = (i_6 \wedge i_7 \wedge \neg i_2) \vee (i_7 \wedge i_9 \wedge \neg i_1) \vee (i_8 \wedge i_9 \wedge \neg i_1)$

$o_3 = (i_1 \wedge i_6) \vee (i_4 \wedge i_9)$

$o_4 = i_1$

$o_5 = 1$

$o_6 = (i_1 \wedge i_2 \wedge \neg i_6) \vee (i_1 \wedge i_3 \wedge \neg i_6) \vee (i_1 \wedge i_7 \wedge \neg i_6) \vee (i_1 \wedge i_8 \wedge \neg i_6)$
$\qquad \vee (i_1 \wedge i_9 \wedge \neg i_6)$

$o_7 = (i_2 \wedge i_6 \wedge \neg i_7) \vee (i_6 \wedge i_8 \wedge \neg i_7) \vee (i_6 \wedge i_9 \wedge \neg i_7) \vee (i_9 \wedge i_2 \wedge \neg i_1)$

$o_8 = i_9 \wedge i_7 \wedge \neg i_4$

$o_9 = (i_7 \wedge i_8 \wedge \neg i_4) \vee (i_4 \wedge i_2 \wedge \neg i_9) \vee (i_4 \wedge i_3 \wedge \neg i_9) \vee (i_4 \wedge i_6 \wedge \neg i_9)$
$\qquad \vee (i_4 \wedge i_7 \wedge \neg i_9) \vee (i_4 \wedge i_8 \wedge \neg i_9)$

of constructing them in practice, beyond *Gedankenmoleküle*, such as those of Hiratsuka *(139)*. Deoxyribozyme logic provides a systematic method for such a construction, and recurrent circuits, including flip-flops and oscillators, have been designed *in silico* on the basis of it *(140,141)*.

4. WORD DESIGN FOR DNA COMPUTING

Most DNA computation models assume that computation is error-free. (Although we describe most of the constraints in terms of DNA, RNA computers also exist [for an example, *see* ref. *64*], and all of the constraints described here are also relevant to RNA.) For example, Adleman *(43)* and Lipton *(45)* used randomly generated DNA strings in their experiments because they assumed that errors due to false positives were rare. However, it has been experimentally shown that randomly generated codes are inadequate for accurate DNA computation as the size of the problem grows *(68)*, because a poorly chosen set of DNA strands can cause hybridization errors. Therefore, for many types of DNA computers, it may be practical or even necessary to create a "library" or "pool" of DNA word codes suitable for computation.

There are three steps to constructing a library. First, rules or constraints must be defined that specify whether a given set of molecules will cause errors; these constraints can be complex because they are subject to the laws of biochemistry as well as the specific algorithm and computation style. Second, an

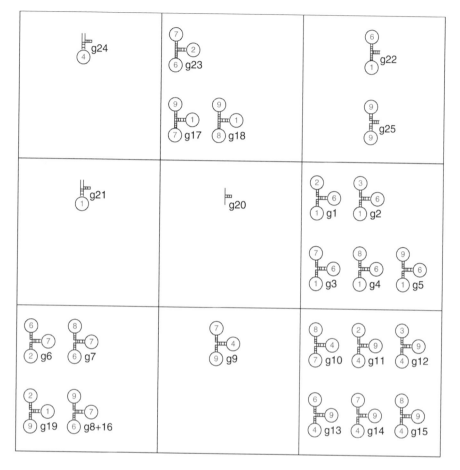

Fig. 11. Realizing a tic-tac-toe automaton using deoxyribozyme logic. The center well contains a consitutively active deoxyribozyme. Each of the eight remaining wells contains a number of deoxyribozyme logic gates as indicated.

algorithm must be found that either generates or finds such a set of molecules; the solution space is large because the number of candidate molecules grows exponentially in the length of the DNA string. Third, it must be proved that the final set of molecules correctly implements the DNA algorithm; for some problem instances, proving this is NP-hard *(142)*. Correspondingly, we define three problems in library design. Given an algorithm for a type of DNA computer, the DNA Code *Constraint* Problem is to find a set of constraints that the DNA strands must satisfy to minimize the number of errors due to the choice of DNA strands. Given a set of constraints, the DNA Code *Design* Problem is to find the largest set of DNA strands which satisfy the given constraints or to

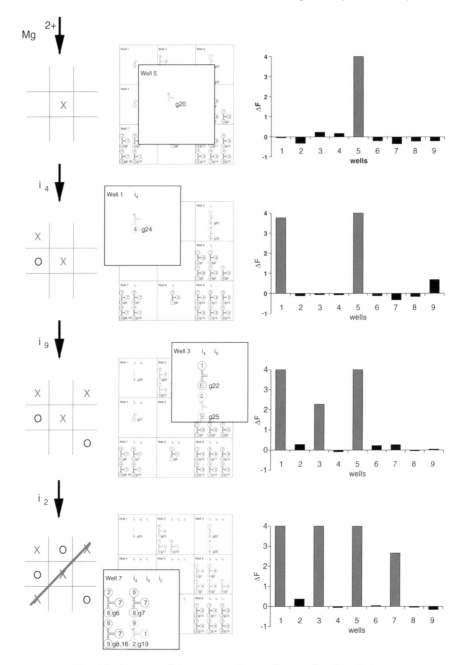

Fig. 12. A game of tic-tac-toe. See main text for description.

find a set of DNA strands of a given size that best satisfy a given set of constraints. The DNA Code *Evaluation* Problem is to evaluate how accurate a set of DNA strands is for implementing a DNA algorithm.

4.1. DNA Code Constraint Problem

A properly constructed library will help minimize errors so that DNA computation is more practical, reliable, scalable, and less costly in terms of materials and laboratory time. (For an overview of library design, *see* ref. *67.*) However, the construction of a library is nontrivial for two reasons. First, there are 4^N unique DNA strings of length N; thus, the number of candidate molecules grows exponentially in the length of the DNA string. Second, the constraints used to find a library are complex because they are subject to the laws of biochemistry as well as the specific algorithm and computation style.

4.1.1. Positive and Negative Design

Although there are many types of DNA computers, most share similar biochemical requirements because they use the same fundamental biochemical processes for computation. The fundamental computation step for most DNA computers occurs through the bonding (hybridization) and unbonding (denaturation) of oligonucleotides (short strands of DNA).

Creating an error-free library typically requires that planned hybridizations and denaturations (between a word and its Watson-Crick complement) occur and unplanned hybridizations and denaturations (between all other combinations of code words and their complements) not occur. The former situation is referred to as the *positive design problem* while the latter is referred to as the *negative design problem (143,144).*

The positive design problem requires that there exist a sequence of reactions that produces the desired outputs, starting from the given inputs. Thus, positive design attempts to "optimize affinity for the target structure" *(144).* These reactions must occur within a reasonable amount of time for feasible concentrations. Usually, the strands must satisfy a specified secondary structure criterion (e.g., the strand must have a desired secondary structure or have no secondary structure at all). Because a strand is typically identified by hybridization with its perfect Watson-Crick complement, the positive design problem requires that each Watson-Crick duplex be stable. In addition, for computation styles that use denaturation, the positive design problem often requires all of the strands in the library to have similar melting temperatures, or melting temperatures above some threshold. In short, positive design tries to maximize hybridization between perfect complements.

The negative design problem requires that: (1) no strand have an undesired secondary structure such as hairpin loops (*see* Fig. 13), (2) no string in the

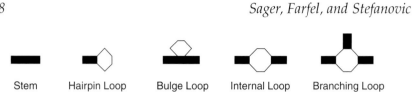

Fig. 13. DNA loops. Solid areas represent double stranded sections. Lines represent single stranded sections.

library hybridize with any other string in the library, and (3) no string in the library hybridize with the complement of any other string in the library. Thus, negative design attempts to "optimize specificity for the target structure" *(144)*. Unplanned hybridizations can cause two types of potential errors: false positives and false negatives. False negatives occur when all (except an undetectable amount) of DNA that encodes a solution is hybridized in unproductive mismatches. Because mismatched strands are generally less stable than perfectly matched strands, false negatives can be controlled by adjusting strand concentrations. Deaton experimentally verified the occurrence of false positives, which happen when a mismatched hybridization causes a strand to be incorrectly identified as a solution *(68)*. False positives can be prevented by ensuring that all unplanned hybridizations are unstable. In short, the negative design problem tries to minimize nonspecific hybridization.

Positive design often uses guanine-cytosine (GC) content and energy minimization as heuristics (*see* below). Negative design uses combinatorial methods (such as Hamming distance, reverse complement Hamming distance, shifted Hamming distance, and sequence symmetry minimization), and thermodynamic methods (such as minimum free energy). Constraints that incorporate both positive and negative designs are probability, average incorrect nucleotides, energy gap, probability gap, and energy minimization in combination with sequence-symmetry minimization. The best-performing models for designing single-strand secondary structure use simultaneous positive and negative design, and significantly outperform either method alone; however, kinetic constraints must be considered separately because low free energy does not necessarily imply fast folding *(144)*. We believe that this same principle holds for designing hybridizations between multiple strands.

4.1.2. Secondary Structure of Single Strands

Most DNA computation styles need strands with no secondary structure (i.e., no tendency to hybridize with itself). There are, on the other hand, cases where specific secondary structures are desired, such as for deoxyribozyme logic gates *(112)*; Fig. 14 shows the desired structure. Even there, structures different from the desired ones must be eliminated.

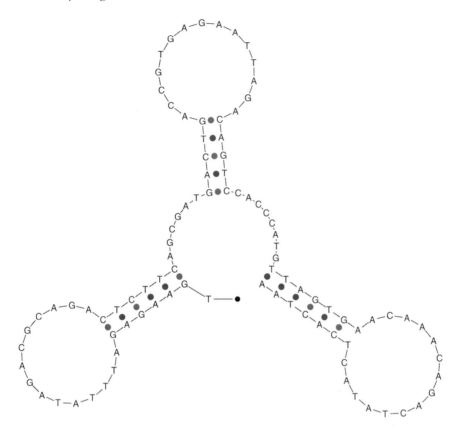

Fig. 14. Example of secondary structure in Stojanovic and Stefanovic's DNA automaton *(112)* as computed by mfold *(174,180,232)* using 140 m*M* Na⁺, 2 m*M* Mg²⁺, and 25°C. The strand has three hairpin loops, which is the desired secondary structure. ΔG is −12.3 kcal/mol.

There are several heuristics that are used to prevent secondary structure. Sometimes, repeated substrings and complementary substrings within a single strand that are nonoverlapping and longer than some minimum length are forbidden in order to prevent stem formation. This heuristic is often called *sequence symmetry minimization (144,145)* or *substring uniqueness (146)*. Another heuristic is to forbid particular substrings; these *forbidden substrings* are usually strings known to have undesired secondary structure. For example, sequences containing GGGGG should be avoided because they may form the four-stranded G4-DNA structure *(147,148)*. (For more information about alternative base-pairing structures, *see* ref. *97*.) Alternatively, strands are designed using only a *three-letter alphabet* (A, C, T for DNA and A, C, U for RNA) to eliminate the potential for GC pairs, which could cause unwanted secondary structure *(149)*.

In order to design a strand with a desired secondary structure (inverse secondary structure prediction), the nucleotides at positions that bond together must be complementary. This simple approach can be improved by also requiring the strands to satisfy some free-energy-based criteria, such as those described below from Dirks et al. *(144)*.

The *minimum free energy* constraint, which can be calculated in $O(N^3)$ time for structures with no pseudoknots *(150)*, is used to choose sequences such that the target structure has the minimum free energy. However, because this method is a negative design, it does not ensure the absence of other structures that the sequence is likely to form. Algorithms also exist to determine whether a set of strands are structure-free, where a set of sequences is considered to be structure-free if the minimum free energy of every strand in the set is greater than or equal to zero *(151–153)*. It has also been suggested that sequences be chosen so that the difference between the free energy of the desired structure and undesired structures is maximal *(67)*.

The *energy minimization* constraint is used to choose sequences that have a low free energy in the target structure, but not necessarily the minimum free energy. To design strands with this constraint, first generate a random string *s* that satisfies the complementary requirements of the desired secondary structure. For each step (Dirks used 10^6 steps), choose a random one-point mutation. Let *s′* be the sequence with this random one-point mutation (and a mutation in the corresponding base required by the structure constraint, if any). Accept the mutation by replacing *s* with *s′* if:

$$e^{-\frac{\Delta G(s')-\Delta G(s)}{RT}} \geq \rho$$

where $\rho \in [0, 1]$ is a random number drawn from a uniform distribution, $\Delta G(s)$ is the free energy of the sequence in secondary structure *s*, and $\Delta G(s')$ is the free energy of the sequence in secondary structure *s′* [the free energy of a given structure can be calculated in $O(N)$ time]. Thus, this equation always accepts any mutations that result in no change or a decrease in free energy, and accepts with some probability any mutations that increase the free energy.

Sequences can also be chosen that maximize the *probability* of sampling the target structure. The probability $p(s)$ that every nucleotide in the sequence exactly matches the target structure *s* at thermodynamic equilibrium is calculated by:

$$p(s) = \frac{1}{Q} e^{-\frac{\Delta G(s)}{RT}}$$

where $\Delta G(s)$ is the free energy of the sequence in secondary structure s. The partition function, Q, is:

$$Q = \sum_{s \in \Omega} e^{-\frac{\Delta G(s)}{RT}}$$

where Ω is the set of all secondary structures that the sequence can form in equilibrium. If s^* is the target secondary structure and $p(s^*) \approx 1$, then the sequence has a high affinity and high specificity for s^*. An optimal dynamic programming algorithm calculates $p(s^*)$ for structures with no pseudoknots in $O(N^3)$ time *(154)*, whereas $p(s^*)$ for secondary structures with pseudoknots can be calculated in $O(N^5)$ time *(155)*.

Additionally, sequences can be chosen to minimize the *average number of incorrect nucleotides*, $n(s)$, ΩS_s for a given sequence of length N in structure s is:

$S_s[i, j] = 1$, if base i is paired with base j in s; 0,otherwise

$S_s[i, N + 1] = 1$, if base i is unpaired in s; 0, otherwise

where $1 \leq i \leq N$ and $1 \leq j \leq N$. The probability matrix P_s is:

$$P_s[i,j] = \sum_{s \in \Omega} p(s) S_s[i,j]$$

where $1 \leq i \leq N$ and $1 \leq j \leq N + 1$. When $1 \leq j \leq N$, $P_s[i, j]$ is the probability of forming a base pair between the nucleotides at position i and j (i.e., the sum of the probabilities of each structure where i and j are paired). $P_s[i, N + 1]$ is the probability that base i is unpaired. Let $n(s)$ be the average number of incorrect nucleotides over the equilibrium ensemble of secondary structures Ω. If s^* is the target structure then:

$$n(s^*) = N - \sum_{i=1}^{N} \sum_{j=1}^{N+1} P_s[i,j] \, S_{s^*}[i,j]$$

where $n(s^*)$ can be calculated in $O(N^3)$ time in structures with no pseudoknots and $O(N^5)$ in structures with pseudoknots.

Dirks and Pierce *(144)* determined that the best-performing models are probability, average incorrect nucleotides, and energy minimization in combination with sequence symmetry minimization for the substrings that are not constrained by the desired secondary structure. The models with medium performance are the negative design methods (minimum free energy, and sequence symmetry minimization alone). The worst-performing model is energy minimization (a positive design method). Surprisingly, minimum free energy performs similarly to sequence symmetry minimization; these results show that free energy measurements do not guarantee good design. An effective search must use both positive and negative design methods.

4.1.3. Secondary Structure of Multiple Strands

The way in which DNA folds in nature is not necessarily how computers should fold DNA strands to obtain the structure, because nature has the advantage of parallel processing and the proximity of the molecules in space. The strength of a perfectly matched duplex, a positive constraint, is often estimated by either: (1) the type of hydrogen bonds, AT vs GC, expressed as the percentage of nucleotides that are G and C bases in a strand or duplex, which is known as *GC content*; or (2) the amount of free energy released from the formation of the hydrogen bonds and the phosphodiester bonds that hold together adjacent nucleotides in a strand. The latter model is known as the nearest-neighbor model.

Because GC base pairs are held together by three hydrogen bonds whereas AT base pairs are held together by only two hydrogen bonds, dsDNA with a high GC content is *often* more stable than DNA with a high AT content. Many DNA library searches require each strand to have a 50% GC content to make the thermodynamic stability of perfectly matched duplexes similar. The GC-content heuristic is simple to calculate; only the length and the number of GC bases are needed, where the length refers to the number of nucleotide base pairs. However, the nearest-neighbor heuristic is more accurate than the GC-content heuristic because the nearest neighbor base-stacking energies account for more of the change in free energy than the energy of the hydrogen bonding between nucleotide bases.

Requiring all pairs of strings in the library to have at least a given minimum *Hamming distance* (i.e., the number of characters in corresponding places which differ between two strings) is intended to satisfy the negative requirement that no pair of strings in the library should hybridize. A variation of this idea is the *reverse complement Hamming distance*, which is the number of corresponding positions that differ in the complement of s_1 and the reverse of s_2. This constraint is used to reduce the false positives that occur from hybridization between a word and the reverse of another word in the library.

The advantage of Hamming distance (and its variations) is its theoretical simplicity and the vast body of extant work in coding theory. Many bounds have been calculated on the optimal size of codes with various Hamming-distance-based constraints *(156)*. Many early DNA library search algorithms used Hamming distance as a constraint to develop combinatorial algorithms based on the results from coding theory. However, Hamming distance alone is an insufficient constraint.

One problem with Hamming-distance-based heuristics is that this measure assumes that position *i* of the first string is aligned with position *i* of the second string. However, because duplexes can be formed with dangling ends and loops, this is not the only possible alignment. Various *Hamming distance slides*, *substring uniqueness (146)*, partial words *(157)*, and H-measure *(71,158)* constraints have been developed to fix the alignment problem. Similarly, many of the previously mentioned constraints (such as GC content and Hamming distance) have also been applied to windows and pairs of windows, which are substrings of a given length. Another problem with heuristics based on Hamming distance is that the percentage of matching base pairs necessary to form a duplex is not necessarily known. Melting temperature can be used to approximate what the minimum Hamming distance should be; however, for a given temperature and word set, there can be significant variation in the required minimum distance.

Now that accurate free-energy information is available for all but the most complicated secondary structures (e.g., branching loops), the nearest-neighbor model is a much more accurate method to use than the constraints based on Hamming distance. It has also been experimentally determined for a sequence *A* of length *n* and a sequence *B* of length *m* that minimum free energy is a superior constraint to *BP*, where

$$BP = min(n, m) - min_{-m<k<n} H(A, \sigma^k(\bar{B}))$$

where $H(*,*)$ is the Hamming distance, $\bar{B}$ is the reverse complement of *B*, and σ^k is the shift rightward when $k > 0$ or leftward when $k < 0$ *(147)* (*BP* is equivalent to the H-measure constraint if $n = m$). One way of using free-energy-based calculations as a constraint to prevent mismatched duplexes is to maximize the gap between the free energy of the weakest specific hybridization and the free energy of strongest nonspecific hybridization, which we refer to as the *energy gap*; this approach was used by Penchovsky *(159)*. A metric also exists that calculates the maximum number of stacked base pairs in any secondary structure; a thermodynamic weighting of this metric gives an upper bound on the free energy of duplex formation *(160)*. The probability, $p(s*)$, measurement could also be applied to duplexes. A reasonable heuristic would be to maximize the gap between the lowest probability of the desired specific hybridizations and the highest probability of undesired nonspecific hybridizations, which we refer to as the *probability gap*. Algorithms exist that calculate the probability, $p(s*)$, for all possible combinations of single- and double-stranded foldings between a pair of strands *(161)*. Various equilibrium thermodynamic approaches have been used *(162–166)*. Computational incoherence predicts the probability of an error hybridization per hybridization event based on statistical thermodynamics *(158,162,167)*.

The physically based models can be divided into categories based on the level of chemical detail *(168)*. Techniques that model single molecules include molecular mechanics models such as Monte Carlo minimum free energy simulations and molecular dynamics, which models the change of the system with time. Techniques that average system behavior, or mass action approaches, are less accurate but more computationally feasible. Molecular mechanics (which models the movement of the system to the lowest energy), chemical kinetics, melting temperature, and statistical thermodynamics are all mass action approaches.

Thermodynamics are best at predicting DNA structure. However, calculating these measures can be costly. According to the requirements mentioned for the negative design problem, checking that a library of size M meets specifications requires $O(M^2)$ string comparisons, where each comparison of a pair of strings of length N is potentially polynomial in N. Thus, the weaker combinatorial and heuristic predictors could be used to quickly filter a candidate set of library molecules, and then the free energy model could be used to more accurately check this set. If this approach is adopted, the correlation between these alternative heuristics and free energy measurements should be explored. Alternatively, free energy or probability approximation algorithms could be used. This approach has the advantage that techniques from randomized algorithm analysis could be used to prove the correctness of the approximation.

4.1.4. Melting Temperature

Melting temperature is typically used as a constraint in DNA paradigms that use multiple hybridization and denaturation steps to identify the answer (for an example, *see* ref. *64*). When DNA is heated, the hydrogen bonds that bind two bases together tend to break apart, and the strands tend to separate from each other. The probability that a bond will break increases with temperature. This probability can be described by the melting temperature, which is the temperature in equilibrium at which 50% of the oligonucleotides are hybridized and 50% of the oligonucleotides are separated. Because temperature control is often used to help denature the strands in intermediate steps, it is advantageous for these paradigms to require all of the strands in the library to have similar melting temperatures, or melting temperatures above some threshold.

The melting temperature of a perfectly matched duplex can be roughly estimated from the 2–4 rule *(67)*, which predicts the melting temperature as twice the number of AT base pairs plus four times the number of GC base pairs.

Another rough estimate of the change in melting temperature due to mismatched duplexes can also be obtained by decreasing the melting temperature of a corresponding matched duplex by 1°C per 1% mismatch; unfortunately, the inaccuracy is typically greater than 10°C *(169)*. Neither method is recommended. A better method is to use the nearest-neighbor model regardless of whether the duplex is perfectly matched or mismatched. This method produces more accurate results because melting temperature is closely related to free energy. Melting temperature has been used to characterize the hybridization potential of a duplex *(170,171)*, but this measure cannot be used to predict whether two strands are bound at a given temperature because the melting temperatures of different duplexes do not necessarily correspond to relative rankings of stability.

4.1.5. Reaction Rates

Once the structure of candidate strands is known, the next logical question to ask is how fast these reactions occur and what concentration is needed. Kinetics deals with the rate of change of reactions. For some implementations of DNA computers, the rate of the reaction could be an additional search constraint. System-level simulation software has been described for this purpose *(172)*.

4.1.6. DNA Prediction Software

There exist many software packages that predict DNA/RNA structure, thermodynamics, or kinetics. A few well known structure prediction software packages are: Dynalign *(173)*, mfold *(174)*, NUPACK *(155,175)*, RNAsoft *(176)*, RNAstructure *(177)*, and the Vienna Package *(178)*. RNA free energy nearest-neighbor parameters are available from the Turner Group *(177)*. Some software packages that calculate thermodynamics are: HyTher *(179–181)*, BIND *(170)*, MELTING *(182)*, MELTSIM *(183)*, and MeltWin *(184)*. Kinfold *(185)* simulates kinetics. EdnaCo *(158)* and Visual OMP (Oligonucleotide Modeling Platform; DNA Software Inc.) *(186)* simulate biochemical protocols *in silico*. In addition, there are many library design software packages such as DNA Design Toolbox *(187)*, DNASequenceCompiler *(146)*, DNASequenceGenerator *(146)*, NACST/Seq *(188)*, NucleicPark *(166)*, PERMUTE *(64)*, PUNCH *(189)*, SCAN *(171)*, SEQUIN *(145)*, SynDCode *(160,190,191)*, and TileSoft *(192)*.

4.2. DNA Code Design Problem

Once the desired constraints are known, how should one design a sequence generator to find strings that satisfy those constraints? A good generator should be reliable, extensible, efficient, and scalable. Ideally, the generator should find as large a set as possible, work for multiple problems, and should allow constraints to be easily added and removed. However, comparisons of sequence-generation algorithms are difficult because the algorithms are usually written and tested for a specific DNA computation problem and specific set of constraints; an algorithm that does well on one constraint set may not do well on another constraint set. Thus in this section we briefly explain several approaches to give the flavor of possible solutions to the DNA Code Design Problem; *see also* ref. *143*.

Early algorithms to find DNA word sets focused on the Hamming-distance constraint or variations thereof to achieve a theoretical abstraction of the constraints, which allowed the use of combinatorial algorithms (e.g., *69*) and proofs of completeness (i.e., that the size of the pool is optimal or near optimal) *(156)*. However, in the process the constraints are simplified so much that they no longer accurately predict DNA structure. Current algorithms tend to use a more complex combination of the constraints. However, because these constraints are difficult to abstract, more recent programs resort to genetic algorithms, random search, exhaustive search, and local stochastic search algorithms. For a survey of algorithms that have been used to solve the DNA/RNA Code Design Problem, *see* ref. *143*.

4.2.1. Combinatoric Algorithms

Because of the association between DNA code design and coding theory, early algorithms tended to focus on finding optimal code sizes. Many proofs have been found which bound the size of optimal codes for simple combinations of constraints based on Hamming distance and reverse complement Hamming distance *(156)*. These proofs can be used to evaluate the optimality of a solution to the DNA Code Design Problem. Algebraic properties, formal language theory, and coding theory have also been used to show properties of DNA-compliant languages *(193)*. However, the tradeoff is that many of these proofs are extremely difficult to extend to complex combinations of constraints that model the physical world more realistically. As a result, these algorithms tended to be deterministic, combinatorial, and specific to the DNA computer that they were designed for.

For example, the "template-map" strategy *(69)* obtains a large number of dissimilar word sequences from a significantly smaller number of templates and maps *using theoritical proofs*, where a template is a string chosen from the alphabet $\{A, C\}$ and a map is a string of the same length chosen from the alphabet $\{0, 1\}$. When a map m is applied to a template t, a character in the template, t_i, is replaced with its complement if the corresponding character in the map, m_i, is 1; if m_i is 0 then there is no change to t_i (e.g., when the map 10100101 is applied on template AACCACCA, it produces the string

TAGCAGCT.) Because each template and map pair uniquely describes a string from the alphabet {A, C, G, T}, additional constraints are needed to prevent nonspecific hybridizations. The templates are also required to be "conflict-free," where two templates are considered to be conflict-free if they generate two strings that have a Hamming distance and reverse complement Hamming distance of at least 4 when paired with any two maps. In addition, the template and map pairs are also required to generate strings with a 50% GC content. The obvious limitation of this method is with respect to extensibility and scalability.

4.2.2. Randomized Algorithms

Later algorithms have tended to focus on being extensible to a variety of problems and constraints and also on accurately modeling the physical world; this trend can be seen in the current discussions about defining a standard for biomolecular computing simulation software *(194)*. Because the search space is large and the constraints are complex, most of the randomized algorithms used for DNA code design tend to be Las Vegas algorithms (algorithms which vary in run time) and not Monte Carlo algorithms (algorithms which sometimes produce incorrect answers); thus the efficiency with which a randomized algorithm finds or converges to a solution is an important consideration for evaluating these types of algorithms. In addition, these algorithms may also vary in solution quality from run to run, so the quality of the solution is also important.

The PERMUTE program *(64)* is an example of a simple randomized algorithm. It generates random nucleotides from the three-letter alphabet {A, C, U} and then permutes the sequence until the constraints are satisfied. If no permutation produces a valid string, then a new random string is generated. A simple variation on this idea is to generate a random candidate string, add the string to the pool only if it satisfies the constraints, and repeat *(195)*. These types of "generate-and-test" algorithms perform well in situations where the search process does not tend to get stuck in local minima. However, the constraints must be set appropriately before algorithm executes and the generator cannot suggest whether it is possible to find better sets which satisfy the same constraints.

The DNA SequenceGenerator *(146)* is an example of a slightly more complicated randomized algorithm. This algorithm generates a pool of n_b-unique sequences from a directed graph whose nodes are labeled with sequences of length n_b, which are referred to as "base strands". A directed edge, (u, v), connects nodes u and v if the last $n_b - 1$ characters of base strand u are the same as the first $n_b - 1$ characters of base strand v. Thus a string of length n_s is represented by a path of length $(n_s - n_b + 1)$; the set of paths of length $(n_s - n_b + 1)$ that do not share any nodes corresponds to a set of n_b unique sequences of length n_s. The nodes of certain base strands (such as self-complementary substrings, forbidden substrings, substrings containing two consecutive GG or CC bases, substrings containing specified GC content, etc.) can be restricted by marking their corresponding nodes as forbidden or by removing them from the graph. In each iteration, the algorithm randomly chooses a start node and performs a random walk to find a path of length $(n_s - n_b + 1)$ that does not contain forbidden nodes, nodes used in other paths, or the reverse complement of nodes used in other paths. If a complete path that satisfies the constraints (such as melting temperature and GC content) is found, the sequence is added to the library of strings, otherwise the walk backtracks and attempts to find another path. A limitation of this algorithm is that a large amount of memory may be needed to store the graph.

Most current research in DNA word design falls in the category of stochastic local search algorithms (which includes the evolutionary algorithms described below). Stochastic local search (SLS) algorithms are the subset of randomized algorithms that make use of the previous randomized choices when generating or selecting new candidate solutions. More specifically, "the local search process is started by selecting an initial candidate solution, and then proceeds by iteratively moving from one candidate solution to a neighboring candidate solution, where the decision on each search step is based on a limited amount of local information only. In stochastic local search algorithms, these decisions as well as the initial search initialization can be randomized" *(196)*. Many SLS algorithms have parameters that must be set manually. The comparison of these algorithms can be misleading when the parameter settings are unevenly optimized; thus, care must be taken to ensure that the parameters are equally optimized or that at least the same amount of effort is spent on each algorithm to optimize the parameters if the optimal settings are uncertain.

Given a set of individual and pairwise constraints on strands (e.g., Hamming distance, reverse Hamming distance, GC content, or thermodynamics), the SLS-THC algorithm *(196–199)* begins with a randomly chosen pool of strings of size N, where each string is of length n and each string satisfies any constraints specified on individual strings. To obtain good performance, the algorithm stores the results of the calculations for the pairwise constraints in a table; thus, modifying a word in the pool requires only $\Omega(N)$ calculations. In each iteration, the algorithm picks a pair of words (uniformly at random) that has a conflict (a violation of a pairwise search constraint) and modifies one of the words. All single-base mutations to each string in the conflicting pair that satisfies the individual constraints (the 1-mutation neighborhood) are considered modifications. With constant probability q, a modification in the 1-mutation neighborhood is chosen at random, otherwise a modification is chosen that maximally reduces the number of pair conflicts in the pool. Empirical analysis of the run-time distributions of the algorithm on hard design problems indicates that the search performance is compromised by stagnation; this problem can be overcome by the occasional random replacement of a small fraction of the strings in the pool *(197)*. The algorithm terminates if S has no conflicts or if a specified number of iterations have been completed. If the algorithm terminates before it finds a valid set of size N, then a word in a conflicting pair is randomly deleted from the pool until no conflicts remain.

The SLS-THC algorithm is a more sophisticated search than the previous randomized algorithms because it utilizes local information in its search process. The search process can be thought of as a conflict-directed random walk. As the algorithm runs, at any given time there may be pair-wise conflicts in the pool; allowing these conflicts to remain may help the algorithm overcome local minima because the decision of which conflicting string to remove is delayed. Because every conflicting pair has the same probability of being mutated in each iteration, there is a high probably that strings that create minor conflicts will be resolved by only a few mutations and a high probability that strings which prevent the pool size from growing (local minima) will be mutated greatly or even replaced. It has been empirically demonstrated that the SLS-THC algorithm matches or improves upon the pool sizes obtained from the best-known theoretical constructions for several different combinations of Hamming distance, reverse Hamming distance, and GC-content constraints *(196)*.

Evolutionary algorithms (EAs) are a subset of SLS algorithms that use techniques inspired by biological evolution. The solution pool is represented by a population of "individuals" or "chromosomes." EAs use selection, mutation, and recombination on the population to utilize local information and prevent local minima in order to efficiently optimize the population. There are several types of evolutionary algorithms, such as genetic algorithms (GAs) *(200)*, evolution strategies *(201,202)*, and evolutionary programming *(203)*. However, because current work often blends concepts from many styles of EAs, we do not emphasize the differences between the types of EAs.

The goal of a GAs is to minimize or maximize a measure of fitness; this concept corresponds to the biological concept of "selection of the fittest." For example, in some GA implementations of the DNA word design problem, the fitness is based on the Hamming distance between strings *(68,204)* or based on the partition function *(205)*. Other GAs have used a single fitness function that incorporates multiple constraints *(195)*; as a result, several experimental runs may be required to decide how to set the parameters. When the constraints are mutually independent, the parameter values can be determined independently. However, in the DNA word design problem, it is often the case that optimizing one constraint causes a relative tradeoff in the optimality of another constraint (e.g., the chance of nonspecific hybridizations can be reduced by using only the three bases A, T, and C, but this technique also increases the similarity of the strings). When the parameters are not mutually exclusive, finding the optimal parameter settings can be difficult *(196)*. It has been suggested that as the number of design constraints is increased, a single fitness measure that incorporates all of the design constraints may not be appropriate for the DNA word design problem because the relative importance of each constraint is often unknown *(195)*. Some more recent GAs, such as NACST/Seq *(188,206,207)*, attempt to resolve these problems using a multi-objective GA.

4.3. DNA Code Evaluation Problem

Of the heuristics previously mentioned, the most appropriate method for obtaining an estimate of the absolute or relative rate of hybridization error is thermodynamics and statistical thermodynamics. For example, $p(s^*)$, $n(s^*)$, pair probabilities, and free energy have been used to evaluate whether a singly stranded sequence will form a desired secondary structure, s^* *(144)*. Statistical thermodynamics (the partition function of all hybridized configurations) have been used to predict the error rate in the set of strands used in Adleman's original Hamiltonian Path problem *(205)*. Computational incoherence *(162,167)*, xi, could also be used for evaluation. In addition, the energy gap or probability gap could be used for evaluation *(199)*. The most significant evaluation criterion is how the strands perform in the laboratory, because this is what the library is ultimately designed for.

Research in DNA libraries has two main goals: (1) to further understand DNA chemistry, and (2) to understand search techniques useful for constructing sets of DNA codes. Although there is a growing consensus that DNA computers will never be as practical or as fast as conventional computers, biological computers have the advantage that their style of computation is closer to natural processes. Deaton states that the process of converting an algorithm into a biomolecular system "is as difficult [i.e., NP-hard or harder] as the combinatorial optimization problems they are intended to solve" *(142)*. However, successful research in DNA libraries will help to reduce errors in DNA computation and may unearth new information about how DNA interacts with itself. Although current DNA computers are simplistic in comparison to natural biochemical processes, DNA computation may help to develop alternative theories for how cells work or could have evolved *(208)*. In addition, research in DNA design also pertains to DNA nanotechnology, PCR-based applications, and DNA arrays. Breakthroughs in this field will add to the current knowledge of DNA chemistry as well as DNA computers.

4.4. *Exploiting Inexact Matching*

In the preceding, we assumed that the applications to which the designed word sets will be put require exact matching for correctness of operation. This is indeed true of combinatorial DNA computing, to avoid false positives, i.e., spurious solutions, and it is somewhat true in enzymatic DNA computing, to minimize cross-talk between signals. On the other hand, there can be an array of applications that inherently allow modest amounts of error. Such is the case with signal-processing applications, where the input data are noisy. It is preferable in such situations to allow imperfect matches, i.e., to build the possibility of imperfect matches directly into the design of the word set.

Tsaftaris *(209,210)* considers a hypothetical scenario in which a database of signals is stored as a pool of DNA. Each signal is represented as a double-stranded section of DNA. The database allows matching queries, in which one asks if a given (short) probe signal is approximately equal to some portion of one of the stored (target) signals; the target signal and the position of the match are identified. To run the matching query, a sample of the database is denatured, the probe is represented as the complementary oligonucleotide, hybridization is allowed to take place, and then the result is isolated. In such a setting, it is explicitly advantageous to allow some degree of hybridization errors between strands that encode *adjacent* signal levels. The word design problem is then not just that of choosing some N oligonucleotides of a given length, but of assigning them to the N discrete signal levels in such a way that for signal levels that are close to one another, the likelihood of a stable mismatch is inversely proportional to the level difference, and for signal levels exceeding some threshold, that likelihood is negligible. This is called the *noise tolerance constraint*, and is imposed in addition to the usual combinatorial constraints. A stochastic algorithm that builds upon thermodynamic models of SantaLucia *(180)* is proposed by Tsaftaris and demonstrated for $N = 128$ and 10-nt oligonucleotides *(209)*.

5. CONCLUSION

This review focuses on a few selected topics in nanocomputing. Meanwhile, the literature is growing by the day. For combinatorial approaches, which predominated at the outset of the DNA computing research era, consult, e.g., ref. *99*. For state-machine-based approaches, predicted at least as early as in the work of Manin, initiated by Rothemund, and forcefully demonstrated by Benenson *(211,212)*, in which finite control is achieved using collections of customized enzymes, consult, e.g., ref. *213*. For cell and membrane computing, consult refs. *85,90,214–216*. For recent achievements in self-assembly, in particular assembly of almost arbitrary

planar shapes, see ref. *217*. For recent achievements in enzymatic computing, see ref. *218*. For architectural advances, spearheaded by dyed-in-the
wool computer scientists, see, e.g., refs. *219, 220*.

REFERENCES

1. International Technology Roadmap for Semiconductors 2003, http://public. itrs.net.
2. Bennett CH. The thermodynamics of computation—a review. Int J Theor Phys 1982;21:905–940.
3. Zhirnov VV, Cavin RK III, Hutchby JA, Bourianoff GI. Limits to binary logic switching—a gedanken model. Proceedings of the IEEE 2003;91: 1934–1939.
4. Ball P. Chemistry meets computing. Nature 2000;406:118–120.
5. Reed M, Tour JM. Computing with molecules. Scientific American 2000; 2000:86–93.
6. Zhirnov VV, Herr DJC. New frontiers: self-assembly and nanoelectronics. IEEE Computer 2001;2001:34.
7. Hutchby JA, Bourianoff GI, Zhirnov VV, Brewer JE. Extending the road beyond CMOS. IEEE Circuits and Devices Magazine 2002;18:28–41.
8. Bourianoff G. The future of nanocomputing. IEEE Computer 2003;2003:36.
9. Stone NJ, Ahmed H. Silicon single electron memory cell. Appl Phys Lett 1998;73:2134–2136.
10. Mahapatra S, Vish V, Wasshuber C, Banerjee K, Ionescu AM. Analytical modeling of single electron transistor for hybrid CMOS-SET analog IC design. IEEE Transactions on Electron Devices 2004;51:1772–1782.
11. Bachtold A, Hadley P, Nakanishi T, Dekker C. Logic circuits with carbon nanotube transistors. Science 2001;294:1317–1320.
12. Chen Y, Ohlberg DAA, Medeiros-Ribeiro G, Chang YA, Williams RS. Self-assembled growth of epitaxial erbium disilicide nanowires. Appl Phys Lett 2000;76:4004–4006.
13. Cui Y, Wei Q, Park H, Lieber CM. Nanowire nanosensors for highly sensitive and selective detection of biological and chemical species. Science 2001;293:1289–1292.
14. Huang Y, Duan X, Cui Y, Lauhon LJ, Kim KH, Lieber CM. Logic gates and computation from assembled nanowire building blocks. Science 2001;294:1313–1317.
15. Reed MA, Zhou C, Muller CJ, Burgin TP. Conductance of a molecular junction. Science 1997;278:252–254.
16. Reed M, Chen J, Rawlett AM, Price DW, Tour JM. Molecular random access memory cell. Appl Phys Lett 2001;78:3735–3737.
17. Chen Y, Ohlberg DAA, Li X, et al. Nanoscale molecular-switch devices fabricated by imprint lithography. Appl Phys Lett 2003;82:1610–1612.
18. Csaba G, Imre A, Bernstein GH, Porod W, Metlushko V. Nanocomputing by field-coupled nanomagnets. IEEE Transactions on Nanotechnology 2002;1: 209–213.

19. Porod W, Lent CS, Bernstein GH, et al. Quantum-dot cellular automata: computing with coupled quantum dots. International Journal of Electronics 1999;86:549–590.
20. Heath JR, Kuekes PJ, Snider GS, Williams RS. A defect-tolerant computer architecture: opportunities for nanotechnology. Science 1998;280:1716–1721.
21. Collier CP, Wong EW, Belohradsky M, et al. Electronically configurable molecular-based logic gates. Science 1999;285:391–394.
22. Metzger RM. Electrical rectification by a molecule: the advent of unimolecular electronic devices. Accounts Chem Res 1999;32:950–957.
23. Ellenbogen JC, Love JC. Architectures for molecular electronic computers: 1. Logic structures and an adder built from molecular electronic diodes. Proceedings of the IEEE 2000;88:386–426.
24. Joachim C, Gimzewski JK, Aviram A. Electronics using hybrid-molecular and monomolecular devices. Nature 2000;408:541–548.
25. Donhauser ZJ, Mantooth BA, Kelly KF, et al. Conductance switching in single molecules through conformational changes. Science 2001;292:2303–2307.
26. Goldstein SC, Budiu M. NanoFabrics: spatial computing using molecular electronics. In: Proceedings of the 28th International Symposium on Computer Architecture. New York: ACM Press, 2001:178–191.
27. Pease AR, Jeppesen JO, Stoddart JF, Luo Y, Collier CP, Heath JR. Switching devices based on interlocked molecules. Accounts Chem Res 2001;34: 433–444.
28. Postma HWC, Teepen T, Yao Z, Grifoni M, Dekker C. Carbon nanotube single-electron transistors at room temperature. Science 2001;293:76–79.
29. Mishra M, Goldstein SC. Scalable defect tolerance for molecular electronics. In: 1st Workshop on Non-Silicon Computing. Cambridge, MA, 2002.
30. de Silva AP, Gunaratne HQN, McCoy CP. A molecular photoionic AND gate based on fluorescent signalling. Nature 1993;364:42–44.
31. de Silva AP, Gunaratne HQN, McCoy CP. Molecular photoionic AND logic gates with bright fluorescence and "off-on" digital action. J Am Chem Soc 1997;119:7891–7892.
32. Credi A, Balzani V, Langford SJ, Stoddart JF. Logic operations at the molecular level. An XOR gate based on a molecular machine. J Am Chem Soc 1997;119:2679–2681.
33. Pina F, Melo MJ, Maestri M, Passaniti P, Balzani V. Artificial chemical systems capable of mimicking some elementary properties of neurons. J Am Chem Soc 2000;122:4496–4498.
34. de Silva AP, McClenaghan ND. Proof-of-principle of molecular-scale arithmetic. J Am Chem Soc 2000;122:3965–3966.
35. Banerjee K, Soukri SJ, Kapur P, Saraswat K. 3-D ICs: A novel chip design for improving deep-submicrometer interconnect performance and systems-on-chip integration. Proceedings of the IEEE 2001;89:602–633.
36. Bilardi G, Preparata FP. Horizons of parallel computation. Tech. Rep. CS-93-20, Department of Computer Science, Brown University, 1993.
37. Abelson H, Allen D, Coore D, et al. Amorphous computing. Communications of the ACM 2000;43:74–82.

38. Goldstein SC, Rosewater D. Digital logic using molecular electronics. In: IEEE International Solid-State Circuits Conference. San Francisco, CA, 2002;12:5.
39. Gruau F, Malbos P. The blob: a basic topological concept for hardware-free distributed computation. In: Calude C, Dinneen MJ, Peper F, eds. Unconventional Models of Computation, Third International Conference Proceedings, Lecture Notes in Computer Science, vol. 2509. Berlin, Heidelberg: Springer, 2002:151–163.
40. Gruau F, Lhuillier Y, Reitz P, Temam O. BLOB computing. In: Vassiliadis S, Gaudiot J-L, Piuri V, eds. Proceedings of the First Conference on Computing Frontiers. ACM SIGMICRO, 2004:125–139.
41. Durbeck LJK, Macias NJ. The cell matrix: an architecture for nanocomputing. Nanotechnology 2001;12:217–230.
42. Munakata T, Sinha S, Ditto WL. Chaos computing: implementation of fundamental logic gates by chaotic elements. IEEE Transactions on Circuits and Systems—I: Fundamental Theory and Applications 2002;49:1629–1633.
43. Adleman LM. Molecular computation of solutions to combinatorial problems. Science 1994;266:1021–1024.
44. Deaton RJ, Garzon M, Rose JA, Franceschetti DR, Stevens SE Jr. DNA computing: a review. Fundamenta Informaticae 1998;35:231–245.
45. Lipton RJ. DNA solution of hard computational problems. Science 1995; 268:542–545.
46. Ruben AJ, Landweber LF. Timeline: the past, present and future of molecular computing. Nat Rev Mol Cell Biol 2000;1:69–72.
47. Wang L, Liu Q, Corn RM, Condon AE, Smith LM. Multiple word DNA computing on surfaces. J Am Chem Soc 2000;122:7435–7440.
48. Winfree E. On the computational power of DNA annealing and ligation. In: Lipton RJ, Baum EB, eds. DNA Based Computers, DIMACS Workshop 1995, vol. 27 of Series in Discrete Mathematics and Theoretical Computer Science. Princeton University: American Mathematical Society, 1996:199–221.
49. Winfree E. Complexity of restricted and unrestricted models of molecular computation. In: Lipton RJ, Baum EB, eds. DNA Based Computers, DIMACS Workshop 1995, vol. 27 of Series in Discrete Mathematics and Theoretical Computer Science. Princeton University: American Mathematical Society, 1996:187–198.
50. Watson J, Crick FHC. A structure for deoxyribose nucleic acid. Nature 1953;171:737.
51. LaBean TH, Yan H, Kopatsch J, et al. Construction, analysis, ligation, and self-assembly of DNA triple crossover complexes. J Am Chem Soc 2000; 122:1848–1860.
52. Watson JD, Hopkins NH, Roberts JW, Steitz JA, Weiner AM. Molecular Biology of the Gene, 4th ed. Menlo Park, CA: Benjamin/Cummings, 1988.
53. Winfree E, Liu F, Wenzler LA, Seeman NC. Design and self-assembly of two-dimensional DNA crystals. Nature 1998;394:539–544.
54. Wang L, Hall JG, Lu M, Liu Q, Smith LM. A DNA computing readout operation based on structure-specific cleavage. Nat Biotechnol 2001;19:1053–1059.

55. Braich RS, Chelyapov N, Johnson C, Rothemund PWK, Adleman L. Solution of a 20variable 3-SAT problem on a DNA computer. Science 2002;296:499–502.
56. Morimoto N, Arita M, Suyama A. Solid phase DNA solution to the Hamiltonian path problem. In: Rubin H, Wood DH, eds. DNA Based Computers III, DIMACS Workshop 1997, vol. 48 of Series in Discrete Mathematics and Theoretical Computer Science. University of Pennsylvania: American Mathematical Society, 1999:193–206.
57. Ouyang Q, Kaplan PD, Liu S, Libchaber A. DNA solution of the maximal clique problem. Science 1997;278:446–449.
58. Pirrung MC, Connors RV, Odenbaugh AL, Montague-Smith MP, Walcott NG, Tollett JJ. The arrayed primer extension method for DNA microchip analysis. Molecular computation of satisfaction problems. J Am Chem Soc 2000;122:1873–1882.
59. Garzon M, Gao Y, Rose JA, et al. In vitro implementation of finite-state machines. In: Proceedings 2nd International Workshop on Implementing Automata WIA'97, Lecture Notes in Computer Science, vol. 1436. Berlin, Heidelberg: Springer Verlag, 1998:56–74.
60. Guarnieri F, Fliss M, Bancroft C. Making DNA add. Science 1996;273: 220–223.
61. Hug H, Schuler R. DNA-based parallel computation of simple arithmetic. In: Jonoska N, Seeman NC, eds. DNA Computing: 7th International Workshop on DNA-Based Computers, DNA 2001, Lecture Notes in Computer Science, vol. 2340. Berlin, Heidelberg: Springer, 2002.
62. Mao C, LaBean TH, Reif JH, Seeman NC. Logical computation using algorithmic self-assembly of DNA triple-crossover molecules. Nature 2000;407:493-496. Erratum, Nature 2000;408:750.
63. Rothemund PWK, Winfree E. The program-size complexity of self-assembled squares. In: The Thirty-Second Annual ACM Symposium on the Theory of Computing, 2000:459–468.
64. Faulhammer D, Cukras AR, Lipton RJ, Landweber LF. Molecular computation: RNA solutions to chess problems. Proc Natl Acad Sci USA 2000;97: 1385–1389. The PERMUTE Program is available at http://www.pnas.org/cgi/content/full/97/4/1385/DC1.
65. Hartmanis J. On the weight of computation. Bulletin of the EATCS 1995;55:136–138.
66. Baum EB. DNA sequences useful for computation. In: Landweber LF, Baum EB, eds. DNA Based Computers II, DIMACS Workshop 1996, vol. 44 of Series in Discrete Mathematics and Theoretical Computer Science. University of Pennsylvania: American Mathematical Society, 1999:235–241.
67. Brenneman A, Condon AE. Strand design for bio-molecular computation. Tech. Rep., University of British Columbia, 2001.
68. Deaton RJ, Murphy RC, Garzon M, Franceschetti DR, Stevens SE Jr. Good encodings for DNA-based solutions to combinatorial problems. In: Landweber LF, Baum EB, eds. DNA Based Computers II, DIMACS Workshop 1996, vol. 44 of Series in Discrete Mathematics and Theoretical Computer Science. University of Pennsylvania: American Mathematical Society, 1999:247–258.

69. Frutos AG, Liu Q, Thiel AJ, et al. Demonstration of a word design strategy for DNA computing on surfaces. Nucl Acids Res 1997;25:4748–4757.

70. Garzon M, Deaton RJ, Niño LF, Stevens E, Wittner M. Encoding genomes for DNA computing. In: Gemetic Programming 1998: Proceedings 3rd Genetic Programming Conference. Morgan Kaufmann, 1998:684–690.

71. Garzon M, Neathery P, Deaton RJ, Murphy RC, Franceschetti DR, Stevens SE Jr. A new metric for DNA computing. In: Proceedings 2nd Genetic Programming Conference, 1997:472–478.

72. Marathe A, Condon AE, Corn RM. On combinatorial DNA word design. In: Winfree E, Gifford DK, eds. DNA Based Computers V, DIMACS Workshop 1999, vol. 54 of Series in Discrete Mathematics and Theoretical Computer Science. MIT: American Mathematical Society, 2000:75–89.

73. Reinert G, Schbath S, Waterman MS. Probabilistic and statistical properties of words: an overview. Journal of Computational Biology 2000;7:1–46.

74. Feldkamp U, Banzhaf W, Rauhe H. A DNA sequence compiler. Tech. Rep., University of Dortmund, 2000.

75. Hug H, Schuler R. Strategies for the development of a peptide computer. Bioinformatics 2001;17:364–368.

76. Sakamoto K, Gouzu H, Komiya K, et al. Molecular computation by DNA hairpin formation. Science 2000;288:1223–1226.

77. Winfree E. Simulations of computing by self-assembly. In: Kari L, Rubin H, Wood DH, eds. DNA Based Computers IV, DIMACS Workshop 1998, Biosystems, vol. 52, issues 1–3. Elsevier, 1999:213–242.

78. Basu S, Karig D, Weiss R. Engineering signal processing in cells: towards molecular concentration band detection. In: Hagiya M, Ohuchi A, eds. DNA Computing: 8th International Workshop on DNA-Based Computers, DNA 2002, Lecture Notes in Computer Science, vol. 2568. Berlin, Heidelberg: Springer, 2003.

79. Conrad M. On design principles for a molecular computer. Communications of the ACM 1985;28:464–480.

80. Guet CC, Elowitz MB, Wang W, Leibler S. Combinatorial synthesis of genetic networks. Science 2002;296:1466–1470.

81. Hayes B. Computing comes to life. American Scientist 2001;89:204–208.

82. Ji S. The cell as the smallest DNA-based molecular computer. BioSystems 1999;52:123–133.

83. Knight TF Jr, Sussman GJ. Cellular gate technology. In: Proceedings UMC98, First International Conference on Unconventional Models of Computation, 1998.

84. LaBean TH, Winfree E, Reif JH. Experimental progress in computation by self-assembly of DNA tilings. In: Winfree E, Gifford DK, eds. DNA Based Computers V, DIMACS Workshop 1999, vol. 54 of Series in Discrete Mathematics and Theoretical Computer Science. American Mathematical Society, 2000:123–140.

85. Landweber LF, Kari L. The evolution of cellular computing: nature's solution to a computational problem. BioSystems 1999;52:3–13.

86. Landweber LF, Kuo TC, Curtis EA. Evolution and assembly of an extremely scrambled gene. Proc Natl Acad Sci USA 2000;97:3298–3303.

87. Reif JH. Parallel biomolecular computation. In: Rubin H, Wood DH, eds. DNA Based Computers III, DIMACS Workshop 1997, vol. 48 of Series in Discrete Mathematics and Theoretical Computer Science. American Mathematical Society, 1999:217–254.

88. Saylor G. Construction of genetic logic gates for biocomputing. In: 101st General Meeting of the American Society for Microbiology, 2001.

89. Weiss R. Cellular Computation and Communication using Engineered Genetic Regulatory Networks. Ph.D. thesis, Massachusetts Institute of Technology, 2001.

90. Weiss R, Basu S. The device physics of cellular logic gates. In: First Workshop on Non-Silicon Computing, 2002.

91. Weiss R, Homsy G, Nagpal R. Programming biological cells. Tech. Rep., MIT Laboratory for Computer Science and Artificial Intelligence, 1998.

92. Weiss R, Homsy GE, Knight TF Jr. Towards *in vivo* digital circuits. In: DIMACS Workshop on Evolution as Computation, 1999.

93. Winfree E, Yang X, Seeman NC. Universal computation via self-assembly of DNA: some theory and experiments. In: Landweber LF, Baum EB, eds. DNA Based Computers II, DIMACS Workshop 1996, vol. 44 of Series in Discrete Mathematics and Theoretical Computer Science. American Mathematical Society, 1999:191–213. Errata: http://www.dna.caltech.edu/Papers/self-assem.errata.

94. Cox JC, Ellington AD. DNA computation function. Curr Biol 2001;11:R336.

95. Yurke B, Mills Jr AP, Cheng SL. DNA implementation of addition in which the input strands are separate from the operator strands. BioSystems 1999;52:165–174.

96. Reif JH. DNA lattices: a method for molecular scale patterning and computation. Computer and Scientific Engineering Magazine 2002;4:32–41.

97. Seeman NC. It started with Watson and Crick, but it sure didn't end there: pitfalls and possibilities beyond the classic double helix. Natural Computing: an international journal 2002;1:53–84.

98. Wang H. Proving theorems by pattern recognition I. Commun ACM 1960;3:220–234.

99. Jonoska N, Kephard DE, Lefevre J. Trends in computing with DNA. J Comput Sci Technol 2004;19:98.

100. Carbone A, Mao C, Constantinou PE, et al. 3D fractal DNA assembly from coding, geometry and protection. Natural Computing 2004;3:235–252.

101. Barish RD, Rothemund PWK, Winfree E. Two computational primitives for algorithmic self-assembly: copying and counting. Nano Lett 2005;5:2586–2592.

102. Winfree E. DNA computing by self-assembly. National Academy of Engineering's The Bridge 2003;33:31–38.

103. Schulman R, Winfree E. Programmable control of nucleation for algorithmic self-assembly. In: Ferretti C, Mauri G, Zandron C, eds. DNA Computing: 10th International Workshop on DNA-Based Computers, DNA 2004, Lecture Notes in Computer Science, vol. 3384. Berlin: Springer, 2005:319–328.

104. Chen HL, Goel A. Error free self-assembly using error prone tiles. In: Ferretti C, Mauri G, Zandron C, eds. DNA Computing: 10th International Workshop

on DNA-Based Computers, DNA 2004, Lecture Notes in Computer Science, vol. 3384. Berlin: Springer, 2005:62–75.
105. Winfree E, Bekbolatov R. Proofreading tile sets: Error-correction for algorithmic self-assembly. In: Chen J, Reif JH, eds. DNA Computing: 9th International Workshop on DNA-Based Computers, DNA 2003, Lecture Notes in Computer Science, vol. 2943. Berlin: Springer, 2004:126–144.
106. Reif JH, Sahu S, Yin P. Compact error-resilient computational DNA tiling assemblies. In: Ferretti C, Mauri G, Zandron C, eds. DNA Computing: 10th International Workshop on DNA-Based Computers, DNA 2004, Lecture Notes in Computer Science, vol. 3384. Berlin: Springer, 2005:293–307.
107. Stojanovic MN, de Prada P, Landry DW. Catalytic molecular beacons. ChemBioChem 2001;2:411–415.
108. Stojanovic MN, Mitchell TE, Stefanovic D. Deoxyribozyme-based logic gates. J Am Chem Soc 2002;124:3555–3561.
109. Stojanovic MN, Kolpashchikov D. Modular aptameric sensors. J Am Chem Soc 2004;126:9266–9270.
110. Stojanovic MN, Semova S, Kolpashchikov D, Morgan C, Stefanovic D. Deoxyribozymebased ligase logic gates and their initial circuits. J Am Chem Soc 2005;127:6914–6915.
111. Stojanovic MN, Stefanovic D. Deoxyribozyme-based half adder. J Am Chem Soc 2003;125:6673–6676.
112. Stojanovic MN, Stefanovic D. A deoxyribozyme-based molecular automaton. Nature Biotechnology 2003;21:1069–1074.
113. Andrews B. Games, Strategies, and Boolean Formula Manipulation. Master's thesis, University of New Mexico, 2005.
114. Epstein IR, Pojman JA. An Introduction to Nonlinear Chemical Dynamics. New York: Oxford University Press, 1998.
115. Field RJ, Körös E, Noyes R. Oscillations in chemical systems. II. Thorough analysis of temporal oscillation in the bromate-cerium-malonic acid system. J Am Chem Soc 1972;94:8649–8664.
116. Noyes R, Field RJ, Körös E. Oscillations in chemical systems. I. Detailed mechanism in a system showing temporal oscillations. J Am Chem Soc 1972; 94:1394–1395.
117. Tyson JJ. The Belousov-Zhabotinskii Reaction. In: Lecture Notes in Biomathematics, vol. 10. Berlin: Springer-Verlag, 1976.
118. Hjelmfelt A, Ross J. Chemical implementation and thermodynamics of collective neural networks. Proc Natl Acad Sci USA 1992;89:388–391.
119. Hjelmfelt A, Ross J. Pattern recognition, chaos, and multiplicity in neural networks of excitable systems. Proc Natl Acad Sci USA 1994;91:63–67.
120. Hjelmfelt A, Schneider FW, Ross J. Pattern recognition in coupled chemical kinetic systems. Science 1993;260:335–337.
121. Hjelmfelt A, Weinberger ED, Ross J. Chemical implementation of neural networks and Turing machines. Proc Natl Acad Sci USA 1991;88:10,983–10,987.
122. Hjelmfelt A, Weinberger ED, Ross J. Chemical implementation of finite-state machines. Proc Natl Acad Sci USA 1992;89:383–387.

123. Laplante JP, Pemberton M, Hjelmfelt A, Ross J. Experiments on pattern recognition by chemical kinetics. J Phys Chem 1995;99:10,063–10,065.
124. Rössler OE. A principle for chemical multivibration. J Theor Biol 1972;36:413–417.
125. Rössler OE, Seelig FF. A Rashevsky-Turing system as a two-cellular flip-flop. Zeitschrift für Naturforschung 1972;27b:1444–1448.
126. Seelig FF, Rössler OE. Model of a chemical reaction flip-flop with one unique switching input. Zeitschrift für Naturforschung 1972;27b:1441–1444.
127. Szilard L. Über die Entropieverminderung in einem thermodynamischen System bei Eingriffen intelligenter Wesen. Zeitschrift für Physik 1929;53:840–856.
128. Matías MA, Güémez J. On the effects of molecular fluctuations on models of chemical chaos. J Chem Phys 1995;102:1597–1606.
129. Moore C. Unpredictability and undecidability in dynamical systems. Phys Rev Lett 1990;64:2354–2357.
130. Wolfram S. Undecidability and intractability in theoretical physics. Phys Rev Lett 1985;54:735–738.
131. Winfree AT. Spiral waves of chemical activity. Science 1972;175:634–635.
132. Steinbock O, Kettunen P, Showalter K. Anisotropy and spiral organizing centers in patterned excitable media. Science 1995;269:1857–1860.
133. Steinbock O, Kettunen P, Showalter K. Chemical wave logic gates. J Phys Chem 1996;100:18,970–18,975.
134. Steinbock O, Toth A, Showalter K. Navigating complex labyrinths: optimal paths from chemical waves. Science 1995;267:868–871.
135. Yurke B, Turberfield AJ, Mills AP Jr, Neumann JL. A molecular machine made of and powered by DNA. In: The 2000 March Meeting of the American Physical Society, 2000.
136. Magnasco MO. Molecular combustion motors. Phys Rev Lett 1994;72:2656–2659.
137. Magnasco MO. Chemical kinetics is Turing universal. Phys Rev Lett 1997;78:1190–1193.
138. Homsy GE. Performance limits on biochemical computation. Tech. Rep., MIT Artificial Intelligence Laboratory, 2000.
139. Hiratsuka M, Aoki T, Higuchi T. Enzyme transistor circuits for reaction-diffusion computing. IEEE Transactions on Circuits and Systems—I: Fundamental Theory and Applications 1999;46:294–303.
140. Morgan C, Stefanovic D, Moore C, Stojanovic MN. Building the components for a biomolecular computer. In: Ferretti C, Mauri G, Zandron C, eds. Preliminary Proceedings of the 10th International Workshop on DNA-Based Computers, DNA 2004, 2004.
141. Farfel J, Stefanovic D. Towards practical biomolecular computers using microfluidic deoxyribozyme logic gate networks. In: Carbone A, Daley M, Kari L, McQuillan I, Pierce N, eds. Preliminary Proceedings of the 11th International Workshop on DNA-Based Computers, DNA 2005, 221–232.

142. Deaton RJ, Garzon M. Thermodynamic constraints on DNA-based computing. In: Paun G, ed. Computing with Bio-Molecules. Singapore: Springer-Verlag, 1998:138–152.

143. Mauri G, Ferretti C. Word design for molecular computing: a survey. In: Chen J, Reif JH, eds. DNA Computing: 9th International Workshop on DNA-Based Computers, Lecture Notes in Computer Science, vol. 2943. Berlin, Heidelberg: Springer, 2004:37–47.

144. Dirks RM, Lin M, Winfree E, Pierce NA. Paradigms for computational nucleic acid design. Nucl Acids Res 2004;32:1392–1403.

145. Seeman NC. *De novo* design of sequences for nucleic acid structural engineering. J Biomolecular Structure & Dynamics 1990;8:573–581.

146. Feldkamp U, Rauhe H, Banzhaf W. Software tools for DNA sequence design. Genetic Programming and Evolvable Machines 2003;4:153–171.

147. Tanaka F, Kameda A, Yamamoto M, Ohuchi A. Specificity of hybridization between DNA sequences based on free energy. In: Carbone A, Daley M, Kari L, McQuillan I, Pierce N, eds. Preliminary Proceedings of the 11th International Workshop on DNA-Based Computers, DNA, 2005:366–375.

148. Sen D, Gilbert W. Formation of parallel four-stranded complexes by guanine-rich motifs in DNA and its implications for meiosis. Nature 1988;334: 364–366.

149. Mir KU. A restricted genetic alphabet for DNA computing. In: Landweber LF, Baum EB, eds. DNA Based Computers II, DIMACS Workshop 1996, vol. 44 of Series in Discrete Mathematics and Theoretical Computer Science. American Mathematical Society, 1999.

150. Zuker M, Stiegler P. Optimal computer folding of large RNA sequences using thermodynamics and auxiliary information. Nucl Acids Res 1981;9: 133–148.

151. Andronescu M, Dees D, Slaybaugh L, et al. Algorithms for testing that sets of DNA word designs avoid unwanted secondary structure. In: Hagiya M, Ohuchi A, eds. DNA Computing: 8th International Workshop on DNA-Based Computers, Lecture Notes in Computer Science, vol. 2568. Berlin, Heidelberg: Springer, 2003:182–195.

152. Kobayashi S. Testing structure freeness of regular sets of biomolecular sequences (extended abstract). In: Ferretti C, Mauri G, Zandron C, eds. DNA Computing: 10th International Workshop on DNA-Based Computers, Lecture Notes in Computer Science, vol. 3384. Berlin, Heidelberg:Springer, 2005:192–201.

153. Kijima A, Kobayashi S. Efficient algorithm for testing structure freeness of finite set of biomolecular sequences. In: Carbone A, Daley M, Kari L, McQuillan I, Pierce N, eds. Preliminary Proceedings of the 11th International Workshop on DNA-Based Computers, DNA, 2005:278–288.

154. McCaskill JS. The equilibrium partition function and base pair binding probabilities for RNA secondary structure. Biopolymers 1990;29:1105–1119.

155. Dirks RM, Pierce NA. A partition function algorithm for nucleic acid secondary structure including pseudoknots. Journal of Computational Chemistry 2003;24:1664–1677. NUPACK is available at http://www.acm.caltech.edu/~niles/software.html.

156. Marathe A, Condon AE, Corn RM. On combinatorial DNA word design. Journal of Computational Biology 2001;8:201–220.

157. Leupold P. Partial words for DNA coding. In: Ferretti C, Mauri G, Zandron C, eds. Preliminary Proceedings of the 10th International Workshop on DNA-Based Computers, DNA, 2004.

158. Garzon M, Deaton RJ, Rose JA, Lu L, Franceschetti DR. Soft molecular computing. In: Winfree E, Gifford DK, eds. DNA Based Computers V, DIMACS Workshop 1999, vol. 54 of Series in Discrete Mathematics and Theoretical Computer Science. American Mathematical Society, 2000:91–100. EdnaCo is available at http://zorro.cs.memphis.edu/~cswebadm/csweb/research/pages/bmc/ or http://engronline.ee.memphis.edu/molec/demos.htm.

159. Penchovsky R, Ackermann J. DNA library design for molecular computation. Journal of Computational Biology 2003;10:215–229.

160. D'yachkov AG, Macula AJ, Pogozelski WK, Renz TE, Rykov VV, Torney DC. A weighted insertion-deletion stacked pair thermodynamic metric. In: Ferretti C, Mauri G, Zandron C, eds. DNA Computing: 10th International Workshop on DNA-Based Computers, Lecture Notes in Computer Science, vol. 3384. Berlin, Heidelberg: Springer, 2005:90–103. Syn-DCode is available at http://cluster.ds.geneseo.edu:8080/ParallelDNA/.

161. Dimitrov RA, Zuker M. Prediction of hybridization and melting for double-stranded nucleic acids. Biophysical Journal 2004;87:215–226.

162. Rose JA, Deaton RJ, Franceschetti DR, Garzon M, Stevens SE Jr. A statistical mechanical treatment of error in the annealing biostep of DNA computation. In: Special Program in GECCO-99. 1999:1829–1834.

163. Rose JA, Deaton RJ. The fidelity of annealing-ligation: a theoretical analysis. In: Condon A, Rozenberg G, eds., DNA Computing: 6th International Workshop on DNA-Based Computers, DNA 2000, Lecture Notes in Computer Science, vol. 2054. Springer, 2001.

164. Rose JA, Deaton RJ, Hagiya M, Suyama A. The fidelity of the tag-antitag system. In: Jonoska N, Seeman NC, eds. DNA Computing: 7th International Workshop on DNA-Based Computers, DNA 2001, Lecture Notes in Computer Science, vol. 2340. Berlin, Heidelberg: Springer, 2002.

165. Rose JA, Deaton RJ, Hagiya M, Suyama A. An equilibrium analysis of the efficiency of an autonomous molecular computer. Physical Review E 2002;65.

166. Rose JA, Hagiya M, Suyama A. The fidelity of the tag-antitag system 2: identifying the regime of stringency. In: Sarker R, Reynolds R, Abbass H, et al., eds, Proceedings of the 2003 Congress on Evolutionary Computation CEC2003. New Jersey: IEEE Press, 2003:2740–2747. NucleicPark is available at http://hagi.is.s.u-tokyo.ac.jp/johnrose/ and http://engronline.ee.memphis. edu/molec/demos.htm.

167. Rose JA, Deaton RJ, Franceschetti DR, Garzon M, Stevens SE Jr. Hybridization error for DNA mixtures of N species, 1999. http://engronline.ee.memphis.edu/molec/Misc/ci.pdf.

168. Rose JA, Suyama A. Physical modeling of biomolecular computers: models, limitations, and experimental validation. Natural Computing 2004;3:411–426.

169. SantaLucia J Jr, Hicks D. The thermodynamics of DNA structural motifs. Annual Review of Biophysics Biomolecular Structure 2004;33:415–440.

170. Hartemink AJ, Gifford DK. Thermodynamic simulation of deoxyoligonu-cleotide hybridization for DNA computation. In: Rubin H, Wood DH, eds. Preliminary Proceedings of DNA Based Computers III, DIMACS Workshop, 1997:15–25.

171. Hartemink AJ, Gifford DK, Khodor J. Automated constraint-based nucleotide sequence selection for DNA computation. In: Kari L, Rubin H, Wood DH, eds. DNA Based Computers IV, DIMACS Workshop 1998, Biosystems, vol. 52, issues 1–3. Elsevier, 1999:227–235.

172. Nishikawa A, Yamamura M, Hagiya M. DNA computation simulator based on abstract bases. Soft Computing 2001;5:25–38.

173. Mathews DH, Turner DH. Dynalign: An algorithm for finding the secondary structure common to two RNA sequences. J Mol Biol 2002;317:191–203.

174. Zuker M. Mfold web server for nucleic acid folding and hybridization prediction. Nucl Acids Res 2003;31:3406–3415. Mfold is available at http://www.bioinfo.rpi.edu/applications/mfold.

175. Dirks RM, Pierce NA. An algorithm for computing nucleic acid base-pairing probabilities including pseudoknots. Journal of Computational Chemistry 2004;25:1295–1304.

176. Andronescu M, Aguirre-Hernandez R, Condon A, Hoos HH. RNAsoft: a suite of RNA secondary structure prediction and design software tools. Nucl Acids Res 2003;31:3416–3422. RNAsoft is available at http://www.rnasoft.ca/.

177. Mathews DH, Disney MD, Childs JL, Schroeder SJ, Zucker M, Turner DH. Incorporating chemical modification constraints into a dynamic program-ming algorithm for prediction of RNA secondary structure. Proc Natl Acad Sci USA 2004;101:7287–7292. The free energy nearest neighbor parameters are available at http://rna.chem.rochester.edu/, RNAstructure is available at http://128.151.176.70/RNAstructure.html.

178. Hofacker IL. Vienna RNA secondary structure server. Nucl Acids Res 2003;31:3429–3431. Vienna Package is available at http://www.tbi.univie.ac.at/~ivo/RNA/.

179. Peyret N, Saro P, SantaLucia J Jr. HyTher server. HyTher Version 1.0 is avail-able at http: //ozone2.chem.wayne.edu/.

180. SantaLucia J Jr. A unified view of polymer, dumbbell, and oligonucleotide DNA nearest-neighbor thermodynamics. Proc Natl Acad Sci USA 1998;95:1460–1465.

181. Peyret N, Seneviratne PA, Allawi HT, SantaLucia J Jr. Nearest-neighbor ther-modynamics and NMR of DNA sequences with internal A-A, C-C, G-G, and T-T mismatches. Biochemistry 1999;38:3468–3477.

182. Novère NL. MELTING, computing the melting temperature of nucleic acid duplex. Bioinformatics 2001;17:1226–1227. Melting is available at http://www.ebi.ac.uk/~lenov/meltinghome.html.

183. Blake RD, Bizzaro JW, Blake JD, et al. Statistical mechanical simulation of polymeric DNA melting with MELTSIM. Bioinformatics 1999;15: 370–375.

184. McDowell JA. MeltWin. MeltWin is available at http://www.meltwin.com/.

185. Flamm C, Fontana W, Hofacker IL, Schuster P. RNA folding at elementary step resolution. RNA 2000;6:325–338. Kinfold is available at http://www.tbi.univie.ac.at/~xtof/RNA/Kinfold/.

186. Visual OMP (Oligonucleotide Modeling Platform), DNA Software, Inc. Visual OMP is available at http://www.dnasoftware.com.
187. The DNA and Natural Algorithms Group. DNA design toolbox. DNA Design Toolbox is available at http://www.dna.caltech.edu/DNAdesign/.
188. Kim D, Shin SY, Lee IH, Zhang BT. NACST/Seq: A sequence design system with multiobjective optimization. In: Hagiya M, Ohuchi A, eds. DNA Computing: 8th International Workshop on DNA-Based Computers, DNA 2002, Lecture Notes in Computer Science, vol. 2568. Berlin: Springer, 2003:242–251.
189. Ruben AJ, Freeland SJ, Landweber LF. PUNCH: An evolutionary algorithm for optimizing bit set selection. In: Jonoska N, Seeman NC, eds. DNA Computing: 7th International Workshop on DNA-Based Computers, DNA 2001, Lecture Notes in Computer Science, vol. 2340. Berlin: Springer, 2002:150–160.
190. Bishop M, Macula AJ, Pogozelski WK, Renz TE, Rykov VV. SynDCode: Cooperative DNA code generating software. In: Carbone A, Daley M, Kari L, McQuillan I, Pierce N, eds. Preliminary Proceedings of the 11th International Workshop on DNA-Based Computers, DNA, 2005:391.
191. Pogozelski WK, Bernard MP, Priore SF, Macula AJ. Experimental validation of DNA sequences for DNA computing: Use of a SYBR green assay. In: Carbone A, Daley M, Kari L, McQuillan I, Pierce N, eds. Preliminary Proceedings of the 11th International Workshop on DNA-Based Computers, DNA, 2005:322–331.
192. Yin P, Guo B, Belmore C, et al. Tilesoft: Sequence optimization software for designing DNA secondary structures, 2004. http://www.cs.duke.edu/~reif/paper/peng/TileSoft/TileSoft.pdf.
193. Kari L, Kitto R, Thierrin G. Codes, involutions and DNA encodings. In: Lecture Notes in Computer Science, vol. 2300. Berlin, Heidelberg: Springer, 2002:376.
194. Blain DR, Garzon M, Shin SY, et al. Development, evaluation and benchmarking of simulation software for biomolecule-based computing. Natural Computing 2004;3:427–442.
195. Arita M, Nishikawa A, Hagiya M, Komiya K, Gouzu H, Sakamoto K. Improving sequence design for DNA computing. Proceedings of the Genetic and Evolutionary Computation Conference (GECCO 2000), 2000:875–882.
196. Hoos HH, Stutzle T. Stochastic Local Search: Foundations and Applications. Morgan Kaufmann, 2004.
197. Tulpan DC, Hoos HH, Condon A. Stochastic local search algorithms for DNA word design. In: DNA Computing: 8th International Workshop on DNA-Based Computers, DNA 2002, Lecture Notes in Computer Science, vol. 2568. Berlin: Springer, 2003:229–241.
198. Tulpan DC, Hoos HH. Hybrid randomised neighbourhoods improve stochastic local search for DNA code design. In: Canadian Conference on AI 2003, Lecture Notes in Computer Science, vol. 2671. Berlin: Springer-Verlag, 2003:418–433.
199. Tulpan D, Andronescu M, Change SB, et al. Thermodynamically based DNA strand design. Nucl Acids Res 2005;33:4951–4964.
200. Holland JH. Adaptation in Natural and Artificial Systems: An Introductory Analysis with Applications to Biology, Control and Artificial Intelligence. Cambridge, MA: MIT Press, 1992.

201. Rechenberg I. Evolutionsstrategie—Optimierung technischer Systeme nach Prinzipien der biologischen Information. Freiburg, Germany: Fromman Verlag, 1973.

202. Schwefel HP. Numerical Optimization of Computer Models. New York: John Wiley & Sons, Inc., 1981.

203. Fogel LJ, Owens AJ, Walsh MJ. Artifical Intelligence Through Simulated Evolution. New York: John Wiley & Sons, 1966.

204. Deaton RJ, Murphy RC, Garzon M, Franceschetti DR, Stevens SE Jr. Genetic search of reliable encodings for DNA-based computation. In: First Genetic Programming Conference. Stanford University, 1996.

205. Deaton RJ, Rose JA. Simulations of statistical mechanical estimates of hybridization error. In: Condon A, Rozenberg G, eds. Preliminary Proceedings of the 6th International Workshop on DNA-Based Computers, DNA, 2000:251–252.

206. Shin SY, Kim DM, Lee IH, Zhang BT. Evolutionary sequence generation for reliable DNA computing. In: Proceedings of the 2002 Congress on Evolutionary Computation (CEC2002), vol. 1. 2002:79–84.

207. Shin SY, Kim DM, Lee IH, Zhang BT. Multiobjective evolutionary algorithms to design error-preventing dna sequences. Tech. Rep. BI-02-003, Biointelligence Lab (BI), School of Computer Science & Engineering, Seoul National University, 2002.

208. Smith WD. DNA computers in vitro and vivo. In: Lipton RJ, Baum EB, eds. DNA Based Computers, DIMACS Workshop 1995, vol. 27 of Series in Discrete Mathematics and Theoretical Computer Science. American Mathematical Society, 1995:121–185.

209. Tsaftaris SA, Katsaggelos AK, Pappas TN, Papoutsakis ET. DNA-based matching of digital signals. In: International Conference on Acoustics, Speech, and Signal Processing, vol. 5. Montreal, Quebec, Canada, 2004.

210. Tsaftaris SA, Katsaggelos AK, Pappas TN, Papoutsakis ET. How can DNA computing be applied to digital signal processing? IEEE Signal Processing Magazine 2004;21.

211. Benenson Y, Paz-Elizur T, Adar R, Keinan E, Livneh Z, Shapiro E. Programmable and autonomous computing machine made of biomolecules. Nature 2001;414:430–434.

212. Benenson Y, Adar R, Paz-Elizur T, Livneh Z, Shapiro E. DNA molecule provides a computing machine with both data and fuel. Proc Natl Acad Sci USA 2003;100:2191–2196.

213. Stojanovic MN, Stefanovic D, LaBean T, Yan H. Computing with nucleic acids. In: Willner I, Katz E, eds. Bioelectronics: From Theory to Applications. Wiley-VCH, 2005.

214. Calude CS, Paun G. Computing with Cells and Atoms. London: Taylor & Francis, 2001.

215. Ehrenfeucht A, Harju T, Petre I, Prescott DM, Rozenberg G. Computation in Living Cells. Berlin: Springer-Verlag, 2004.

216. Paun G. Computing with Bio-Molecules. Singapore: Springer-Verlag, 1998.

217. Rothemund PWK. Folding DNA to create nanoscale shapes and patterns. Nature 2006;440:297–302.

218. Baron R, Lioubashevski O, Katz E, Niazov T, Willner I. Elementary arithmetic operations by enzymes: A model for metabolic pathway based computing. Angewandte Chemie International Edition 2006;45:1572–1576.
219. Pistol C, Lebeck AR, Dwyer C. Design automation for DNA self-assembled nanostructures. In: Design Automation Conference (DAC), 2006.
220. Patwardhan J, Johri V, Dwyer C, Lebeck AR. A defect tolerant self-organizing nanoscale simd architecture. In: Proceedings of the Twelth International Conference on Architectural Support for Programming Languages and Operating Systems (ASPLOS XII), 2006.
221. Lipton RJ, Baum EB. DNA Based Computers, DIMACS Workshop 1995, vol. 27 of Series in Discrete Mathematics and Theoretical Computer Science. American Mathematical Society, 1996.
222. Rubin H, Wood DH. DNA Based Computers III, DIMACS Workshop 1997, vol. 48 of Series in Discrete Mathematics and Theoretical Computer Science. American Mathematical Society, 1999.
223. Jonoska N, Seeman NC. DNA Computing: 7th International Workshop on DNA-Based Computers, DNA 2001, Lecture Notes in Computer Science vol. 2340. Berlin, Heidelberg: Springer, 2002.
224. Landweber LF, Baum EB. DNA Based Computers II, DIMACS Workshop 1996, vol. 44 of Series in Discrete Mathematics and Theoretical Computer Science. American Mathematical Society, 1999.
225. Winfree E, Gifford DK. DNA Based Computers V, DIMACS Workshop 1999, vol. 54 of Series in Discrete Mathematics and Theoretical Computer Science. American Mathematical Society, 2000.
226. Kari L, Rubin H, Wood DH. DNA Based Computers IV, DIMACS Workshop 1998, Biosystems, vol. 52, issues 1–3. Elsevier, 1999.
227. Hagiya M, Ohuchi A. DNA Computing: 8th International Workshop on DNA-Based Computers, DNA 2002, Lecture Notes in Computer Science, vol. 2568. Berlin, Heidelberg: Springer, 2003.
228. Ferretti C, Mauri G, Zandron C. DNA Computing: 10th International Workshop on DNA-Based Computers, DNA 2004, Lecture Notes in Computer Science, vol. 3384. Berlin, Heidelberg: Springer, 2005.
229. Chen J, Reif JH. DNA Computing: 9th International Workshop on DNA-Based Computers, DNA 2003, Lecture Notes in Computer Science, vol. 2943. Berlin, Heidelberg: Springer, 2004.
230. Ferretti C, Mauri G, Zandron C, eds. Preliminary Proceedings of the 10th International Workshop on DNA-Based Computers, DNA 2004.
231. Carbone A, Daley M, Kari L, McQuillan I, Pierce N. Preliminary Proceedings of the 11th International Workshop on DNA-Based Computers, DNA 2005.
232. Peyret N. Prediction of Nucleic Acid Hybridization: Parameters and Algorithms. Ph.D. thesis, Wayne State University, Dept. of Chemistry, 2000.

11

Biomolecular Automata

Nataša Jonoska

Summary

   As an emerging new research area, DNA nanoengineering and computation extends into other fields such as nanotechnology and material design, and is developing into a new subdiscipline of science and engineering. This chapter provides a brief overview of the design and development of computational devices by DNA. In particular, two approaches for biomolecular models of automata are described. The first model is based on using the action of restriction endonucleases and the second is based on DNA self-assembly and DNA nanomolecular devices.

Key Words: DNA computing; biomolecular computing; DNA automata; DNA devices; DNA transducers.

1. INTRODUCTION

   Half a century after the discovery of the structure of DNA, extraordinary advances in genetics and biotechnology are being made on a daily basis. At the same time, we are witnessing research developments that employ DNA in a completely new way, treating this molecule of life as a nanomaterial for computation. The first conception of using DNA for computation ripened in the mid 1980s with the first theoretical model of splicing systems introduced by Head *(1)*. These ideas came to full fruition with Adleman's seminal experiment *(2)*, which solved a small instant of a combinatorial problem using solely DNA molecules and biomolecular laboratory techniques. The impact of these first ideas on many researchers can be observed by the numerous theoretical results and innovative experimental solutions that followed. Results of these studies are considered rather significant in that much of the research has been published by leading scientific journals such as *Science* and *Nature*. The research has spurred new scientific interactions and opened connections between mathematics and computer sciences from one side, and molecular biology, nanotechnology, and biotechnology from the

From: *NanoBioTechonology: BioInspired Devices and Materials of the Future*
Edited by: Oded Shoseyov and Ilan Levy © Humana Press Inc., Totowa, NJ

other. We mention some developments, both experimental and theoretical, that use DNA to obtain three-dimensional (3D) nanostructures, for computation and as a material for nanodevices.

1.1. Nanostructures

The inherently informational character of DNA (as a sequence of nucleotides) and the Watson-Crick complementarity of the bases make it an attractive molecule for use in applications that entail targeted assembly. Genetic engineers have used the specificity of sticky-ended cohesion, guided by the complementarity of the bases, to direct the construction of plasmids and other vectors. Naturally occurring DNA is a linear molecule in the sense that its helix axis can be considered a 1D curve in space. Linear DNA molecules are not well suited to serve as components of complex nanomaterials, but it has proven to not be diffcult to construct DNA molecules with stable branch points *(3)*. Synthetic molecules have been designed and shown to assemble into branched species *(4,5)*, and more complex motifs that entail the lateral fusion of DNA double helices *(6)*, such as DNA double crossover (DX) molecules *(7)*, triple crossover (TX) molecules *(8)*, or paranemic crossover (PX) molecules. DX and TX molecules have been used as tiles and building blocks for large nanoscale arrays *(9,10)*. In addition, 3D structures such as a cube *(11)*, a truncated octahedron *(12)*, and arbitrary graphs *(13,14)* have been constructed from DNA duplex and junction molecules. More recently, DX and PX have been employed in the construction of an octahedron *(15)*.

1.2. Computation

Theoretically, it has been shown that 2D arrays can simulate the dynamics of a bounded 1D cellular automaton and so are capable of potentially performing computations as a Universal Turing Machine *(9)*. Several successful experiments performing computation have been reported, most notably the initial successful experiment by Adleman *(2)* and the recent one from the same group solving an instance of satisfiability (SAT) with 20 variables *(16,17)*. Successful experiments with confirmed computation such as the binary addition (simulation of exclusive disjunction [XOR]) using TX molecules (tiles) have been reported in *(18)*. In ref. *19*, a 9-bit instance of the "knight problem" has been solved using RNA and in ref. *20*, a small instance of the maximal clique problem has been solved using plasmids. Theoretically, it has been shown that by self-assembly of 3D graph structures, many hard computational problems can be solved in one (constant) biostep operation *(13,21)*.

1.3. Nanodevices

Based on the B-Z transition of DNA, a nanomechanical device was introduced in ref. *22*. Soon after, "DNA fuel" strands were used to produce devices whose activity is controlled by DNA strands *(23,24)*. The PX-JX$_2$ device introduced in ref. *23* has two distinct structural states, differing by a half-rotation; each state is obtained by addition of a pair of DNA strands that hybridizes with the device such that the molecule is in either the JX$_2$ state or in the PX state. Some of these devices are described under Subheading 4.2. Most recently, these devices have been utilized in the construction and assembly of autonomous nanorobots. A theoretical model of a "walker" and a "crawler" was introduced in ref. *25*, but the first real implementation came from the Seeman's laboratory *(26)*, followed by similar designs *(27,28)*.

The research in biomolecular computing has already taken many different paths (theoretical and experimental), such that it is diffcult to cover all of those aspects. This chapter concentrates on DNA implementation of finite state automata, which, in the author's opinion, have brought about significant advancement in the field. The reader is advised to consult the Proceedings of the Annual Meetings on DNA-Based Computers, currently in its 11th year, where many of the researchers present their results (*see* refs. *29–38*).

This chapter is based mostly on the experimental and theoretical results in *(21,24,39–47)*. In these highly interdisciplinary studies, theory and experiments are tightly interlaced and it is assumed that the reader is familiar with basic biomolecular techniques. For a reference, the first section gives a brief introduction to the notions used in the following sections. It is also assumed that the reader is familiar with mathematical writing as well as with the basic ideas of theoretical computer science, which are used in describing the theoretical background.

The chapter starts with a rather brief introduction to some notions from molecular biology and biotechnology. A short description of DNA molecules used in successful construction of nanostructures and nanomechanical devices is given under Subheading 2.2. The main ideas exploited with the first experimental success are described under Subheading 2.3. As finite-state automata are the main subject of the chapter, the theoretical model concerning finite state automata is described under Subheading 3, and Subheading 4 deals with the implementation. A description of an autonomous finite state machine that employs a restriction enzyme coupled with clever encoding that "reads" the input and recognizes, "accepts" certain encoded strings is covered under Subheading 4.1. The same idea is suggested for developing a more powerful computational device, the so-called push-down automaton. We also briefiy describe the potential of such automata in genetics and medicine. The section on nanodevices, Subheading 4.3, describes three very significant experimental results. The first one is a DNA-based mechanical device that uses DNA "fuel" strands to change from one state to another. The second is a DNA-based "switch" that uses the idea of fuel strands to switch between two positions. This device is proposed as a base for developing a finite-state automaton with programmable input. These ideas as well as the experimental results are described under Subheading 4.2.

The chapter ends with a few concluding remarks.

2. BIOTECHNOLOGY USED: BRIEF OVERVIEW

Briefly, we recall the basic DNA structure and the actions of several types of enzymes. A more thorough and not very technical description of the structure of DNA and the operations performed by enzymes can be found in refs. *48,49* (*see* also ref. *50*). Detailed laboratory protocols can be found in ref. *51*.

Information in a DNA molecule is stored in a sequence of nucleotides, also called by their chemical group, the bases A, G, C, T *(adenine, guanine, cytosine, and thymine)*, joined together by phosphodiester bonds. In the case of RNA, the thymine is substituted with *uracil* (U). A single strand of DNA, i.e., a chain of nucleotides, has also a "beginning" (usually denoted by 5′) and an "end" (denoted by 3′), and so the molecule is oriented. A chain of nucleotides is called *oligonucleotide* or simply just *oligo*. By the well-known Watson-Crick complementarity, A is complementary to T and C is complementary to G. A double-stranded DNA is formed by establishing hydrogen bonds between the complementary bases of two single-stranded molecules that have opposite orientation. This process is usually called *hybridization* or *annealing*.

2.1. Enzymes and DNA Operations

Polymerase (used in the operation of "amplifying" or "detecting"): DNA polymerases are enzymes that synthesize DNA. With these enzymes, a DNA strand can be duplicated or extended. One of the commonly used protocols in molecular biology is the so-called *polymerase chain reaction* (PCR). This reaction detects certain DNA sequences and synthesizes a large number of such molecules from an existing pool of molecules. This method is used to detect (extract) a certain sequence within a large mix of molecules or to amplify it.

Restriction enzymes (endonucleases) (used in the operation of "separating" or "cutting"): such an enzyme recognizes a specific sequence of nucleotides in a double-stranded DNA molecule and cuts the molecule into two pieces by destroying the phosphodiester bonds at specific places in the two strands. Different enzymes recognize different sequences of nucleotides, and even if they recognize the same sequence, they may cut the molecule in a different way.

Ligase (used in the operation of "gluing"): It is said that a double-stranded DNA molecule has a *nick* if the phosphodiester bond between two consecutive nucleotides within one of the strands is broken. A ligase is an enzyme that closes the nicks, i.e., recovers the broken phosphodiester bonds in a double-stranded DNA.

Length (weight) selection (used in the operation of "separating" or "detecting"): a technique called "gel electrophoresis" separates DNA molecules by their weight. The DNA is negatively charged and after being placed in a small well of a gel in an active electric field, it slowly moves toward the positive side. Larger (heavier) molecules move slower and smaller molecules move faster. The portion of the gel that contains molecules with the desired length can be cut out of the gel, DNA-purified, and then used in subsequent experiments.

2.2. DNA Molecules Used in Nanostructures

In nature, DNA appears as a linear double-stranded molecule (in eukaryotes), but it can also be in a circular form (mostly in viruses and bacteria, prokaryotes). Circular DNA can also be obtained in a laboratory by joining (ligating) the ends of a linear DNA. Such molecules are used in several models of DNA-based computers (*see*, for example, ref. *49*, Chapter 9). Circular molecules have been used as building blocks for DNA knots. In theory, virtually any knot can be constructed using right-handed B-DNA for negative crossings and left-handed Z-DNA for positive crossings (*52–55*). Many catenas and linkages of DNA molecules are known, but just recently, the first Borromean DNA rings were assembled using B- and Z-DNA forms (*47*).

In DNA-based computing, the complementarity of the nucleotides (A $\leftrightarrow$ T, G $\leftrightarrow$ C) is one of the basic properties used for encoding information as a tool for computation, as well as for obtaining DNA nanostructures. Besides the linear duplex DNA, there are two additional DNA building blocks frequently used, junction molecules and DNA prototiles made of DX or TX molecules.

Junction molecules are fairly well understood. These molecules have been used in the construction of DNA polyhedra, a quadrilateral, a truncated octahedron, and a cube (*11,12,54*). The *k*-armed branched molecules (*k* is a natural number ≥2) seem to be suitable for graph construction. An example of a four-armed branched molecule is presented in Fig. 1 to the left. In this figure, the double helix of the molecule is not presented. Hydrogen bonds between the anti-parallel, complementary Watson-Crick bonds are depicted as dotted segments between the strands. Polarity of the DNA strands is indicated with arrowheads being placed at the 3' end. The angles between the "arms" are known to be flexible. If we allow each "arm" to be over 200 or 300 base pairs (bp) long, then the "arms" of this molecule become rather

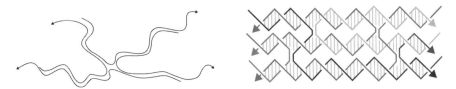

Fig. 1. A four-armed branched junction molecule, and a DNA tile made of a triple crossover molecule *(61)*.

fiexible and we deliberately show them curved. On the other hand, short arms of two or three helical turns (one helical turn employs about 10.5 bases) would provide a quite rigid structure. Such rigid structures show high potential for constructing 3D crystals. Various 2D arrays made of rigid structures have been reported (e.g., ref. *56*). The 3′ ends can be extended such that each arm ends with a single-stranded portion. This single-stranded part, also called "sticky end," can anneal to its Watson-Crick complement once placed in a test tube. Construction and properties of these molecules are fairly well understood *(3,45,57)* such that their potential for utilizing them in more complex structures is becoming rather feasible. Recently, general nonregular graphs have been successfully constructed by using junction molecules for the vertices and duplex molecules for the edges *(58–60)*.

Another very important step toward constructing 3D DNA crystals was made by designing and assembling a 2D array made of DNA tiles *(10)*. The construction of these arrays was enabled by the use of DX and TX molecules that act as tiles. These molecules are double or triple duplex molecules (two or three double helices) such that DNA strands interchange between different helices. An example of a TX molecule is presented in Fig. 1 to the right. The 3′ ends are indicated with an arrow, and they may be extended to be used as sticky ends such that connecting TX molecules in a 2D array is possible *(8)*. These DNA tiles made of DX and TX molecules were initially designed in Seeman's laboratory at New York University and now are used by several groups in Caltech, Duke, University of Southern California, and others.

2.3. The First Experiments

In his seminal experiment, Adleman *(2)* solved a small instant of a combinatorial problem known as the Hamiltonian Path Problem (HPP) for a directed graph. A graph is a structure made of points called *vertices* (one can think of them as cities or towns) and arrows called *edges* connecting the vertices (one can think of the edges as one-way roads connecting the towns). The HPP asks whether for a given graph G there is a path from one vertex (town), denoted v_{in}, to another vertex, denoted v_{out}, that visits every other vertex

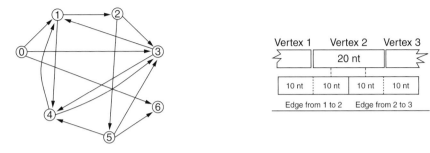

Fig. 2. Schematic representation of Adelman's experiment *(2)*. Set-up of the Hamiltonian Path Problem.

exactly once. If such a path exists, then it is called Hamiltonian. The graph used in the experiment is depicted in Fig. 2 to the left. The vertex labeled 0 represented v_{in} and the one labeled 6 represented v_{out}. A Hamiltonian path follows the following vertex order: $0 \rightarrow 3 \rightarrow 4 \rightarrow 1 \rightarrow 2 \rightarrow 5 \rightarrow 5$.

In Adleman's experiment, the edges of G are represented by single-stranded DNA oligos made of 20 (randomly chosen) nucleotides. The vertices in the graph are also oligos of 20 nucleotides having the first 10 nucleotides complementary to the last 10 of the incoming edge, and the other 10 being complementary to the first 10 nucleotides of the outgoing edge (*see* Fig. 2 to the right).

A path $e_1 \ldots e_k$ of length k in G is represented by a double-stranded DNA molecule of length 20 Kbp with 10 nucleotides (single-stranded) overhanging from each end. So, if a Hamiltonian path exists, then it must be represented by a double-stranded molecule of length 20 n, where n is the number of vertices. The experiment that solved the problem had the following key elements:

1. Encode the information about the graph into DNA strands.
2. Use self-assembly of the molecules led by Watson-Crick complementarity to generate a large library of strands that encode paths in the graph.
3. Use known biomolecular techniques (ligation, gel electrophoresis, PCR, affinity separation) to extract the right solution, i.e., to extract the molecules that represent paths from v_{in} to v_{out} and visit every other vertex exactly once.
4. Provide a way (gel electrophoresis) to read out the output.

The problem that was chosen for this experiment was a well known non-deterministic polynomial-time (NP)-complete problem that is generally "intractable" in the sense that, for a relatively modest size of graph, with any known algorithm, an impractical amount of computer time is needed for its solution. Adleman's approach to this problem is not much different from a brute-force search, but the way in which it was encoded and solved, and the use of *massive parallelism, self-assembly,* and *nondeterminism* made it very novel, inspiring, and a basis for innovative ideas, both experimental and theoretical.

Several theoretical models based on the lab protocols used in Adleman's experiment can be found in the literature (*see*, for example, refs. *16,49,42*). They all use more or less the same set of operations: *merge, separate, detect, amplify, etc.*, and they all can feasibly be executed by a robotic system. Lipton *(62)* showed that using these operations, the satisfiability problem for propositional formulas could be solved and consequently, a large set of problems could be solved by DNA. In ref. *63*, it is shown how these operations can be used to break the Data Encryption Standard (DES). Approximately 1 g of DNA is needed and using robotic arms (assuming each operation to last 1 min), breaking the DES is estimated to take 5 d. The most significant point in the analysis for breaking the DES is that success is quite likely, even with the at this point unavoidable number of errors within the lab protocols.

The big drawback in Adleman's and Lipton's approach is the need for a very large pool of initial molecules that have to be generated in order to ensure the correct solution to the problem. For a larger graph, say a modest size of 200 vertices, one needs "DNA more than the weight of the Earth" (*see* ref. *64*). Subsequent studies have concentrated on developing algorithms such that not necessarily all of the potential solutions are constructed at once (for example, refs. *65,66*). However, scaling up the models to larger and practically significant problems remains one of the problems in using DNA for computation. Other approaches, not necessarily solving large search problems, but constructing nanodevices, turn out to be more feasible and with attractive potentials. Simulating finite-state automata is one such approach, described in the following.

3. FINITE STATE MACHINES—THEORETICAL MODELS

3.1. Automata with Finite Memory

Finite-state automata as well as Turing machines are one of the basic concepts of computing theory and as such, have been used as a basis for designing various computational biomolecular models. We chose to concentrate on the models that are not just theoretical, but have, at least initially, successful experimental components, bringing them closer to reality.

In this section, we define finite-state automata (with or without) output. Because these automata, besides their states, do not use additional memory for computation (i.e., an additional tape), they are also known as automata with finite memory. Successful experimental results with completely different approaches have been reported for both types of automata, with or without output. We briefly recall the definition of a transducer, finite-state machine with output. This notion is well known in automaton theory and an introduction to transducers (Mealy machines) can be found in ref. *67*. The notion of finite-state automata without output is essentially the same, except, the automaton does not produce an output at the end of computation.

A finite-state machine with output or *a transducer* $\mathcal{T} = (\Sigma, \Sigma', Q, \delta, s_0, F)$ is a construct specified by six elements: Σ and Σ' are finite alphabets, Q is a finite set of states, δ is the transition function, $s_0 \in Q$ is the initial state, and $F \subseteq Q$ is the set of final or terminal states. The alphabet Σ is the input alphabet and the alphabet Σ' is the output alphabet. We denote with Σ^* the set of all words over the alphabet Σ. This includes the word with "no symbols," the empty word denoted with λ. For a word $w = a_1 \ldots a_k$ where a_i is a symbol in Σ, the length of w, denoted by $|w|$, is k. For the empty word λ, we have $|\lambda| = 0$.

The transition operation δ consists of four-tuples (q,a,a',q'), where q,q' are states from Q and a,a' are symbols from Σ and Σ', respectively. The transitions form a subset of $Q \times \Sigma \times \Sigma' \times Q$. An element (q,a,a',q') of δ is also denoted with $(q,a) \xrightarrow{\delta} (a',q')$, meaning that when $\mathcal{T}$ is in state q and scans input symbol a, then $\mathcal{T}$ changes into state q' and gives output symbol a'. In the case of *deterministic* transducers, δ is a function $\delta: Q \times \Sigma \to \Sigma' \times Q$, i.e., at a given state reading a given input symbol, there is a unique output state and an output symbol. Usually, the states of the transducer are presented as vertices of a graph and the transitions defined with δ are presented as directed edges with input/output symbols as labels. If there is no edge from a vertex q in the graph that has input label a, we assume that there is an additional "junk" state $\bar{q}$ where all such transitions end. This state is usually omitted from the graph because it is not essential for the computation. The transducer is said to recognize a string (or a word) w over alphabet Σ if there is a path in the graph from the initial state s_0 to a terminal state in F with input label w. The set of all words recognized by a transducer $\mathcal{T}$ is denoted with $L(\mathcal{T})$ and is called a *language* recognized by $\mathcal{T}$. It is well known that finite-state transducers recognize the class of regular languages.

We concentrate on deterministic transducers. In this case, the transition function δ maps the input word $w \in \Sigma^*$ to a word $w' \in (\Sigma')^*$. So the transducer $\mathcal{T}$ can be considered to be a function from $L(\mathcal{T})$ to $(\Sigma')^*$, i.e., $\mathcal{T}: L(\mathcal{T}) \to (\Sigma')^*$.

Examples:

1. The transducer $\mathcal{T}_1$ presented in Fig. 3A has initial and terminal state s_0. The input alphabet is $\{0,1\}$ and the output alphabet is $\Sigma' = \varnothing$, i.e., it is a finite-state automaton without output. It recognizes the set of binary strings that represent numbers divisible by 3. The states s_0, s_1, s_2 represent the remainders 0, 1 and 2, respectively, of the division of the input string by 3. For example, string 110 in binary represents the number 6 which is divisible by 3. A path that starts at s_0 labeled by 110 ends at state s_0 as well. On the other hand, the string 1010 in binary represents the number 10, which has a remainder 1 after dividing by 3. The corresponding path that starts at s_0 ends at state s_1.
2. The transducer $\mathcal{T}_2$ presented in Fig. 3B is essentially the same as $\mathcal{T}_1$ except that now the output alphabet is also $\{0,1\}$. The output in this case is the result of the division of the binary string with three. On input 10101 (21 in decimal) the transducer gives the output 00111 (7 in decimal).

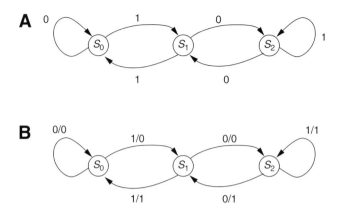

Fig. 3. (A) Finite-state machine that accepts binary strings that are divisible by 3. The machine in **(B)** outputs the result of dividing the binary string with 3 in binary.

3. Our next example refers to encoders. As a result of manufacturing constraints of magnetic storage devices, the binary data cannot be stored verbatim on the disk drive. One method of storing binary data on a disk drive is by using the modified frequency modulation (MFM) scheme currently used on many disk drives. The MFM scheme inserts a 0 between each pair of data bits, unless both data bits are 0, in which case it inserts a 1. The finite-state machine, transducer, which provides this simple scheme is represented in Fig. 4A. In this case the output alphabet is $\Sigma' = \{00,01,10\}$. If we consider rewriting of the symbols with $00 \to \alpha$, $01 \to \beta$ and $10 \to \gamma$ we have the transducer in Fig. 4B.

4. A transducer that performs binary addition is presented in Fig. 4C. The input alphabet is $\Sigma = \{00,01,10,11\}$ representing a pair of digits to be added, i.e., if $x = x_1 \dots x_k$ and $y = y_1 \dots y_k$ are two numbers written in binary $(x_i,y_i = \{0,1\})$, the input for the transducer is written in the form $[x_k y_k] [x_{k-1} y_{k-1}] \dots [x_1 y_1]$. The output of the transducer is the sum of those numbers. The state s_1 is the "carry," s_0 is the initial state and all states are terminal. In ref. *18*, essentially the same transducer was simulated by gradually connecting TX molecules.

For a given $\mathcal{T} = (\Sigma,\Sigma',Q,\delta,s_0,F)$, the transition $(q,a) \overset{\delta}{\to} (a',q')$ can be schematically represented with a square as shown in Fig. 5. Such a square can be considered a Wang tile *(68)* with colored edges, such that left and right we have the state colors encoding the input and output states of the transition and down and up we have colors encoding input and output symbols. Then a computation with $\mathcal{T}$ is obtained by a process of assembling the tiles such that the abutting edges are of the same color. Only translation, and no rotations of the tiles are allowed. We describe this process in more detail below.

3.2. Finite-State Machines with Tile Assembly

A Wang tile is a unit square with colored edges. A finite set of distinct unit squares with colored edges are called *Wang prototiles*. We assume that from

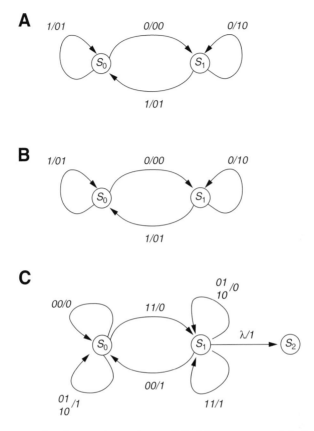

Fig. 4. An encoder that produces the modified frequency modulation code in **A** and **B**. Transducer performing addition modulo 2 in **C**.

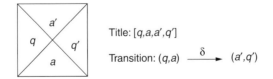

Fig. 5. A computational tile for a transducer.

each prototile, there are an arbitrarily large number of copies that we call *tiles*. A tile τ with left edge colored l, bottom edge colored b, top edge colored t, and right edge colored r is denoted with $\tau = [l,b,t,r]$. No rotation of the tiles is allowed. Two tiles $\tau = [l,b,t,r]$ and $\tau' = [l',b',t',r']$ can be placed next to each other, τ to the left of τ' if $r = l'$, and τ' on top of τ if $t = b'$. In other words, two tiles sit next to each other if their abutting sides have the same colors.

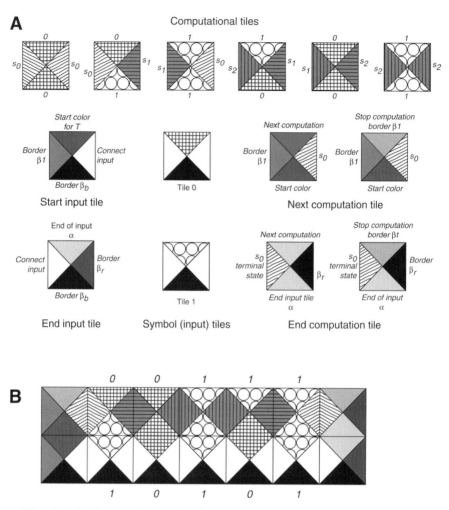

Fig. 6. (A) All prototiles needed for performing the computation of the transducer presented in Fig. 3; **(B)** a simple computation over the string 10101 with result 00111. The last row with output tiles is not included.

For simulation of transducers, we need several types of tiles: computational, input and some special start and end tiles. We refer the reader to ref. *21* for a detailed description of these tiles; here we provide just a short description.

- *Computational tiles.* For a transducer $\mathcal{T}$ with a transition of form $(q,a) \rightarrow (a',q')$, we associate a prototile $[q,a,a',q']$ as presented in Fig. 5. If there are m transitions in the transducer, we associate m such prototiles. Computational tiles for the transducer in Fig. 3B are listed in the top row of Fig. 6.

- *Input and output tiles.* Additional border colors are added to the set of colors. These colors will not be part of the assembly, but will represent the boundary of the computational assembly. Hence, one could take the left border to be distinct from the right border. We assume that each input word ends with α where α is a new symbol "end of input" that does not belong to Σ. For each symbol, we associate a prototile. The input tiles have the same left and right colors such that they can be placed in a sequence. The top color represents the input symbol, and the bottom color is the color of the bottom border. The output tiles are essentially the same as the input tiles, except that they have the top color representing the top border and the bottom color representing the symbols of the output. For DNA implementation, β_i may be represented with a set of different motifs that will facilitate the "readout" of the result. With these sets of input and output tiles, every computation with $\mathcal{T}$ is obtained as a tiled rectangle surrounded by boundary colors (*see* bottom of Fig. 6).
- *Start tiles and accepting (end) tiles.* There are additional tiles that start the computation and the input. These tiles have their colors such that they trigger the assembly of the input and the simulation of the transducer (*see* Fig. 6).

The set of tiles for executing a computation for transducer $\mathcal{T}_2$ that performs division by 3 (*see* Fig. 3B) is depicted in Fig. 6A.

Computation: The simulation is performed by first assembling the input starting with a starting tile and a sequence of input tiles ending with "end of input tile." The computation of the transducer starts by assembling the computation tiles according to the input state (to the left) and the input symbol (at the bottom). The computation ends by assembling the end tile, which can lie next to both the last input tile and the last computational tile if and only if it ends with a terminal state. The output result will be read from the sequence of the output colors assembled with the second row of tiles and application of the output tiles. In this way, one computation is obtained with a tiled $3 \times n$ rectangle ($n > 2$) such that the sides of the rectangle are colored with boundary colors.

The tile computation of $\mathcal{T}_2$ from Fig. 3B for the input string 10101 is shown in Fig. 6B. The output tiles are not included.

It has been shown that the computational power of tile self-assembly and composition of transducers is equivalent to the computational power of a Universal Turing Machine *(9,21)*. This means that all computational functions can be performed in this way.

3.3. Finite-State Automata with Unbounded Memory

Push-down automata with one stack are the simplest example of finite-state automata that use additional infinite tape (stack) as a memory, and as such can be considered automata with infinite memory. In this section, we recall the definition of push-down automata, and some well known theoretical results. It is mainly based on the material found in the classical automata theory book *(67)*.

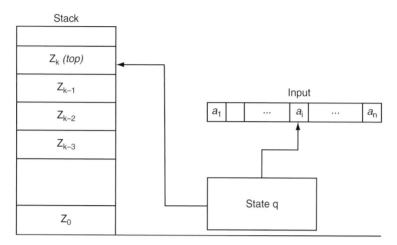

Fig. 7. A push-down automaton.

Although the experimental results for implementation of this type of automata are not reported here, the main idea for implementation as presented in ref. *44* is very similar to the experiments reported in refs. *17,39,41,43.*

A push-down automaton (PDA) is a finite-state automaton with a stack memory (called simply, stack). The class of languages recognized (accepted) by PDA is the class of context-free languages that strictly includes the class of regular languages (recognized by finite-state automata).

The PDA has control of both an input tape and a stack (*see* Fig. 7). The stack is the memory of the machine and it works as a "first in-last out" list. That is, symbols may be entered or removed only at the top of the list such that a symbol that is entered (pushed) at the top pushes the rest of the symbols on the stack one step "down." Similarly, when a symbol is removed (popped) from the top of the list, the remaining symbols on the stack move one step up.

Informally, a transition of a PDA is defined in the following way: at each step, an input symbol and the stack symbol at the top of the stack are read. According to these symbols and the current state, the PDA changes its state and updates the stack, i.e., it either adds a stack-symbol at the top of the stack, removes one stack-symbol from the top of the stack, or leaves the stack unchanged. The computation stops when no more transitions can be applied. The input is accepted if (and only if) it has been entirely read and the PDA is in a final state (similar to the case of Stack finite-state automata). Well-known examples of languages recognized by PDA are the language of *palindrome words* and the set of all words starting with a sequence of a's followed by an equal number of b's. $\{a^n b^n \mid n \in N\}$. Formally (*see* refs. *67,69*), a PDA M is a structure $(Q, \Sigma, \Gamma, \delta, q_0, Z_0, F)$ where Q is a finite set of states, Σ is an input alphabet (its elements are called input-symbols), Γ is a stack alphabet (its elements are called stack-symbols), q_0 in Q is the initial state, Z_0 in Γ is a particular stack-symbol called the start symbol, $F \subseteq Q$ is the set of final (terminal) states, and δ is the transition mapping from $Q \times (\Sigma \cup \{\varepsilon\}) \times \Gamma$ to finite subsets of $Q \times \Gamma^*$. The interpretation of one move (transition):

$$(q,a,z) \xrightarrow{\delta} (p,\gamma)$$

where q and p_i are states, a is in Σ, Z is a stack-symbol, and γ in Γ^* is the following. The PDA in state q, reading an input-symbol a with Z as the top stack-symbol, can enter state p, replace symbol Z by string γ, and advance the input-head one symbol. The acceptance of a language by a PDA is defined as in the case of finite-state automata. For a PDA $M = (Q,\Sigma,\Gamma,\delta,q_0,Z_0,F)$ we define $L(M)$, the *language accepted by M*, to be the set of all w such that the PDA enters a terminal state after starting in state s_0 and reading all of w.

The PDA as a computational device is more powerful than the finite-state automaton without output, as it accepts the class of so-called context-free languages which is a larger class than the class of regular languages. Context-free languages are most often used in compilers translating a higher-order programming language into a machine code.

4. BIOMOLECULAR IMPLEMENTATION

4.1. Implementation by Linear DNA and Endonucleases

4.1.1. Implementation of Finite-State Automata

One of the recent significant accomplishments in the area concerning use of DNA for computation was obtained by a collaboration between a computer scientist E. Shapiro and a biochemist E. Keinan *(39)*. They treated a type II restriction endonuclease as a tool to change states in a finite-state machine, such that together with a rather clever encoding of the states and the symbols, a successful implementation of the finite-state automaton was obtained. They used three main ideas: (a) use a restriction endonuclease *Fok*I, (b) encode a pair (state, symbol) with both the sequence and the length of the segment, and (c) use accepting sequence for final readout.

The enzyme that was chosen, *Fok*I, has a recognition site GGATG, but it cuts a double-stranded molecule 9 and 13 bases away from the recognition site, leaving a 5′ overhang (*see* Fig. 8A). The sequence of bases between the recognition site and the cutting position is completely irrelevant for the action of the enzyme. This is exactly the place where encoding of the symbols and the states of the automaton could take the whole advantage.

The authors demonstrated simulations for several automata with two states. One of these automata consists of two states, s_0 and s_1, with the following

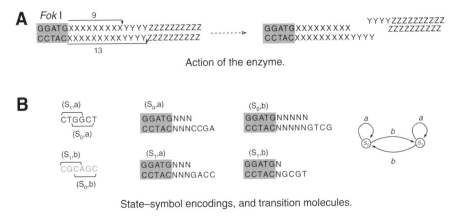

Fig. 8. (A) Action of the enzyme. **(B)** State-symbol encodings, and transition molecules.

transitions: $(s_0,a) \to s_0$, $(s_1,a) \to s_1$, $(s_0,b) \to s_1$ and $(s_1,b) \to s_0$, with s_0 being the initial state and the terminal state. This automaton recognizes (accepts) all words that have an even number of b's. The (two) symbols are encoded with sequences of six nucleotides such that the first four of the sequence encode the state s_1 (i.e., being in state s_1 reading the encoded symbol) and the last four encode the state s_0 (Fig. 8B). The state transitions are encoded with four short transition molecules, each starts with the recognition site of *Fok*I, followed by a sequence of computationally irrelevant base pairs. These transition molecules end with a 5′ overhang of four bases complementary to the encoding of the pair (*state, symbol*). The transitions for (s_0,a) and (s_1,a) have three base pairs which allow cuts of the input molecule to be at the same position of the six base pairs encoding an input symbol. The encoding of (s_0,b) has five irrelevant base pairs, and this "moves" the cut of the enzyme to the left, leaving a 5′ overhang with the first four nucleotides of the six encoding an input symbol. This moves the reading of the next input sequence to encodings of s_1. The encoding of (s_1,b) has only one irrelevant base pair which "moves" the cut of the enzyme to the right, reading the last four nucleotides (as are needed for state s_0). The input to the automaton is a strand that contains a restriction site for *Fok*I, seven (irrelevant) base pairs, a sequence of base pairs encoding a word with symbols a and b. At the end there is a terminal sequence which can anneal to the transition molecule that encodes a terminal state. The "automaton" changes its states and reads the input without outside mediation, and solely by the use of the enzyme. Figure 9 shows several transitions (computational steps) of the finite-state machine. The initial

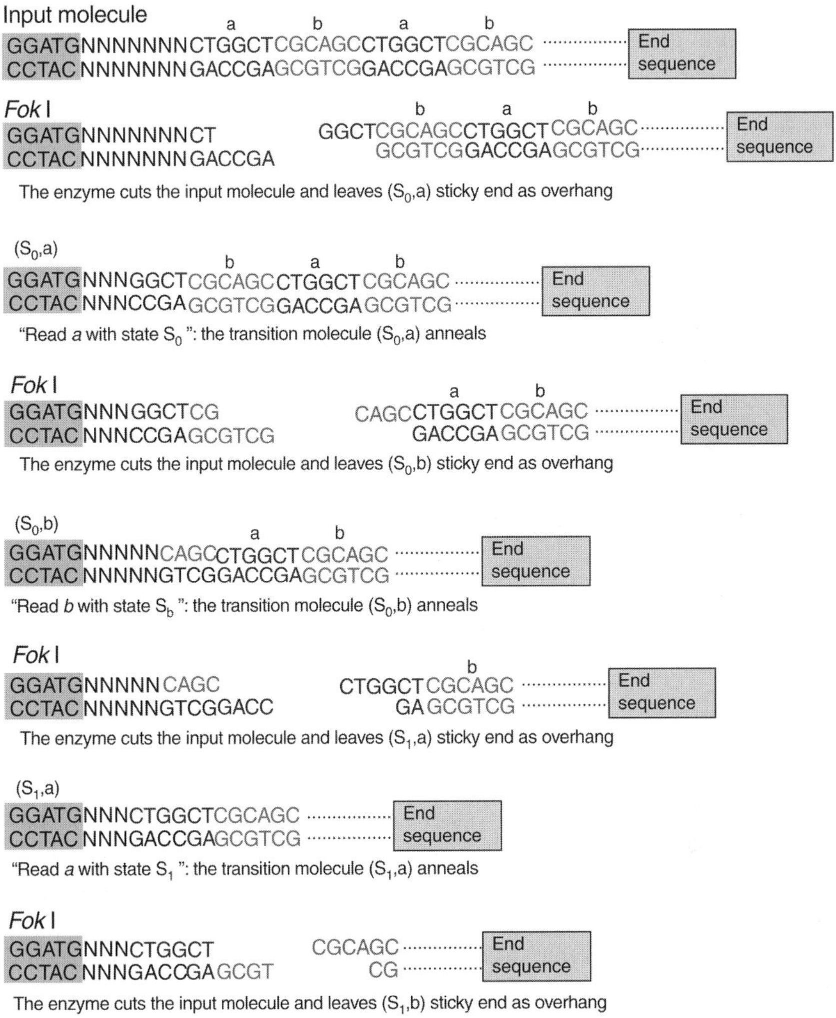

Fig. 9. Several computational steps of the finite-state machine.

experiment reported in ref. *39* employed a ligase to "glue" the transition molecules to the remainder of the input, but their subsequent study showed that this is not necessary *(41,42)*.

This autonomous computational device is one significant advance toward the ultimate goal of achieving a biomolecular computer. Even a much more complex automaton with three states over a three-symbol alphabet has been successfully achieved with this idea *(43)*. It shows that with proper coding and use of the appropriate enzyme, a computational device is possible

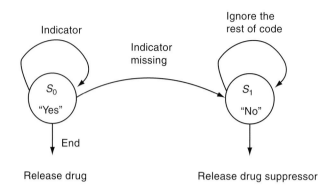

Fig. 10. Automaton that checks whether all disease-indicator genes are present and releases a drug or a suppressor for the drug accordingly.

without any outside mediation. This opens up the door for using such devices not just for performing computation, but potentially also in genetics and medicine, as seen further on.

4.1.2. Using Automata for Logical Control of Gene Expression
(see ref. 6 for details)

The above ideas were shown (at least in vitro) to have potential application in the control of gene expression. As a proof of principle, the automaton described under Subheading 4.1.1 was modified to accept as input an mRNA containing disease-related genes (associated with lung and prostate cancer) *(40)*. For example, the diagnostic rule for prostate cancer checks whether genes PPAP2B and GSTP1 are underexposed and genes PIM1 and HPN are overexposed. In that case (if all four indicators are present), a single-stranded DNA molecule that inhibits the synthesis of a given protein (in this case MDM2) is released, such that this single-stranded DNA molecule binds with the mRNA of that protein. The finite-state automaton that makes this diagnosis is presented in Fig. 10. The transition molecules for this automaton are very similar to those presented in Fig. 8B, except that the sequence that encodes symbols is substituted with a sequence that encodes the given indicator. The input molecule consists of a molecule that encodes all of the indicators that are to be checked. One can think of each symbol *a* or *b* as presented in Figs. 8 and 9, as being a code for one of the indicators. In this case, the "end sequence" (shown in Fig. 9) consists of a hairpin that encodes the single-stranded DNA molecule. This single-stranded DNA should be released once an accepting state is reached. The state S_0 is the start and terminal state for the automaton. The automaton will remain in this state if all of the indicators are present, in which case, the encoding of the drug is released (from the end

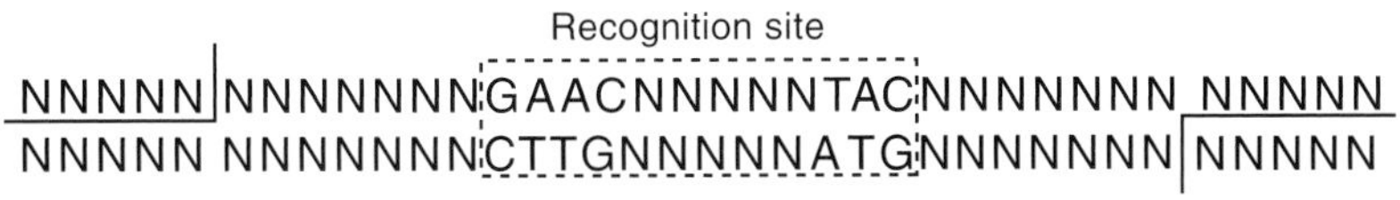

Recognition site

```
 NNNNN|NNNNNNN'GAACNNNNNTAC'NNNNNNN NNNNN
 NNNNN NNNNNNN'CTTGNNNNNNATG'NNNNNNN|NNNNN
```

Fig. 11. The cleavage of enzyme PsrI.

sequence). Otherwise, the automaton ends in state S_1, indicating that at least one of the indicators is missing and at that point a drug suppressant is released.

This "proof-of-principle" experiment shows the potential of this approach for designing an *autonomous biomolecular device* for a wide variety of applications in biochemistry, genetic engineering and even in medical diagnosis.

4.1.3. Implementation of PDA

Implementing a PDA can be achieved in the above-described fashion, using the enzyme *PsrI* together with circular molecules containing the information for the stack and the tape, and linear DNA strands for the transitions. A similar idea was proposed much earlier by Paul Rothemund *(70)* for Turing machines. The enzyme *PsrI* cleaves as depicted in Fig. 11 (for further specifics, *see* ref. *71*). Details describing the implementation of the PDA and their inclusion in a 2D array of DNA tiles can be found in ref. *11*. Experimental implementation of this idea has not been reported to date. The sketch of the idea follows.

Consider a PDA with two input symbols $\{a,b\}$, a stack symbol Z, and five transitions as presented in Fig. 13. This PDA accepts words of the form $a^n b^n$, a language that cannot be recognized by a finite-state machine. The automaton pushes a Z on the stack for every a it reads, keeping track of the number of a's. Then it erases a Z from the stack for each b that is read. If the input ends at the same time as the stack is emptied, the word is accepted.

For the implementation, the input letters are encoded as $a = TTC$ and $b = AAC$. Codes of the stack symbols, using strings of five nucleotides, can be chosen $Z = TCCAG$ and $\# = CAAAC$ for the end of the stack symbol.

The initial circular DNA strand corresponds to the initial configuration of the PDA: it contains the initial configuration of the stack and the input to be read. This is "codified" as follows. The input is written such that any pair of input-symbols are separated by GC, which is also added in front of the first symbol and after the last symbol. A stop-sequence (the end-of-input), $CAGGC$, follows the input. The sequence GC allows "moves" between different states.

This coding allows for *three states* reading the same symbol to be encoded as shown in Fig. 12.

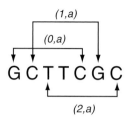

Fig. 12. Coding used to simulate three different states.

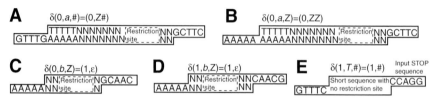

Fig. 13. Transitions of the push-down automaton.

The circular DNA strand representing the initial configuration of the PDA is depicted in Fig. 14A. The first part, *CAAAC*, represents the initial configuration of the stack (containing only symbol #; virtually an empty stack). The middle portion, *GAACNNNNNNTAC*, is the restriction site for the enzyme *PsrI*. The final part is the input described above.

Together with the circular molecule that represents the initial configuration of the PDA, five linear DNA strands that encode the transitions of the PDA are also needed: one strand for each transition. The linear strands corresponding to the transitions in our example are depicted in Fig. 13. These molecules are added in the solution together with the circular molecules corresponding to the initial configuration of the PDA. The enzyme cuts the circular molecule and the transition molecules are allowed to connect to the circular molecule, after which the enzyme cleaves again and the process is repeated (*see* Fig. 14).

4.2. Implementation by DNA Tiles

4.2.1. Transducers: Automata with Output

Winfree et al. *(9)* have shown that 2D arrays made of DX molecules can simulate dynamics of 1D cellular automata (CA). This provides a way to simulate a Turing Machine, the ultimate computer model, since at one time-step the configuration of a 1D CA can represent one instance of the Turing Machine. In the case of tiling representations of finite-state machines (transducers), each tile represents a transition, such that the result of the computation with a transducer is obtained within one row of tiles. Transducers can be

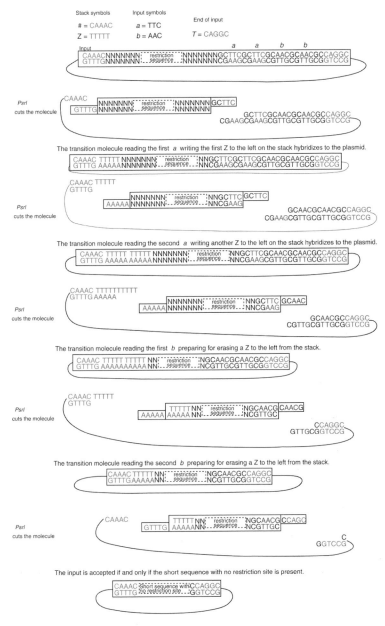

Fig. 14. Accepting of the string aabb.

seen as Turing machines without left movement. Hence, the Wang tile simulation gives the result of a computation with such a machine within one row of tiles, which is not the case in the simulation of CA. A single tile simulating a transducer transition requires more information and DX molecules are not

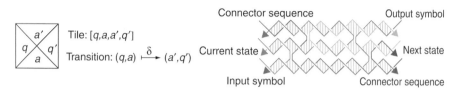

Fig. 15. A computational tile for a fluorescent speckle microscope and a proposed computational triple-crossover molecule tile.

suitable for this simulation. Here, we propose to use tiles made of TX DNA molecules with sticky ends corresponding to one side of the Wang tile. An example of such a molecule is presented in Fig. 15B, with 3' ends indicated with arrowheads. The connector is a sticky part that does not contain any information but it is necessary for the correct assembly of the tiles (*see* below). The TX tile representing transitions for a transducer has the middle duplex longer than the other two (*see* the schematic presentation in Fig. 21, discussed later). It has been shown that such TX molecules can self-assemble in an array *(8)*, and they have been linearly assembled such that a cumulative XOR operation has been executed *(18)*. However, we have yet to demonstrate that they can be assembled reliably in a programmable way in a 2D array. Progress toward using DX molecules for assembly of a 2D array that generates the Sierpinski triangle was recently reported by the Winfree lab *(72)*.

Controlling the right assembly can be done via two approaches: (1) by regulating the temperature of the assembly such that only tiles that can bind three sticky ends at a time hybridize, and those with less than three do not, or (2) by including competitive imperfect hairpins that will extend to proper sticky ends only when all three sites are paired properly.

An iterated computation of the machine can be obtained by allowing third, fourth, etc. rows of assembly. The input for this task is a combination of DX and TX molecules as presented in Fig. 16. The top TX duplex (not connected to the neighboring DX) will have the right-end sticky part encoding one of the input symbols and the left sticky end will be used as a connector. The left (right) boundary of the assembly is obtained with TX molecules that have the left (right) sides of their duplexes ending with hairpins instead of sticky ends. For a two-symbol alphabet, the output tile for one symbol may contain a motif that acts as a topographic marker, and the other not. In this way, the output can be detectable by atomic force microscopy.

4.3. Programmable Computations with DNA Devices

The relatively predictable results from the hybridization of two complementary DNA strands is one of the appealing reasons for using nucleic acids

Fig. 16. Input for the computational assembly.

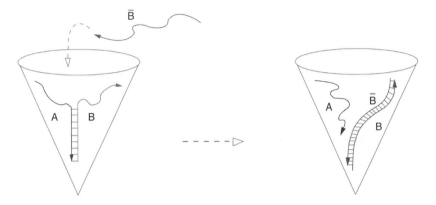

Fig. 17. Assume a tube contains partially hybridized molecules A and B such that B has a portion that is single-stranded and not annealed (left). If a complete complement to B, strand $\bar{B}$ is introduced, then the hybridization between B and $\bar{B}$ prevails and strand A is released.

in nanotechnology. It has a minuscule size (about 2 nm in diameter), and the single-stranded parts can be considered "sticky parts" that anneal to its complement with the formation of a double helix as a final result. The sticky parts provide predictable intermolecular interaction and the double helical structure provides a predictable final geometrical structure. Moreover, nature has provided unique tools in the form of enzymes that allow us to have a tractable and controllable system. These properties are rather attractive for use in nanotechnology and for construction of nanodevices. We mention some recent developments in using DNA as a tool for constructing nanomechanical devices or as a model for a molecular finite-state automaton.

4.3.1. DNA Actuator

Yurke *(24,73)* realized that, in a solution that contains DNA strands that are partially hybridized, when a strand that is completely complementary to one of the strands is introduced, then the hybridization of the complementary strands overcomes the partial hybridization (*see* Fig. 17). This new complementary strand can be used as "fuel" to "move" strands from one hybridization to another and with this, to change the geometry of the self-assembled structures in the tube.

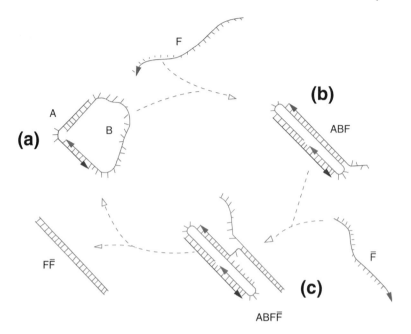

Fig. 18. DNA-fueled actuator-like nanodevice.

The actuator-like nanodevice uses DNA strands as fuel and operates with the same principles. Two strands are assembled, strand A and strand B. The second strand, B, has a sequence length approximately double the length of strand A (*see* Fig. 18A). Strand A has both 3′ and 5′ ends complementary to the corresponding 5′ and 3′ ends of strand B except for a few bases in the center. The double-stranded regions are less than 100 bp and as such are rather stiff, whereas the single-stranded region is much more flexible. In this initial assembly, the double-stranded stiff regions are bent due to the central nucleotides that are free.

Motion on the actuator-like device is induced by employing DNA fuel strands, F, that are complementary to the single-stranded portion of strand B. This makes a rigid type of DX molecule made of two duplex molecules. In this case, the device is straight (Fig. 18B). The fuel strand is a bit longer than the single-stranded portion of strand B and it has a nonhybridized single-stranded part. The original (relaxed) state of the device is obtained by introduction of a complementary strand $\bar{F}$ to the strand F. The free single-stranded portion of F anneals to $\bar{F}$ and then $\bar{F}$ starts to compete with the complex AB for binding with F (Fig. 18C). Because $\bar{F}$ is a full complement of F and is firmly attached to its free single-stranded portion, going through the process of three-strand

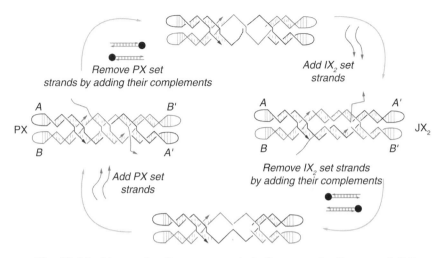

Fig. 19. Machine cycle of two-state switch (for more details, *see* ref. *74*).

branch migration, $\bar{F}$ wins over the complex AB. It produces a waste product $F\bar{F}$ and relaxes the complex AB.

Yurke and his collaborators have developed several similar DNA-fueled nanomechanical devices *(46,47)*. One can envision such devices incorporated into 2D DNA arrays such that the whole surface could fold and open up. This combination has great potential in developing nanomaterials and nanotemplates for circuit developments.

4.3.2. Two-State Switch, Automata with Programmable Input

Using the same idea of fuel DNA strands, Seeman's laboratory developed a two-state switch *(23)*. This device is a combination of a PX and JX_2 molecule that flip-flops between these two states as different fuel strands are added or removed *(see* Fig. 19). One can consider a PX molecule as a double helix made of two DNA duplex molecules. This robust device, whose machine cycle is shown in Fig. 19, is directed by the addition of set strands to the solution that forms its environment.

The set strands, drawn in green and purple, establish which of the two states the device will assume. They differ by a half-turn rotation in the bottom parts of their structures. The sequence-driven nature of the device means that many different devices can be constructed, each of which is individually addressable; this is done by changing the sequences of the red and blue strands where the green or purple strands pair with them. As was the case with the fuel strands of the actuator, the green and purple strands have short, single-stranded extensions. The state of the device is changed by first binding full complements

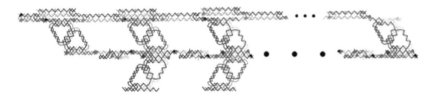

Fig. 20. Linear array of a series of devices to set up the input.

of green or purple strands (fuel strands) and then removing them from solution (strands are biotin-tailed and can be removed by magnetic streptavidin
beads). Adding the other strands changes the state of the device. Large trapezoids have been prototyped and atomic force microscopy has shown that the
device can change the orientation of such large DNA trapezoids *(23)*.

This device has two very significant advantages. First, it is robust, and
second, it is sequence-dependent. One can envision several such devices,
each addressable by different sets of strands used in a single nanostructure.

Linear arrays of a series of PX-JX$_2$ devices can be adapted to set the input
of a fluorescent speckle microscope (FSM). This is presented in Fig. 20, where
we have replaced the trapezoids with double-triangle double-diamonds. The
computational set-up is illustrated schematically in Fig. 21, with just two computational tiles made of DNA TX molecules.

Both diamonds and trapezoids are the result of edge-sharing motifs *(75)*. A
key difference between this structure and the previous one is that the devices
between the two illustrated double-diamond structures differ from each other.
Consequently, we can set the states of the two devices independently. The
double-diamond structures contain domains that can bind to DX molecules;
this design will set up the linear array of the input (*see* Fig. 20). Depending
on the state of the devices, one or the other of the double diamonds will be
on the top side. The two DX molecules will then act similarly to bind with
computational TX molecules, in the gap between them, as shown schematically
in Fig. 21. The system is designed to be asymmetric by allowing an extra
diamond for the "start" and "end" tiles (this is necessary in order to distinguish
the top and bottom in atomic force microscopy visualizations). The left side
must contain an initiator (shown blunt at the top) and the right side will contain
a terminator. The bottom assembly can be removed using the same Yurke-type
techniques *(24)* that are used to control the state of the devices for removal of
the set strands. Successive layers can be added to the device by binding to the
initiator on the left, and to the terminator on the right.

The assembly of the central four-diamond patterns has been prototyped,
connecting them with sticky ends, rather than PX-JX$_2$ devices. Moreover a

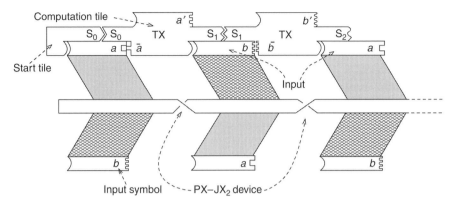

Fig. 21. Schematic view for a couple of first-step computations. The linear array of double-crossover and triple-crossover molecules that sets up the input is substituted only with a rectangle for each of the input symbols.

sequence of two different PX-JX$_2$ devices connecting the diamond motifs has been assembled in a row and programming the input of the transducer has been shown to be possible *(45)*.

5. CONCLUDING REMARKS

There are many other issues concerning biomolecular computations that are rather significant for prosperous development of the scientific discovery that were not included in this short review. They are both theoretical and experimental. When dealing with the use of synthetic DNA, one of the diffcult problems is encoding the bases such that cross-hybridization of noncomplementary strands is minimized. It turns out that this problem is quite complex and many authors have concentrated on developing theoretical coding models *(76,77)*, computer simulations *(78–80)*, and even experimental build-up of coding libraries *(81)*.

From the theoretical point of view, although there are many theoretical models for DNA-based computations, the real task facing theoreticians is to characterize these models within the scope of the experimental limitations.

Aside from Adleman's latest success *(17)*, all other experimental results in biomolecular computing are still at the "toy" level. Although Adleman's experiment solved a complex computational problem (an NP-complete problem), it is now clear that due to the amount of DNA needed to scale up this approach to a larger problem, solutions to large combinatorial search problems via biomolecular protocols will not improve on conventional computers. Further, there are real challenges for the experimentalists to obtain protocols that are suffciently reliable, controllable, and predictable. The standard biomolecular protocols do not have the precision needed for computation. These techniques must undergo many adjustments and sometimes, completely new protocols are necessary in order to improve their yield. The solution of a 20 variable SAT problem was obtained by designing protocols for exquisitely sensitive and error-resistant separation of a small set of molecules. The solution of a 20 variable SAT problem could be considered a "test" for this new protocol. Hence, a search for a DNA solution of such combinatorial problems, although not computationally significant, may prove to be fruitful in developing new technologies.

The real potential of the whole approach seems to arise in assembling DNA in complex 3D structures such as crystals, arrays, or graphs. These structures can serve as scaffolds for other materials such as gold or carbon nanotubes. Further, coupling scaffolding structures together with autonomous automaton devices may prove to be a powerful way of designing and using new nanomaterials. On the other hand, the whole research area is very recent and for such a short time (about 10 yr since the first experimental result), progress has been tremendous. For the past half-century, DNA, RNA, and proteins have been exclusively the province of molecular biologists and medical scientists who have concentrated on understanding their biological impact and properties in living organisms. As these past few years of biomolecular computing have shown, it is most likely that nanoengineers, computer scientists and material scientists will explore these molecules within nonbiological contexts. We now have models and experimental designs for 2D arrays, 3D structures, DNA devices fueled by DNA strands, and autonomous finite-state biomolecular machines, to name a few, all of which are in their infancy. It is not yet clear which of these roads will turn out to be fruitful. That only few of them may be successful is a very real possibility, each in a different field of our scientific community and aspects of life. Whatever the results, this area of research has brought together theoreticians (mathematicians and computer scientists) and experimentalists (molecular biologists and biochemists) into a very successful collaboration. Just the exchange of fresh ideas and discussions among these communities brings excitement, and quite often, provides new lines of development that could not otherwise have been possible.

ACKNOWLEDGMENTS

This work has been partially supported by grants CCF-0523928 and CCF-0432009 from the National Science Foundation, USA. The author thanks Nadrian C. Seeman for many useful comments about the topics covered in this chapter.

REFERENCES

1. Head T. Formal language theory and DNA: an analysis of the generative capacity of specific recombinant behaviors. Bull Math Biol 1987;49:737–759.
2. Adleman L. Molecular computation of solutions of combinatorial problems. Science 1994;266:1021–1024.
3. Seeman NC. Nucleic acid junctions and lattices. J Theor Biol 1982;99:237–247.
4. Kallenbach NR, Ma R-I, Seeman NC. An immobile nucleic acid junction constructed from oligonucleotides. Nature 1983;305:829–831.
5. Wang Y, Mueller JE, Kemper B, Seeman NC. The assembly and characterization of 5-arm and 6-arm DNA junctions. Biochemistry 1991;30:5667–5674.
6. Seeman NC. DNA nicks and nodes and nanotechnology. NanoLett 2001;1:22–26.
7. Fu TJ, Seeman NC. DNA double crossover structures. Biochemistry 1993;32:3211–3220.
8. LaBean TH, Yan H, Kopatsch J, Liu F, Winfree E, Reif JH, Seeman NC. The construction, analysis, ligation and self-assembly of DNA triple crossover complexes. J Am Chem Soc 2000;122:1848–1860.
9. Winfree E, Yang X, Seeman NC. Universal computation via self-assembly of DNA: some theory and experiments. In: Proceedings of the First Annual Meeting, DIMACS Series in Discrete Mathematics and Theoretical Computer Science, vol. 27. Providence, RI: American Mathematical Society, 1996.
10. Winfree E, Liu F, Wenzler L, Seeman NC. Design of self-assembly of two-dimensional crystals. Nature 1998;494:539–544.
11. Chen J, Seeman NC. Synthesis from DNA of a molecule with the connectivity of a cube. Nature 1991;350:631–633.
12. Chen J, Kallenbach NR, Seeman NC. A specific quadrilateral synthesized from DNA branched junctions. J Am Chem Soc 1989;111:6402–6407.
13. Jonoska N, Karl S, Saito M. Three dimensional DNA structures in computing. BioSystems 1999;52:143–153.
14. Jonoska N, Karl S, Saito M. Creating 3-dimensional graph structures with DNA. In: Proceedings of the First Annual Meeting, DIMACS Series in Discrete Mathematics and Theoretical Computer Science, vol. 44. Providence, RI: American Mathematical Society, 1998:123–136.
15. Shihm WM, Quispe JD, Joyce GF. A 1.7-kilobase single-stranded DNA folds into a nano-scale octahedron. Nature 2004;427:618–621.
16. Adleman L. On constructing a molecular computer. In: Lipton R, Baum E, eds. DNA Based Computers. DIMACS: Series in Discrete Mathematics and Theoretical Computer Science, vol. 27. Providence, RI: American Mathematical Society, 1996:1–21.
17. Braich RS, Chelyapov N, Johnson C, Rothemund PWK, Adleman L. Solution of a 20-variable 3-SAT problem on a DNA computer. Science 2002;296:499–502.
18. Mao C, LaBean T, Reif JH, Seeman NC. Logical computation using algorithmic self-assembly of DNA triple crossover molecules. Nature 2000;407:493–496.
19. Faulhammer D, Cukras AR, Lipton RJ, Landweber FL. Molecular computation: RNA solutions to chess problems. Proc Natl Acad Sci USA 2000;97:1385–1389.

20. Head T et al. Computing with DNA by operating on plasmids. BioSystems 2000;57:87–93.
21. Jonoska N, Liao S, Seeman NC. Transducers with programmable input by DNA self-assembly in aspects of molecular computing. In: Jonoska N, Paun Gh, Rozenberg G, eds, LNCS, vol. 2950. Heidelberg: Springer-Verlag, 2004:219–240.
22. Mao C, Sun W, Shen Z, Seeman NC. A nanomechanical device based on the B-Z transition of DNA. Nature 2000;397:144–146.
23. Yan H, Zhang X, Shen Z, Seeman NC. A robust DNA mechanical device controlled by hybridization topology. Nature 2002;415:62–65.
24. Yurke B, Turberfield AJ, Mills AP, Simmel FC Jr. A DNA fueled molecular machine made of DNA. Nature 2000;406:605–608.
25. Yin P, Tuberfield AJ, Reif JH. Design of autonomous unidirectional walking DNA devices. In: Proceedings of the Tenth Annual Meeting, LNCS, vol. 3384. Heidelberg: Springer-Verlag, 2005:410–425.
26. Sherman WB, Seeman NC. A precisely controlled DNA biped walking device. Nano Lett 2004;4:1203–1207.
27. Shin JS, Pierce NA. A synthetic DNA walker for molecular transport. J Am Chem Soc 2004;126:10,834–10,835.
28. Tuberfield AJ et al. DNA fuel for free-running nanomachines. Phys Rev Lett 2003;90:118102.
29. Lipton R, Baum E. DNA based computers. Proceedings of the First Annual Meeting, DIMACS Series in Discrete Mathematics and Theoretical Computer Science, vol. 27. Providence, RI: American Mathematical Society, 1996.
30. Landweber L, Baum E. DNA based computers. Proceedings of the First Annual Meeting, DIMACS Series in Discrete Mathematics and Theoretical Computer Science, vol. 44. Providence, RI: American Mathematical Society, 1998.
31. Rubin H, Wood D. DNA based computers, Proceedings of the Third Annual Meeting DIMACS series in Discrete Mathematics and Theoretical Computer Science, vol. 48. Providence, RI: American Mathematical Society, 1999.
32. Kari L, Wood D. DNA based computers, revised papers. Proceedings of the Fourth Annual Meeting BioSystems (special issues), 1999:52.
33. Winfree E, Gifiord D.K. DNA based computers, Proceedings of the Fifth Annual Meeting, DIMACS Series in Discrete Mathematics and Theoretical Computer Science, vol. 54. Providence, RI: American Mathematical Society, 2000.
34. Condon A, Rozenberg G. DNA based computers. Proceedings of the Sixth Annual Meeting. LNCS, vol. 2054. Heidelberg: Springer-Verlag, 2001.
35. Jonoska N, Seeman NC. DNA based computers. Proceedings of the Seventh Annual Meeting, LNCS, vol. 2340. Heidelberg: Springer-Verlag, 2002.
36. Hagiya M, Ohuchi A. DNA Based computers. Proceedings of the Eighth Annual Meeting, LNCS, vol. 2568. Heidelberg: Springer-Verlag, 2002.
37. Reif J, Chen J. DNA Based Computers, LNCS, vol. 2943. Heidelberg: Springer-Verlag, 2004.
38. Ferretti C, Mauri G, Zandron C. DNA Based computers. Proceedings of the Tenth Annual Meeting, LNCS, vol. 3384. Heidelberg: Springer-Verlag, 2005.
39. Benenson Y, Paz-elizur T, Adar R, Keinan E, Livneh Z, Shapiro E. Programmable and autonomous machine made of biomolecules. Nature 2001; 414:430–434.

40. Benenson Y, Adar R, Shapiro E. An autonomous molecular computer for logical control of gene expression. Nature 2004;429:423–429.
41. Benenson Y, Adar R, Paz-elizur T, Livneh Z, Shapiro E. DNA molecule provides a computing machine with both data and fuel. Proc Natl Acad Sci USA 2003; 100:2191–2196.
42. Benenson Y, Adar R., Paz-elizur T, Livneh Z, Shapiro E. Molecular computing machine uses information as fuel. Preliminary Proceedings of the 8th Int. Meeting on DNA Based Computers, Hokkaido University, June 10–13, 2002:198.
43. Soreni M, Yogev S, Kossoy E, Shoham Y, Keinan E. Parallel biomolecular computation with advanced finite automata. J Am Chem Soc, in press.
44. Cavalere M, Jonoska N, Keinan E, Seeman N. Implementing Computing Devices with Unbounded Memory Using Biomolecules, LNCS. Heidelberg: Springer-Verlag, 2005, in press.
45. Liao S, Seeman NC. Translation of DNA signals into polymer assembly instructions. Science 2004;306:2072–2074.
46. Simmel FC, Yurke B. Using DNA to construct and power a nanoactuator. Phys Rev E 2001;63(041913):1–5.
47. Simmel FC, Yurke B. Operation of a purified DNA nanoactuator. In: Proceedings of the Seventh Annual Meeting, LNCS, vol. 2340. Heidelberg: Springer-Verlag, 2002:248–257.
48. Head T, Paun Gh, Pixton D. Language theory and molecular genetics. In: Rozenberg G, Salomaa A, eds. Handbook of Formal Languages, vol. 2. Heide-lberg: Springer-Verlag, 1997:295–358.
49. Paun Gh, Rozenberg G, Salomaa A. DNA Computing, New Computing Paradigms. Heidelberg: Springer-Verlag, 1998.
50. Kari L. DNA Computing: arrival of biological mathematics. The Mathematical Intelligencer 1997;19(2):9–22.
51. Ausubel FM, Brent R, Kingston RE, et al. Current Protocols in Molecular Biology. New York, NY: Greene Publishing Associates and Wiley-Interscience, 1993.
52. Seeman NC. The design of single-stranded nucleic acid knots. Mol Eng 1992; 2:197–307.
53. Seeman NC et al. The perils of polynucleotides: the experimental gap between the design and assembly of unusual DNA structures. In: Proceedings of the First Annual Meeting, DIMACS Series in Discrete Mathematics and Theoretical Computer Science, vol. 44. Providence, RI: American Mathematical Society, 1998:215–234.
54. Seeman NC, Zhang Y, Du SM, Chen J. Construction of DNA polyhedra and knots through symmetry minimization. In: Siegel JS, ed. Supermolecular Stereochemistry 1995:27–32.
55. Du SM, Wang H, Tse-Dinh YC, Seeman NC. Topological transformations of synthetic DNA knots. Biochemistry 1995;34:673–682.
56. Mao C, Sun W, Seeman N.C. Designed two-dimensional Holliday junction arrays visualised by atomic force microscopy. J Am Chem Soc 1999;121: 5437–5443.
57. Seeman NC et al. Gel electrophoretic analysis of DNA branched junctions. Electrophoresis 1989;10:345–354.

58. Jonoska N, Sa-Ardyen P, Seeman NC. Computation by self-assembly of DNA graphs. J Gen Prog Evolvable Machines 2003;4:123–137.

59. Sa-Ardyen P, Jonoska N. Seeman NC. Self-assembly of graphs represented by DNA Helix Axis Topology. J Am Chem Soc 2004;126(21):6648–6657.

60. Sa-Ardyen P, Jonoska N, Seeman NC. Self-assembling DNA graphs. In: Hagiya M, Ohuchi A, eds. Revised Papers of 8th International Meeting on DNA Based Computers, LNCS, vol. 2568. Heidelberg: Springer-Verlag, 2002:1–9; also in Natural Computing 2003;24:427–438.

61. Seeman NC. Private communication.

62. Lipton R. DNA solution of hard computational problems. Science 1995;268: 542–545.

63. Adleman L, Rothemund PWK, Roweis S, Winfree E. On applying molecular computation to the Data Encryption Standard. In: DIMACS: Series in Discrete Mathematics and Theoretical Computer Science, vol. 44. Providence, RI: American Mathematical Society, 1999:31–44.

64. Hartmanis J. On the weight of a computation, Bull EATCS 1995;55:136–138.

65. Morimoto N, Arita M, Suyama A. Solid phase DNA solution to the Hamiltonian path problem. In: Rubin H, Wood DH, eds. DNA Based Computers III, DIMACS series, vol. 48. Providence, RI: American Mathematical Society, 1999:193–206.

66. Ouyang Q, Kaplan PD, Liu S, Libchaber A. DNA solution to the Maximal Clique problem. Science 1997;278:446–449.

67. Hopcroft JE, Ullman JD. Introduction to Automata Theory, Languages, and Computation. Boston, MA: Addison-Wesley, 1979.

68. Wang H. Notes on a class of tiling problems. Fundamenta Mathematicae 1975;82:295–334.

69. Davis M, Sigal R, Weyuker EJ. Computability, Complexity and Languages. New York, NY: Elsevier, 1994.

70. Rothemund PWK. A DNA and restriction enzyme implementation of Turing machines. In: Proceedings of the First Annual Meeting, DIMACS Series in Discrete Mathematics and Theoretical Computer Science, vol. 27. Providence, RI: American Mathematical Society, 1996;27:75–119.

71. http://rebase.neb.com/rebase/rebase.html.

72. Rothemund PWK, Papadakis N, Winfree E. Algorithmic self-assembly of DNA Sierpinsky triangles. PLoS Biol 2004;2(12):e424.

73. Yurke B, Mills AP Jr. Using DNA to power nanostructures. Gen Prog Evolving Machines 2003;4:111.

74. Seeman NC. DNA in a material world. Nature 2003;421(6921):427–431.

75. Yan H, Seeman NC. Edge-sharing motifs in DNA nanotechnology. J Supramol Chem 2003;1:229–237.

76. Hussini S, Kari L, Konstantinidis S. Coding properties of DNA languages, DNA computing. In: Jonoska N, Seeman NC, eds. Proceedings of the 7th International Meeting on DNA Based Computers, LNCS, vol. 2340. Heidelberg: Springer-Verlag, 2002:57–69.

77. Mahalingam K, Jonoska N. Languages of DNA based code words. In: Chen J, Reif J, eds. Preliminary Proceedings of the 9th International Meeting of DNA Based Computers, Madison, WI, June 1–4, 2003:58–68.
78. Garzon M, Deaton R, Reanult D. Virtual test tubes: a new methodology for computing. Proc. 7th. Int. Symposium on String Processing and Information Retrieval, A Coruna, Spain. IEEE Computing Society Press 2000:116–121.
79. Seeman NC. De novo design of sequences for nucleic acid structural engineering. J Biomolec Struct Dyn 1990;8(3):573–581.
80. Feldkamp U, Saghafi S, Rauhe H. DNASequenceGenerator—a program for the construction of DNA sequences. In: Jonoska N, Seeman NC, eds. DNA Computing, LNCS, vol. 2340. Heidelberg: Springer-Verlag, 2002:23–32.
81. Deaton R, Chen J, Bi H, Garzon M, Rubin H, Wood DH. A PCR-based protocol for in vitro selection of non-crosshybridizing oligonucleotides. Proceedings of the Eighth Annual Meeting, LNCS, vol. 2568. Heidelberg: Springer-Verlag, 2002:196–204.

IV
NANOMEDICINE, NANOPHARMACEUTICALS, AND NANOSENSING

12
Nanomedicine

Kewal K. Jain

Summary

Nanomedicine is the application of nanobiotechnologies in medicine to improve diagnosis as well as therapy. Nanobiotechnologies play an important role in the discovery of biomarkers and molecular diagnostics and facilitate the integration of diagnosis and therapy. Nanopharmaceuticals is the application of nanotechnologies to improve drug discovery and drug delivery. Various nanotechnologies facilitate methods of disease treatment, both medical and surgical, and these include nanoparticles to facilitate imaging, nanoendoscopes, nanolasers, and nanorobotics. This chapter gives examples of applications of nanobiotechnologies to improve management of various diseases and eventually, in the next decade or so, to develop personalized medicine.

Key Words: Biomarkers; cancer; nanobiotechnology; nanodiagnostics; nanoendoscopes; nanomedicine; nanoparticles; nanopharmaceuticals; nanorobotics; personalized medicine.

1. INTRODUCTION

Medicine is constantly evolving, and new technologies are incorporated into the diagnosis and treatment of patients. Various nanobiotechnologies are currently under investigation in medical research, and diagnostics will soon find applications in the practice of medicine *(1)*. Nanobiotechnologies are being used to create and study models of human disease. The introduction of nanobiotechnologies in medicine does not imply that a separate branch of medicine will be created, but simply that diagnosis as well as therapy will be improved and can be referred to as nanomedicine. Relationships of biotechnology, nanotechnology, and medicine are shown in Fig. 1. Some of the areas in which nanomedicine will be applied are shown in Table 1. This chapter will focus on the most promising technologies that are applied in both medicine and surgery to improve the effectiveness and safety of treatment.

From: *NanoBioTechonology: BioInspired Devices and Materials of the Future*
Edited by: Oded Shoseyov and Ilan Levy © Humana Press Inc., Totowa, NJ

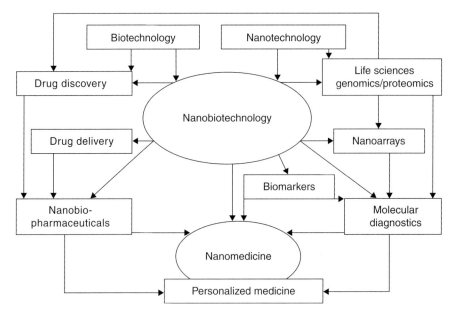

Fig. 1. Relationships of biotechnology, nanotechnology, and medicine.

2. PROMISING NANOBIOTECHNOLOGIES FOR APPLICATIONS IN MEDICINE

2.1. Clinical Molecular Diagnostics

The role of nanotechnology in molecular diagnostics has been described in detail in a special monograph on this topic *(2)*. This will have a tremendous impact on the practice of medicine. Biosensor systems based on nanotechnology could detect emerging disease in the body at a stage that may be curable. This is extremely important in management of infections and cancer. Some of the body functions and responses to treatment will be monitored without cumbersome laboratory equipments. Some examples are a radiotransmitter small enough to put into a cell and acoustical devices to measure and record the noise a heart makes.

2.2. Nanotechnology and Discovery of Biomarkers

Any specific molecular alteration of a cell either on DNA, RNA, or protein level can be referred to as a molecular biomarker. A biomarker is a characteristic that can be objectively measured and evaluated as an indicator of normal biological processes, pathogenic processes, or pharmacological responses to a therapeutic intervention *(3)*. Biomarkers are assuming an important role in health science as a basis for molecular diagnostics, for drug discovery and development (monitoring of clinical trials), and for disease

Table 1
Nanomedicine in the 21st Century

Nanodiagnostics
  Molecular diagnostics
  Nanoendoscopy
  Nanoimaging
Nanotechnology-based drugs
  Drugs with improved methods of delivery
  Therapies combining diagnostics with drugs
Regenerative medicine
  Tissue engineering with nanotechnology
Transplantation medicine
  Exosomes from donor dendritic cells for drug-free organ transplants
Nanorobotic treatments
  Vascular surgery by nanorobots introduced into the vascular system
  Nanorobots for detection and destruction of cancer
Implants
  Bioimplantable sensors that bridge the gap between electronic and neurological
    circuitry
  Durable rejection-resistant artificial tissues and organs
  Implantations of nanocoated stents in coronary arteries to elute drugs and to
    prevent reocclusion
  Implantation of nanopumps for drug delivery
Minimally invasive surgery using catheters
  Miniaturized nanosensors implanted in catheters to provide real-time data to
    surgeons
NanoSurgery by integration of nanoparticles and external energy

JainPharmaBiotech.

prognosis. Nanobiotechnology plays an important role in discovery and use of biomarkers.

Nanomaterials are suitable for biolabeling. Water-soluble, biocompatible, fluorescent, and stable silver/dendrimer nanocomposites have been synthesized that exhibit a potential for labeling cells in vitro as cell biomarkers *(4)*. Efforts to improve the performance of immunoassays and immunosensors by incorporating different kinds of nanostructures have gained considerable momentum over the last decade. Nanoproteomics—application of nanobiotechnology to proteomics—enables identification of low abundant proteins and proteins that can only be isolated from limited source material such as needle biopsies. Nanotechnology offers the possibility of creating devices that can screen for disease biomarkers at very fast rates. The tools will be developed by identifying biomarkers for particular diseases that can then lead to diagnostic tests.

2.3. Nanoparticles for Molecular Imaging

Although developments in molecular imaging have been dominated by nuclear medicine agents in the past, the advent of nanotechnology led to magnetic resonance imaging (MRI) molecular agents that enable detection of sparse biomarkers with a high-resolution imaging. A wide variety of nanoparticulate MRI contrast agents are available, most of which are superparamagnetic iron oxide-based constructs. Perfluorocarbon (PFC) nanoparticulate platform is not only effective as a T1-weighted agent, but also supports 19F magnetic resonance spectroscopy and imaging. The unique capability of 19F permits confirmation and segmentation of MRI images as well as direct quantification of nanoparticle concentrations within a voxel. Ultrasmall superparamagnetic iron oxide (USPIO) is a cell-specific contrast agent for MRI. An open-label phase II study has tested the potential of USPIO-enhanced MRI for macrophage imaging in human cerebral ischemic lesions *(5)*. USPIO-induced signal alterations throughout differed from signatures of conventional gadolinium-enhanced MRI, thus being independent from breakdown of the blood-brain barrier (BBB). Macrophages, as the prevailing inflammatory cell population in stroke, contribute to brain damage. USPIO-enhanced MRI may provide an in vivo surrogate marker of cellular inflammation in stroke and other central nervous system (CNS) pathologies.

2.4. Nanotechnology-Based Integration of Diagnostics with Therapeutics

The current trend in medicine is to base treatment on diagnosis. Integration of diagnosis with therapeutics is an important component of personalized medicine, which simply means selection of the treatment best suited to an individual. Nanobiotechnology will facilitate this integration, as some nanoparticles serve both as diagnostics and therapeutics. The surface groups of dendrimers can be modified and tailored for specific applications. Both therapeutic and diagnostic agents are usually attached to surface groups on dendrimers by chemical modification. The best examples are in cancer and will be elaborated in the section on oncology.

2.5. Nanotechnology-Based Devices for Medical and Surgical Treatments

2.5.1. Application of Nanotechnology in Radiology

X-ray radiation is widely used in medical diagnosis. The basic design of the X-ray tube has not changed significantly in the last century. Now, medical diagnostic X-ray radiation can be generated using carbon nanotube (CNT)-based field emission cathode. The device can readily produce both continuous and pulsed X-rays with programmable wave form and repetition rate. The X-ray intensity is sufficient to image a human extremity. The CNT-based cold-cathode X-ray technology can potentially lead to portable and miniature X-ray sources for medical applications. An X-ray device based on CNTs emits a scanning X-ray beam composed of multiple smaller beams while also remaining stationary *(6)*. As a result, the device can create images of objects from numerous angles and without mechanical motion, which is a distinct advantage for any machine because it increases imaging speed, can reduce the size of the device, and requires less maintenance. This technology can lead to smaller and faster X-ray imaging systems for tomographic medical imaging such as computed tomography (CT) scanners. Other advantages will be that scanners will be cheaper, use less electricity, and produce higher-resolution images.

2.5.2. Nanolasers in Surgery

Scalpel and needle may remain adequate instruments for most surgery work and biological compounds may still be needed to prod cells to certain actions. The introduction of lasers in surgery more than a quarter of century ago has already refined surgery and experimental biological procedures to enable manipulations beyond the capacity of the human hand-held instruments. Laser microsurgery was used both for ablation and repair of tissues *(7)*. Mechanical devices such as microneedles are too large for the cellular scale, whereas biological and chemical tools can only act on the cell as a whole rather than on any one specific mitochondrion or other structure.

Further developments are leading to manipulation of cellular structures at the micrometer and nanometer scale. This is opening up the field of nanoscale laser surgery. Femtosecond (one millionth of a billionth of a second) laser pulses can selectively cut a single strand in a single cell in the worm and selectively knock out the sense of smell. One can target a specific organelle inside a single cell (a mitochondrion, e.g., or a strand on the cytoskeleton) and zap it out of existence without disrupting the rest of the cell. The lasers can neatly zap specific structures without harming the cell or hitting other mitochondria only a few hundred nanometers away. It is possible to carve channels slightly less than 1 micron wide, well within a cell's diameter of 10 to 20 microns. By firing a pulse for only 10 to 15 femtoseconds in beams only 1 micron wide, the amount of photons crammed into each burst becomes incredibly intense: 100 quadrillion watts per square meter, 14 orders of magnitude greater than outdoor sunlight. That searing intensity creates an electric field strong enough to disrupt electrons on the target and create a micro-explosion. But because the pulse is so brief, the actual energy delivered into the cell is only a few nanojoules. To achieve that same intensity with nanosecond or millisecond pulses would require so much more energy that the cell would be destroyed.

That opens the door to researching how cytoskeletons give a cell its shape, or how organelles function independently from each other rather than a whole system. The technology might be scaled up to do surgery without scarring or perhaps to deliver drugs through the skin. Near-infrared femtosecond laser pulses have been applied in a combination of microscopy and nanosurgery on fluorescently labeled structures within living cells *(8)*. Femtolasers are already in use in corneal surgery.

Understanding how nerves regenerate is an important step toward developing treatments for human neurological disease, but investigation has so far been limited to complex organisms (mouse and zebrafish) in the absence of precision techniques for severing axons (axotomy). Femtosecond laser surgery has been used for axotomy in the roundworm *Caenorhabditis elegans*, and these axons functionally regenerated after the operation *(9)*. Femtolaser acts like a pair of tiny "nano-scissors" that is able to cut nano-sized structures like nerve axons. The pulse has a very short length, making the photons in the laser concentrate in one area, delivering much power to a tiny, specific volume without damaging surrounding tissue. Once cut, the axons vaporize and no other tissue is harmed. Application of this precise surgical technique should enable the study of nerve regeneration in vivo.

2.5.3. Nanorobotics

Robotics is already developing for applications in life sciences and medicine. Robots can be programmed to perform routine surgical procedures. Nanobiotechnology introduces another dimension in robotics, leading to the development of nanorobots also referred to as nanobots. Instead of performing procedures from outside the body, nanobots can be miniaturized for introduction into the body through the vascular system or at the end of catheters into various vessels and other cavities in the human body. A surgical nanobot, programmed by a human surgeon, could act as an autonomous on-site surgeon inside the human body. Various functions such as searching for pathology, diagnosis, and removal or correction of the lesion by nanomanipulation can be performed and coordinated by an on-board computer. Such concepts, once science fiction, are now considered to be within the realm of possibility. Nanorobots will be capable of performing precise and refined intracellular surgery, which is beyond the capability of manipulations by the human hand.

A device was developed for facilitating minimally invasive beating-heart intrapericardial interventions *(10)*. This is based on the concept of an endoscopic robotic device that adheres to the epicardium by suction and navigates by crawling like an inchworm to any position on the surface under the control of a surgeon. This approach obviates cardiac stabilization, lung deflation, differential lung ventilation, and reinsertion of laparoscopic tools for accessing different treatment sites, thus offering the possibility of reduced trauma to the patient. The device has a working channel through which various tools can be introduced for treatment. The current prototype demonstrated successful prehension, turning, and locomotion on beating hearts in a limited number of trials in a pig model.

2.5.4. Nanoendoscopes

Video capsule endoscopy is a major innovation that provides high-resolution imaging of the small intestine in its entirety *(11)*. In the 4 yr since its introduction, capsule endoscopy has demonstrated its viability as a first-line investigation in patients with obscure gastrointestinal bleeding after a negative esophagogastroduodenoscopy and colonoscopy, and it has a positive impact on the outcome. Video capsule endoscopy is also useful in the evaluation of inflammatory and neoplastic disorders of the small bowel. Controlling the positioning and movement on a nanoscale will greatly improve the accuracy of this method.

Endoscopic microcapsules or "gutbots," which are in development, are based on nanotechnology including nanosensors and sticking devices. These can be ingested and precisely positioned. A control system allows the capsule to attach to the walls of the digestive tract and move within its lumen. Several different methods are being researched for the attachment of microcapsules, including both dry and wet adhesion as well as mechanical methods such as a set of tripod legs with adhesive on feet. A simple model with surface characteristics similar to those of the digestive tract can be constructed to test these methods. Precisely positioned microcapsules would enable physicians to view any part of the inside lining of the digestive tract in detail, resulting in more efficient, accurate, and less invasive diagnosis. In addition, these capsules could be modified to include treatment mechanisms as well, such as the release of a drug or chemical near an abnormal area. Similar nanorobots are under development for other parts of the body.

2.5.5. Role of Nanopharmaceuticals in Nanomedicine

Nanopharmaceuticals implies drug discovery and development as well drug delivery based on nanobiotechnology. This topic has been dealt with in other chapters in this book. Nanoparticles and various nanodevices play an important role in refinements of drug delivery and facilitate the development of nanomedicine. Nanocarriers solve some of the drug solubility problems and enable targeted drug delivery and controlled release.

3. ROLE OF NANOTECHNOLOGY IN METHODS OF TREATMENT

3.1. Nanobiotechnology Applications in Gene Therapy

Gene therapy can be broadly defined as the transfer of defined genetic material to specific target cells of a patient for the ultimate purpose of preventing or altering a particular disease state. Vectors are usually viral, but several nonviral techniques are being used as well. Genes and DNA are now being introduced without the use of vectors, and various techniques are being used to modify the function of genes in vivo without gene transfer. Nanoparticles and other nanostructures can be used for gene delivery. The success of the gene therapy for clinical applications, in part, would depend on the efficiency of the expression vector as determined by the level as well as the duration of gene expression. Although various cationic polymers and lipid-based systems are being investigated, most of these systems exhibit higher-level but transient gene expression. Most often, the emphasis is on the level of gene expression rather than on the duration of gene expression. In certain disease conditions, a relatively low level of gene expression (therapeutic level) but for a sustained duration may be more effective than higher-level but transient gene expression. Therefore, a gene expression system that can modulate the level as well as the duration of gene expression in the target tissue is desirable. Polymer-based sustained release formulations such as nanoparticles have the potential of developing into such a system.

3.2. Nanobiotechnology for Tissue Engineering

Tissue engineering is a field that applies the principles of engineering and the life sciences to the development of biological substitutes that restore, maintain, or improve tissue function. Tissue engineering is an emerging field between traditional medical devices and regular pharmaceuticals. It faces many challenges, and it is also a field that is interdisciplinary, requiring the efforts of physicians, cell biologists, material scientists, chemical engineers, and chemists. Apart from the use of nanoparticles for diagnostic and therapeutic purposes, nanotechnology has applications for the development of tissue engineering as indicated by some of the studies in life sciences *(12)*.

Microfluidic devices enable the study of methods for patterning cells, topographical control over cells and tissues, and bioreactors. They have not been used extensively in tissue engineering, but major contributions are expected in two areas. The first is the growth of complex tissue, where microfluidic structures ensure a steady blood supply, thereby circumventing the well known problem of providing larger tissue structures with a continuous flow of oxygen as well as nutrition and removal of waste products. The second, and probably more important, function of microfluidics combined with nanotechnology lies in the development of in vitro physiological systems for studying fundamental biological phenomena.

Ideally, the tissue-engineering scaffolds should be analogous to native extracellular matrix (ECM) in terms of both chemical composition and physical structure. The polymeric nanofiber matrix is similar, with its nanoscaled nonwoven fibrous ECM proteins, and thus is a candidate ECM-mimetic material *(13)*. Scaffolds for tissue engineering are typically solid or porous materials with isotropic characteristics, and present regenerative cues such as growth factors or ECM proteins, but these do not explicitly guide tissue regeneration.

3.3. Transplant/Organ Replacement Devices

Several devices are used to repair, replace, or assist the function of damaged organs such as kidneys. The technologies range from those for tissue repair to those for devices to take over or assist the function of the damaged organs. The following sections include a few examples of these applications.

3.3.1. Exosomes for Drug-Free Organ Transplants

A major problem with organ transplantation is rejection. Immunosuppressant drugs used to prevent this have undesirable side effects. Exosomes are nanovesicles shed by dendritic cells. They may hold the key to achieving transplant tolerance, i.e., the long-term acceptance of transplanted organs without the need for drugs *(14)*. Exosomes are no larger than 65–100 nm; yet, each contains a potent reserve of major histocompatibility complex (MHC) molecules—gene products that cells use to determine self from nonself. Millions of exosomes scurry about within the bloodstream, and although their function has been somewhat of a mystery, researchers are beginning to surmise that they play an important role in immune regulation and response.

Because certain dendritic cells have tolerance-enhancing qualities, several approaches under study involve giving recipients donor dendritic cells that have been modified in some way. MHC-rich vesicles, siphoned from donor dendritic cells, are captured by recipient dendritic cells and processed in a manner important for cell-surface recognition. Thus, one can efficiently deliver donor antigen using the exosomes as a magic bullet. This finding is significant because current immunosuppression therapies used in the clinical setting are not able to efficiently prevent T-cell activation via the indirect pathway. The process of internalizing the donor exosomes does not affect maturation of the dendritic cell. Only immature dendritic cells can capture antigens efficiently and are believed to participate in the induction of transplant tolerance. By contrast, once mature, dendritic cells are capable of triggering the T-cell activation that leads to transplant rejection. Additional research will be required to determine whether donor-derived exosomes will enhance the likelihood that an organ transplant from the same donor will be accepted.

3.3.2. Nanotechnology-Based Human Nephron Filter for Renal Failure

Patients with renal failure from end-stage renal disease require treatment through dialysis or renal transplantation. Despite the availability of various forms of renal replacement therapy for nearly four decades, mortality and morbidity is high and patients often have a poor quality of life. A human nephron filter (HNF) development could eventually enable a continuously functioning, portable, or implantable artificial kidney *(15)*. The HNF is the first application in developing a renal replacement therapy to potentially eliminate the need for dialysis or kidney transplantation in end-stage renal disease patients. The HNF utilizes a unique membrane system created through applied nanotechnology. The HNF system, by eliminating dialysate and utilizing a novel membrane system, represents a breakthrough in renal replacement therapy based on the functioning of native kidneys. The enhanced solute removal and wearable design should substantially improve patient outcomes and quality of life.

4. NANOMEDICINE ACCORDING TO THERAPEUTIC AREAS

Nanotechnology has been applied in all therapeutic areas. The most extensive investigations are in cancer.

4.1. Cancer

The application of nanotechnology to cancer can be termed nano-oncology. The major innovations are in diagnostics and drug delivery for cancer. Two nanotechnology-based products are already approved for the treatment of cancer—Doxil (a liposome preparation of doxorubicin) and Abraxane (paclitaxel in nanoparticle formulation). Approximately 150 drugs in development for cancer are based on nanotechnology. Some of the nanotechnologies and their applications in developing cancer therapies are described in this section.

4.1.1. Nanobody-Based Cancer Therapy

A nanobody with subnanomolar affinity for the human tumor-associated carcinoembryonic antigen (CEA) has been identified *(16)*. This nanobody was conjugated to *Enterobacter cloacae* beta-lactamase, and its site-selective anticancer prodrug activation capacity was evaluated. In vitro experiments showed that the nanobody–enzyme conjugate effectively activated the release of phenylenediamine mustard from the cephalosporin nitrogen mustard prodrug 7-(4-carboxybutanamido) cephalosporin mustard at the surface of CEA-expressing LS174T cancer cells. In vivo studies demonstrated that the conjugate had an excellent biodistribution profile and induced regressions and cures of established tumor xenografts. The easy generation and manufacturing yield of nanobody-based conjugates together with their potent antitumor activity make nanobodies promising vehicles for new-generation cancer therapeutics.

4.1.2. Nanoparticles for Targeting Tumors

Nanoparticles can deliver chemotherapy drugs directly to tumor cells and then give off a signal after the cells are destroyed. Drugs delivered this way are 100 times more potent than standard therapies. Gold nanoparticles can help X-rays kill cancerous cells more effectively in mice (17). The technique works because gold, which strongly absorbs X-rays, selectively accumulates in tumors. This increases the amount of energy that is deposited in the tumor compared with nearby normal tissue. Because the gold also shows up on CT and planar X-rays, it can be useful for early imaging and detection of tumors. The technique is being refined to improve targeting of the nanoparticles to tumors for human application.

Efficient conversion of strongly absorbed light by plasmonic gold nanoparticles to heat energy and their easy bioconjugation suggest their use as selective photothermal agents in molecular cancer cell targeting (18). Carcinoma cell lines incubated with anti-epithelial growth factor receptor (EGFR) antibody conjugated gold nanoparticles and then exposed to continuous visible argon ion laser at 514 nm showed that malignant cells required less than half the laser energy to be killed than the benign cells. Au nanoparticles thus offer a novel class of selective photothermal agents using a laser at low powers. The ability of gold nanoparticles to detect cancer has already been demonstrated. Now it will be possible to design an "all-in-one" active agent that can be used to noninvasively find the cancer and then destroy it. This selective technique has a potential in molecularly targeted photothermal therapy in vivo.

4.1.3. Nanoshell-Based Cancer Therapy

Nanoshells may be combined with targeting proteins and used to ablate target cells. This procedure can result in the destruction of solid tumors or possibly metastases not otherwise observable by the oncologist. In addition, Nanoshells can be utilized to reduce angiogenesis present in cancer. Experiments in animals, in vitro and in tissue, demonstrate that cancer cells can be targeted and destroyed by an amount of infrared light that is otherwise not harmful to surrounding tissue. This procedure may be performed using an infrared laser applied externally (outside the body). Prior research has indicated the ability to deliver the appropriate levels of infrared light at depths of up to 15 cm, depending upon the tissue. Photothermal tumor ablation in mice has been achieved by using near infrared-absorbing nanoparticles *(19)*. The advantages of Nanoshell-based tumor cell ablation include:

- Targeting to specific cells and tissues to avoid damage to surrounding tissue.
- Safer side-effect profile than targeted chemotherapeutic agents or photo-dynamic therapy.
- Seamless integration of cancer detection and therapy.
- Repeatability because of:
  - No "tissue memory" as in radiation therapy.
  - Biocompatibility.
  - Ability to treat metastases, and inoperable tumors.

4.1.4. Nanobomb for Cancer

Nanotechnology can used to devise a nanobomb that can literally blow up tumors. Nanobombs are bombs on nanoscale that are selective, localized, and minimally invasive. Nanoclusters (gold nanobombs) can be activated in cancer cells only by confining near-infrared laser pulse energy within the critical mass of the nanoparticles in the nanocluster *(20)*. Another approach to nanobombs is based on CNTs as drug-delivery vehicles. It is possible to trigger microscopic explosions of nanotubes under a wide variety of conditions, such as localized thermal energy imbalance, that occur in tumors. Like cluster bombs, CNTs start exploding one after another once they are exposed to light and the resulting heat. The nanobomb holds great promise as a therapeutic agent for killing cancer cells, particularly breast cancer cells, because its shockwave kills the cancerous cells as well as the biological pathways that carry instructions to generate additional cancerous cells and the small blood vessels that nourish the tumor. Its effect can be spread over a wide area to create structural damage to the surrounding cancer cells. Once the nanobombs are exploded and kill cancer cells, macrophages can effectively clear the cell debris and the exploded nanotube along with it.

4.1.5. PEBBLEs for Brain Tumor Therapy

Probes Encapsulated by Biologically Localized Embedding (PEBBLE) nanosensors consist of sensor molecules entrapped in a chemically inert matrix by a microemulsion polymerization process that produces spherical sensors in the size range of 20 to 200 nm *(21)*. These sensors are capable of real-time inter- and intracellular imaging of ions and molecules and are insensitive to interference from proteins. In human plasma, they demonstrate a robust oxygen-sensing capability, little affected by light scattering and autofluorescence *(22)*. PEBBLE has been developed further as a tool for diagnosis as well as treatment of cancer.

PEBBLEs have been designed to carry a variety of agents on their surface, each with a unique function. This multifunctionality is the major potential advantage of using nanoparticles to treat cancer. One target molecule immobilized on the surface could guide the PEBBLE to a tumor. Another agent could be used to help visualize the target using MRI, while a third agent attached to the PEBBLE could deliver a destructive dose of drug or toxin to cancer cells. All three functions can be combined in a single tiny polymer sphere to make a potent weapon against cancer.

The MRI contrast agent, gadolinium, has been incorporated in the PEBBLEs. When injected into the bloodstream, the nanoparticles wend their way through the bloodstream. But because they can transverse the BBB, and because they have a targeting agent attached, the PEBBLEs accumulate in the brain tumor enabling a clear MRI image within just a few hours. Each PEBBLE carries a photocatalyst. When stimulated by a light source through a micrometer-sized fiber-optic probe inserted into the skull, the photocatalyst converts oxygen into a singlet state, which effectively destroys nearby cells. The PEBBLEs are inert and harmless until the light is switched on. Used in combination with MRI imaging, PEBBLEs kill cancer cells at will, while tracking the effectiveness of the treatment with imaging.

The targeted treatments using nanoparticles offers a number of advantages over traditional chemotherapy. In chemotherapy, the drugs permeate cells throughout the body to damage their DNA and prevent rapid growth, and are only moderately more toxic to cancer cells over normal cells. In contrast, PEBBLEs are highly localized to the cancer target, and do very little damage to surrounding healthy tissue. PEBBLEs and other nanoparticle drugs could also avoid another serious problem occurring in traditional chemotherapy—development of multi-drug resistance, which occurs when cancer cells mutate and begin to pump the chemotherapy drugs back out before they can destroy the cell. The cancer becomes resistant to the drug. But PEBBLEs act on the outside of the cell, and the toxic payload of oxygen that they deliver acts quickly, without giving the cancer much chance to survive and develop resistance. In rat models of brain cancer, 9L-gliosarcoma, PEBBLE-based treatment can significantly increase survival time from 5 d without treatment to 2 mo with MRI image showing elimination of the tumor *(23)*. The investigators hope ultimately to prove the utility and safety of this approach to treating brain cancer in humans.

4.2. Cardiovascular Diseases

Recent rapid advances in nanotechnology and nanoscience offer a wealth of new opportunities for diagnosis and therapy of cardiovascular diseases. To review the challenges and opportunities offered by these nascent fields, the US National Heart, Lung, and Blood Institute convened a Working Group on Nanotechnology. Working Group participants discussed the various aspects of nanotechnology and its applications to heart, lung, and blood disorders and the cardiovascular complications of sleep apnea. An overall recommendation of the Working Group was to focus on translational applications of nanotechnology to solve clinical problems *(24)*. The Working Group recommended the creation of multidisciplinary research centers capable of developing applications of nanotechnology and nanoscience to research and medicine. Centers would also disseminate technology, materials, and resources and train new investigators. Individual investigators outside these centers should be encouraged to conduct research on the application of nanotechnology to biological and clinical problems.

4.2.1. Cardiac Monitoring in Sleep Apnea

Because sleep apnea is a cause of irregular heartbeat, hypertension, heart attack, and stroke, it is important that patients be diagnosed and treated before these highly deleterious sequelae occur. For patients suspected of experiencing sleep apnea, in vivo sensors could constantly monitor blood concentrations of oxygen and cardiac function to detect problems during sleep. In addition, cardio-specific antibodies tagged with nanoparticles may allow physicians to visualize heart movement while a patient experiences sleep apnea to determine both short- and long-term effects of apnea on cardiac function.

4.2.2. Unstable Plaques in the Arteries

The diagnosis and treatment of unstable plaque is an area in which nanotechnology could have an immediate impact. Research is under way using probes targeted to plaque components for noninvasive detection of patients at risk. In an extension of this approach, targeted nanoparticles, multi-functional macromolecules, or nanotechnology-based devices could deliver therapy to a specific site, localized drug release being achieved either passively (by proximity alone) or actively (through supply of energy as ultrasound, near-infrared, or magnetic field). Targeted nanoparticles or devices could also stabilize vulnerable plaque by removing material, e.g., oxidized low-density lipoproteins. Devices able to attach to unstable plaques and warn patients and emergency medical services of plaque rupture would facilitate timely medical intervention.

4.2.3. Tissue Engineering and Regeneration of the Cardiovascular System

Nanotechnology may facilitate repair and replacement of blood vessels, myocardium, and myocardial valves. It also may be used to stimulate regenerative processes such as therapeutic angiogenesis for ischemic heart disease. Cellular function is integrally related to morphology, so the ability to control cell shape in tissue engineering is essential to ensure proper cellular function in final products. Precisely constructed nanoscaffolds and microscaffolds are needed to guide tissue repair and replacement in blood vessels and organs. Nanofiber meshes may enable vascular grafts with superior mechanical properties to avoid patency problems common in synthetic grafts, particularly small-diameter grafts. Cytokines, growth factors, and angiogenic factors can be encapsulated in biodegradable microparticles or nanoparticles and embedded in tissue scaffolds and substrates to enhance tissue regeneration. Scaffolds capable of mimicking cellular matrices should be able to stimulate the growth of new heart tissue and direct revascularization.

4.2.4. Restenosis after Percutaneous Coronary Angioplasty

Restenosis after percutaneous coronary intervention continues to be a serious problem in clinical cardiology. Recent advances in nanoparticle technology have enabled the delivery of NK911, an antiproliferative drug, selectively to the balloon-injured artery for a longer time *(25)*. NK911 is a core-shell nanoparticle of polyoxyethylene glycol (PEG)-based block copolymer encapsulating doxorubicin. It accumulates in vascular lesions with increased permeability. In a balloon injury model of the rat carotid artery, intravenous administration of NK911 significantly inhibited the neointimal formation. The effect of NK911 was due to inhibition of vascular smooth muscle proliferation but not to enhancement of apoptosis or inhibition of inflammatory cell recruitment. NK911 was well tolerated, without any adverse systemic effects. These results suggest that nanoparticle technology is a promising and safe approach to target vascular lesions with increased permeability for the prevention of restenosis after balloon injury.

Currently available stents implanted in arterial lumens have problems with imaging within the stent structure, where potential restenosis can occur. A thin-film nanomagnetic particle coating solution can enable the noninvasive, MRI-based imaging of these devices. Nitric oxide (NO)-eluting nanofibers are being developed for incorporation into drug-eluting stents for antithrombogenic action. NO has vasodilating action as well, which may be beneficial in ischemic heart disease.

4.3. Diseases of Bones and Joints

4.3.1. Reducing Reaction to Orthopedic Implants

In orthopedic implants, biomaterials (usually titanium and/or titanium alloys) often become encapsulated with undesirable soft, fibrous—but not hard—bony tissue. Although possessing intriguing electrical and mechanical properties for neural and orthopedic applications, carbon nanofibers/ nanotubes have not been widely considered for these applications previously. A carbon nanofiber (CN)-reinforced polycarbonate urethane (PU) composite has been developed in an attempt to determine the possibility of using CNs as orthopedic prosthetic devices *(26)*. Mechanical characterization studies show that such composites have properties suitable for orthopedic applications. These materials enhanced osteoblast (bone-forming cell) functions, whereas functions of cells that contribute to fibrous-tissue encapsulation events for bone implants (fibroblasts) decreased on PU composites containing increasing amounts of CNs. In this manner, this study provided the first evidence of the future that CN formulations may have in interacting with bone cells, which is important for the design of successful neural probes and orthopedic implants.

4.3.2. Nanotechnology-Based Scaffolds for Bone Growth

Artificial bone scaffolds have been made from a wide variety of materials, such as polymers or peptide fibers. Their drawbacks include low strength and the potential for rejection in the body. Chemically functionalized single-walled carbon nanotubes (SWNTs) have been used as scaffolds for the growth of artificial bone material *(27)*. The strength, flexibility, and light weight of SWNTs enable them to act as scaffolds to hold up regenerating bone. Bone tissue is a natural composite of collagen fibers and crystalline hydroxyapatite, which is a mineral based on calcium phosphate. SWNTs can mimic the role of collagen as a scaffold for inducing the growth of hydroxyapatite crystals. By chemically treating the nanotubes, it is possible to attract calcium ions and to promote the crystallization process while improving the biocompatibility of the nanotubes by increasing their water solubility. SWNTs may lead to improved flexibility and strength of artificial bone, new types of bone grafts, and to inroads in the treatment of osteoporosis and fractures.

Bone cells can grow and proliferate on a scaffold of CNTs. Because CNTs are not biodegradable, they behave like an inert matrix on which cells can proliferate and deposit new living material, which becomes functional, normal bone *(28)*. CNTs carrying a neutral electric charge sustained the highest cell growth and production of plate-shaped crystals. There was a dramatic change in cell morphology in osteoblasts cultured on multiwalled CNTs, which correlated with changes in plasma membrane functions. CNTs hold promise in the treatment of bone defects in humans associated with the removal of tumors, trauma, and abnormal bone development and in dental implants. More research is needed to determine how the body will interact with CNTs, specifically in its immune response.

One study has assessed bone formation from mesenchymal stem cells (MSCs) on a novel nanofibrous scaffold in a rat model *(29)*. A highly porous, degradable poly(-caprolactone) scaffold with an ECM-like topography was produced by electrostatic fiber spinning. MSCs derived from the bone marrow of neonatal rats were cultured, expanded, and seeded on the scaffolds. The cell–polymer constructs were cultured with osteogenic supplements and maintained the size and shape of the original scaffolds when explanted. Morphologically, the constructs were rigid and had a bone-like appearance. Cells and ECM formation were observed throughout the constructs. In addition, mineralization and type I collagen were also detected. This study establishes the ability to develop bone grafts on electrospun nanofibrous scaffolds in a well vascularized site using MSCs.

4.4. Diseases of the Nervous System

Two important issues in management of diseases of the CNS are regeneration and neuroprotection; both can be facilitated by nanobiotechnology. The ultimate goal is to develop novel technologies that directly or indirectly aid in providing neuroprotection and/or a permissive environment and active signaling cues for guided axon growth.

4.4.1. Application of Nanotechnology for Neuroprotection

Nanobiotechnology is being used to help neuroscientists better understand the physiology of and develop treatments for disorders such as brain injury, spinal cord injury, degenerative retinal disorders, and Alzheimer's disease in order to develop strategies for neuroprotection *(30)*. Quantum dot (QD) technology is used to gather information about how the CNS environment becomes inhospitable to neuronal regeneration following injury or degenerative events by studying the process of reactive gliosis. Glial cells, housekeeping cells for neurons, have their own communication mechanisms that can be triggered to become reactive following injury. QDs are being used to build data capture devices that are easy to use by neuroscientists, and a new protocol has been developed for tracking glial activity. Other research is looking at how QDs might spur the growth of neurites by adding bioactive molecules to the QDs as a means to provide a medium that will encourage this growth in a directed way.

4.4.2. Nanotechnology for Repair of CNS Injuries

One of the major problems CNS injuries is repair of neural tissues: these do not regenerate spontaneously, and there are no satisfactory methods of neural transplants. Nanotechnology-based methods have been useful for growing nerve cells in tissue cultures. Neural progenitor cells have been encapsulated in vitro within a three-dimensional network of nanofibers formed by self-assembly of peptide amphiphile molecules *(31)*. The self-assembly of the nanofiber scaffold was initiated by mixing cell suspensions in media with dilute aqueous solutions of the molecules, and cells survived the growth of the nanofibers around them. These nanofibers were designed to present to cells the neurite-promoting laminin epitope. Relative to laminin or soluble peptide, the artificial nanofiber scaffold induced very rapid differentiation of neural progenitor cells into neurons, while discouraging the development of astrocytes. These new materials, because of their chemical structure, interact with cells of the CNS in ways that may help prevent the formation of scar due to astrocyte proliferation, which is often linked to paralysis resulting from traumatic spinal cord injury.

4.5. Eye Diseases

Some of the strategies for treating eye disorders involve prevention of neovascularization. Examples of how nanotechnology can refine these procedures are as follows.

Photodynamic therapy (PDT) has been used for exudative age-related macular degeneration (AMD). This therapy can be refined by using a supramolecular nanomedical device, i.e., a novel dendritic photosensitizer (DP) encapsulated by a polymeric micelle formulation *(32)*. The characteristic dendritic structure of the DP prevents aggregation of its core sensitizer, thereby inducing a highly effective photochemical reaction. With its highly selective accumulation on choroidal neovascularization (CNV) lesions, this treatment results in a remarkably efficacious CNV occlusion with minimal unfavorable phototoxicity.

Dendrimer glucosamine 6-sulfate has been shown to block fibroblast growth factor (FGF)-2-mediated endothelial cell proliferation and neoangiogenesis in human Matrigel and placental angiogenesis assays *(33)*. When dendrimer glucosamine and dendrimer glucosamine 6-sulfate were used together in a validated and clinically relevant rabbit model of scar tissue formation after glaucoma filtration surgery, they increased the long-term success of the surgery from 30% to 80%. Synthetically engineered dendrimers can be tailored to have defined immuno-modulatory and antiangiogenic properties, and they can be used synergistically to prevent scar tissue formation.

4.6. Infections

Nanobiotechnology has improved the diagnosis of infectious diseases. Even a few viruses or bacteria can be detected. Furthermore, biotechnology provides many opportunities for microbicidal agents.

4.6.1. Nanotubes for Detection and Destruction of Bacteria

A simple molecule has been synthesized from a hydrocarbon and an ammonium compound to produce a unique nanotube structure with antimicrobial capability *(34)*. The quaternary ammonium compound is known for its ability to disrupt cell membranes and causes cell death, whereas the hydrocarbon diacetylene can change colors when appropriately formulated; the resulting molecule would have the desired properties of both a biosensor and a biocide. In the presence of *Escherichia coli*, some strains of which are food-borne pathogens, the nanotubes turned shades of red and pink. Moreover, with the aid of an electron microscope, the tubes were observed to pierce the membranes of the bacteria like a needle being inserted into the cell. Both the polymerized (those that can change color) and the unpolymerized nanotube structures were effective antimicrobials, completely killing all the *E. coli* within an hour's time. The findings have implications for developing products that can simultaneously detect and kill biological weapons.

4.6.2. Nanoemulsions as Microbicidal Agents

The antimicrobial nanoemulsions (NanoBio Inc.) are emulsions that contain water and soya bean oil with uniformly sized droplets in the 200–400 nm range. These droplets are stabilized by surfactant and are responsible for the microbicidal activity. In concentrated form, the nanoemulsions appear as a white milky substance with the taste and consistency of cream. They can be formulated in a variety of carriers allowing for gels, creams, liquid products, etc. In most applications, the nanoemulsions become largely water-based, and in some cases, such as a beverage preservative, comprise 0.01% or less of the resultant mixture. Laboratory results indicate a shelf life of at least 2 yr and virtually no toxicity. The nanoemulsions destroy microbes effectively without toxicity or harmful residual effects. The nanoparticles fuse with the membrane of the microbe and the surfactant disrupts the membrane, killing the microbe. The classes of microbes eradicated are virus (e.g., HIV, herpes), bacteria (e.g., *E. coli*, Salmonella), spores (e.g., anthrax), and fungi (e.g., *Candida albicans*). Clinical trials have shown efficacy in healing cold sores due to herpes simplex virus 1 and toenail fungus. The nanoemulsions also can be formulated to kill only one or two classes of microbes. As a result in large part of the low toxicity profile, the nanoemulsions are a platform technology for any number of topical, oral, vaginal, cutaneous, preservative, decontamination, veterinary, and agricultural antimicrobial applications.

4.6.3. Silver Nanoparticle Coating as Prophylaxis Against Infection

The Institute for New Materials (Saarbrucken, Germany), a research institute specializing in applied nanotechnology applications, has developed a silver nanoparticle surface coating that is deadly to fungi and bacteria. Applications include any surface where microorganisms can gather and possibly endanger public health, including surfaces in hospitals, public buildings, and factories. The coating could be applied to almost any surface that people touch often, such as metal, glass, or plastic, and would eliminate the need for constant cleaning with liquid disinfectants, especially in areas where hygienic conditions are crucial. People who normally cannot use hearing aids that lie inside the ear because of the risk of infection of the auditory canal can safely wear nanocoated appliances. Commercial preparations of silver nanoparticles are available.

4.7. Skin Disorders

4.7.1. Nanoparticles for Improving Targeted Topical Therapy of Skin

Long-term topical glucocorticoid treatment can induce skin atrophy by the inhibition of fibroblasts. Therefore, there is a need for drug carriers that may contribute to a reduction of this risk by an epidermal targeting. Prednicarbate (PC; 0.25%) was incorporated into solid lipid nanoparticles of various compositions, and studies were conducted in which conventional PC cream of 0.25% and ointment served as reference *(35)*. Local tolerability as well as drug penetration and metabolism were studied in excised human skin and reconstructed epidermis. PC incorporation into nanoparticles appeared to induce a localizing effect in the epidermal layer that was pronounced at 6 h and declined later. Dilution of the PC-loaded nanoparticle preparation with cream did not reduce the targeting effect, whereas adding drug-free nanoparticles to PC cream did not induce PC targeting. Therefore, the targeting effect is closely related to the PC nanoparticles and is not a result of either the specific lipid or PC adsorbance to the surface of the formerly drug-free nanoparticles. Lipid nanoparticle-induced epidermal targeting may increase the benefit/risk ratio of topical therapy.

4.7.2. Topical Nanocreams for Inflammatory Disorders of the Skin

Inflammatory skin diseases, including atopic dermatitis and psoriasis, are common. The current treatment is unsatisfactory, although several topical and systemic therapies, including steroids and immunomodulators, are available. Their efficacy is not durable and they are associated with adverse effects. Efforts continue to develop safer alternative treatment for these disorders. Nanocrystalline silver has been demonstrated to have exceptional antimicrobial properties, and has been successfully used in wound healing. Studies conducted by Nucryst Pharmaceuticals have revealed that topical application of Nanocrystalline silver cream (0.5%, and 1%) ointment produced significant suppressive effects on allergic contact dermatitis in a guinea pig model. There was a clear concentration–response relationship to the decrease of inflammation, as lower concentrations were not effective. The effects were equivalent to the immunosuppressant tacrolimus ointment. This study suggests that nanocrystalline silver cream has therapeutic potential for treating inflammatory skin diseases.

4.8. Wound Healing

Several nanotechnology-based products have been used for wound care. Polyurethane membrane, produced via electrospinning (a process by which nanofibers can be produced by an electrostatically driven jet of polymer solution), is particularly useful as a wound dressing because of the following properties: it soaks fluid from the wound so that it does not build up under the covering, and does not cause wound desiccation (36). Water loss by evaporation is controlled, there is excellent oxygen permeability, and exogenous microorganism invasion is inhibited because of the ultra-fine pores size. Histological examination of the wound shows that the rate of epithelialization is increased and the dermis becomes well organized if wounds are covered with an electrospun nanofibrous membrane. This membrane has potential applications for wound dressing.

Silver nanoparticles have been incorporated in preparations for wound care to prevent infection. Acticoat bandages (Smith & Nephew) contain nanocrystal silver, which is highly toxic to pathogens in wounds.

5. SAFETY ISSUES OF NANOPARTICLES IN THE HUMAN BODY

The success of nanomaterials is due to their small size, which enables us to get them into parts of the body where usual inorganic materials cannot enter because of their large particle size. There is an enormous advantage in drug-delivery systems or cancer therapeutics. Current research is trying to find simple ways to control the degree of a particle's toxicity. This control means that the particle will be toxic only under certain desirable circumstances, such as for curing cancer. This also raises questions about unintentional effects of such powerful agents on the human body. This, however, would not be an issue for the use of nanoparticles for in vitro diagnostics.

Effects of particles on human health have been studied previously by toxicologists, as they are present in environments. Effects of larger particles generated by the wearing down of implants in the body and aerosolized particles of all sizes have also been studied. However, there is little information on the health impacts of nano-engineered particles under 20 nm. The main concern will be about particles less than 50 nm, which can enter the cells.

The biological impacts of nanoparticles are dependent on size, chemical composition, surface structure, solubility, shape, and aggregation. These parameters can affect cellular uptake, protein binding, translocation from portal of entry to the target site, and the possibility of causing tissue injury. The effects of nanoparticles depend on the routes of exposure, including gastrointestinal tract, skin, lung, and systemic administration for diagnostic and therapeutic purposes. The interaction of nanoparticles with cells, body fluids, and proteins play a role in their biological effects and ability to distribute themselves throughout the body. Nanoparticle binding to proteins may generate complexes that are more mobile and can enter tissue sites that are normally inaccessible. Accelerated protein denaturation or degradation on the nanoparticle surface may lead to functional and structural changes, including interference in enzyme function. Nanoparticles also encounter a number of defenses that can eliminate, sequester, or dissolve them. There are still many unanswered questions about their fate in the living body. Because of the huge diversity of materials used and the wide range in size of nanoparticles, these effects will vary greatly. It is conceivable that the particular sizes of some materials may turn out to have toxic effects. At this stage, no categorical statement can be made about the safety of nanoparticles, i.e., one cannot say that nanoparticles are entirely safe or that they are dangerous. Further investigations will be needed. Currently available information about toxicity studies of various nanoparticles have been summarized elsewhere *(1)*.

6. FUTURE OF NANOMEDICINE

6.1. Place in Future Healthcare Systems

Remarkable changes are taking place in the practice of medicine in the 21st century. Developments in biotechnology, particularly genomics and proteomics, during the past decade are having a clinical impact. The number of biotechnology-based drugs is increasing. There is a concern about the adverse effects of drugs, and efforts are being made to find safer and more effective drugs. Nanobiotechnology has an assured place in contributing to an increased understanding of the pathomechanism of disease, the design of better treatments, the improvement of drug delivery, and, finally, the personalization of medicine. All of these advances would justify the term "nanomedicine."

6.2. Role of Nanobiotechnology in Development of Personalized Medicine

Personalized medicine is beginning to be recognized, and is expected to become a part of medical practice within the next decade. Nanobiotechnology will facilitate the development of personalized medicine by:

- Improving the understanding of molecular mechanism of disease.
- Refining molecular diagnostics for early detection and monitoring of disease as well as therapy.
- Providing the possibility of integrating diagnostics with therapeutics.

One example of the role of nanobiotechnology in personalized medicine is in cancer, where it will facilitate both diagnosis and drug delivery *(37)*. This would be the ultimate in personalized management of cancer. With current advances in nanotechnology, it is feasible to design a miniature robotic device that can be introduced in the body to locate, identify, and destroy cancer cells while its in vivo activities are monitored by an external device. Such a device could integrate diagnostics and therapeutics by incorporating a nanobiosensor to identify cancer cells and carry a supply of nanoparticulate anticancer substances that could be released on encountering cancer cells.

6.3. Concluding Remarks

Nanotechnology will also provide devices to examine tissue in minute detail. Biosensors that are smaller than a cell would give us an inside look at cellular function. Tissues could be analyzed down to the molecular level, giving a completely detailed "snapshot" of cellular, subcellular, and molecular activities. Such a detailed diagnosis would guide the appropriate treatment.

Nanobiotechnology has greatly improved drug delivery to the desired site of action. This will improve the efficacy and reduce the toxicity of drugs. It is expected that within the next few years, we will have a better understanding of how to coat or chemically alter nanoparticles to reduce their toxicity to the body, which will allow us to broaden their use for disease diagnosis and for drug delivery. Two nanotechnology-based anticancer drugs are already on the market, and more are expected to be approved in the near future.

Disease and other disturbances of function are caused largely by damage at the molecular and cellular level, but current surgical tools are large and crude. Even a fine scalpel is a weapon more suited for tearing and injuring than healing and curing. It would make more sense to operate at the cell level to correct the cause of disease, rather than to cut off large lesions as a result of the disturbances at cell level.

Nanotechnology will enable construction of computer-controlled molecular tools that are much smaller than a human cell and built with the accuracy and precision of drug molecules. Such tools will be used for interventions in a refined and controlled manner at the cellular and molecular levels. They could remove obstructions in the circulatory system, kill cancer cells, or take over the function of subcellular organelles. Instead of transplanting artificial hearts, a surgeon of the future would be transplanting artificial mitochondrion.

REFERENCES

1. Jain KK. A Handbook of Nanomedicine. Totowa, NJ: Humana Press/Springer, 2007 (in press).
2. Jain KK. Nanobiotechnology in Molecular Diagnostics. Norwich, UK: Horizon Scientific Press, 2007:1–608.
3. Jain KK. Biomarkers: Technologies, Markets and Companies. Basel: Jain PharmaBiotech Publications, 2007:1–622.
4. Lesniak W, Bielinska AU, Sun K, et al. Silver/dendrimer nanocomposites as biomarkers: fabrication, characterization, in vitro toxicity, and intracellular detection. Nano Lett 2005;5:2123–2130.
5. Saleh A, Schroeter M, Jonkmanns C, et al. In vivo MRI of brain inflammation in human ischaemic stroke. Brain 2004;127(Pt 7):1670–1677.
6. Zhang J, Yang G, Cheng, et al. Stationary scanning x-ray source based on carbon nanotube field emitters. Appl Phys Lett 2005;86:DOI:10.1063/1.1923750.
7. Jain KK. Handbook of Laser Neurosurgery. Springfield, IL: Charles C. Thomas, 1983.
8. Sacconi L, Tolic-Norrelykke IM, Antolini R, Pavone FS. Combined intracellular three-dimensional imaging and selective nanosurgery by a nonlinear microscope. J Biomed Opt 2005;10:14002.
9. Yanik MF, Cinar H, Cinar HN, et al. Neurosurgery: functional regeneration after laser axotomy. Nature 2004;432:822.
10. Riviere CN, Patronik NA, Zenati MA. Prototype epicardial crawling device for intrapericardial intervention on the beating heart. Heart Surg Forum 2004;7:E639–E643.
11. Raju GS, Nath SK. Capsule endoscopy. Curr Gastroenterol Rep 2005;7:358–364.
12. Emerich DF, Thanos CG. Nanotechnology and medicine. Expert Opin Biol Ther 2003;3:655–663.
13. Ma Z, Kotaki M, Inai R, et al. Potential of nanofiber matrix as tissue-engineering scaffolds. Tissue Eng 2005;11:101–109.
14. Morelli AE, Larregina AT, Shufesky WJ, et al. Endocytosis, intracellular sorting, and processing of exosomes by dendritic cells. Blood 2004;104:3257–3266.
15. Nissenson AR, Ronco C, Pergamit G, et al. The human nephron filter: toward a continuously functioning, implantable artificial nephron system. Blood Purification 2005;23:269–274.
16. Cortez-Retamozo V, Backmann N, Senter PD, et al. Efficient cancer therapy with a nanobody-based conjugate. Cancer Res 2004;64:2853–2857.
17. Hainfeld J, Slatkin DN, Smilowitz HM. The use of gold nanoparticles to enhance radiotherapy in mice. Phys Med Biol 2004;49:N309–N315.
18. El-Sayed IH, Huang X, El-Sayed M. Selective laser photo-thermal therapy of epithelial carcinoma using anti-EGFR antibody conjugated gold nanoparticles. Cancer Lett 2006;239:129–135.
19. O'Neal DP, Hirsch LR, Halas NJ, et al. Photo-thermal tumor ablation in mice using near infrared-absorbing nanoparticles. Cancer Lett 2004;209:171–176.
20. Zharov VP, Galitovskaya EN, Johnson C, Kelly T. Synergistic enhancement of selective nanophotothermolysis with gold nanoclusters: potential for cancer therapy. Lasers Surg Med 2005;37:219–226.

21. Sumner JP, Aylott JW, Monson E, Kopelman R. A fluorescent PEBBLE nanosensor for intracellular free zinc. Analyst 2002;127:11–16.
22. Cao Y, Lee Koo YE, Kopelman R. Poly(decyl methacrylate)-based fluorescent PEBBLE swarm nanosensors for measuring dissolved oxygen in biosamples. Analyst 2004;129:745–750.
23. Kopelman R, Philbert M, Koo YEL, et al. Multifunctional nanoparticle platforms for in vivo MRI enhancement and photodynamic therapy of a rat brain cancer. J Magnetism Magnetic Mater 2005;293:404–410.
24. Buxton DB, Lee SC, Wickline SA, et al. Recommendations of the National Heart, Lung, and Blood Institute Nanotechnology Working Group. Circulation 2003;108:2737–2742.
25. Uwatoku T, Shimokawa H, Abe K, et al. Application of nanoparticle technology for the prevention of restenosis after balloon injury in rats. Circ Res 2003;92: e62–e69.
26. Webster TJ, Waid MC, McKenzie JL, et al. Nano-biotechnology: carbon nanofibres as improved neural and orthopaedic implants. Nanotechnology 2004;15:48–54.
27. Zhao B, Hu H, Mandal SK, Haddon RC. A bone mimic based on the self-assembly of hydroxyapatite on chemically functionalized single-walled carbon nanotubes. Chem Mater 2005;17:3235–3241.
28. Zanello LP, Zhao B, Hu H, Haddon RC. Bone cell proliferation on carbon nanotubes. Nano Lett 2006;6:562–567.
29. Shin M. In vivo bone tissue engineering using mesenchymal stem cells on a novel electrospun nanofibrous scaffold. Tissue Eng 2004;10:33–41.
30. Silva GA. Nanotechnology approaches for the regeneration and neuroprotection of the central nervous system. Surg Neurol 2005;63:301–306.
31. Silva GA, Czeisler C, Niece KL, et al. Selective differentiation of neural progenitor cells by high-epitope density nanofibers. Science 2004;303:1352–1355.
32. Ideta R, Tasaka F, Jang WD, et al. Nanotechnology-based photodynamic therapy for neovascular disease using a supramolecular nanocarrier loaded with a dendritic photosensitizer. Nano Lett 2005;5:2426–2431.
33. Shaunak S, Thomas S, Gianasi E, et al. Polyvalent dendrimer glucosamine conjugates prevent scar tissue formation. Nat Biotech 2004;22:977–984.
34. Lee SB, Koepsel R, Stolz DB, et al. Self-assembly of biocidal nanotubes from a single-chain diacetylene amine salt. J Am Chem Soc 2004;126:13,400–13,405.
35. Santos Maia C, Mehnert W, Schaller M, et al. Drug targeting by solid lipid nanoparticles for dermal use. J Drug Target 2002;10:489–495.
36. Khil MS, Cha DI, Kim HY, et al. Electrospun nanofibrous polyurethane membrane as wound dressing. J Biomed Mater Res B Appl Biomater 2003;67: 675–679.
37. Jain KK. Role of nanobiotechnology in developing personalized medicine for cancer. TCRT 2005;4:407–416.

13
Nano-Sized Carriers for Drug Delivery

Sanjeeb K. Sahoo, Tapan K. Jain, Maram K. Reddy,
and Vinod Labhasetwar

Summary

Drug delivery is an important issue, especially with a new generation of therapeutics, which are either unstable in the biological environment, have poor transport
properties across biological membranes, are insoluble in water, or have very low
bioavailability. Nano-sized drug carriers can address some of the above issues and
enhance their therapeutic efficacy. Different types of nano-sized carriers, such as
nanoparticles, nanowires, nanocages, dendrimers, etc., are being developed for
various drug-delivery applications. The challenge is to determine the therapeutic
dose of the drug formulated in a system, which could be significantly different from
that of the drug nanocarrier. In this regard, a better understanding of the patho-
physiology of the disease condition under consideration is critical so that one can select
and design an appropriate drug carrier system that would deliver a therapeutic dose
of the drug in the target tissue or a body compartment at a rate and for the duration
that is therapeutically effective and can cure the disease.

Key Words: Conjugates; drug therapy; nanomedicine; nanosystems; polymers;
targeting.

1. INTRODUCTION

The National Cancer Institute, under the National Nanotechnology
Initiative Program, recently defined nano-sized drug carriers as those which
are typically 300 nm or smaller in size (1). The main objective of developing nano-sized drug carriers is to enhance the therapeutic potential of drugs
so that they are less toxic and more effective. The goal is to alter the biodistribution of therapeutic agents so that they concentrate more in the target
tissue. The new generation of therapeutics, which have been developed as
a result of research in genomics and proteonomics, have a different set of
drug-delivery issues. Although these molecules are potent and specific in

From: *NanoBioTechonology: BioInspired Devices and Materials of the Future*
Edited by: Oded Shoseyov and Ilan Levy © Humana Press Inc., Totowa, NJ

their action, their instability in the biological environment and, in some instances, their inaccessibility to the target site limit their therapeutic utility. In general, the objectives of drug delivery are to: (1) design nano-sized drug carrier systems that can incorporate different types of therapeutic agents in sufficient doses, (2) identify disease-specific ligands that can be conjugated to drug carrier systems to achieve targeted drug therapy, (3) protect drug molecules from degradation in the body prior to their reaching the target, (4) develop drug carrier systems that can release the drug at the target site and at a desired or controllable rate, and for the duration necessary to elicit the desired pharmacological response, (5) achieve effective intracellular drug delivery for those therapeutic agents whose receptor or site of action is intracellular, and (6) develop nano-sized drug carrier systems that are biocompatible and biodegradable so that these can be used safely in humans. Because of the complex nature of these issues, drug delivery has become an integral part of drug discovery and development. Nano-sized drug carrier systems are expected to play a critical role in overcoming some of the above issues *(2)*. In this book chapter, we review different nano-sized drug carrier systems (Fig. 1) and briefly describe their therapeutic applications in drug delivery.

2. NANO-SIZED DRUG CARRIER SYSTEMS

2.1. Solid Lipid Nanoparticles

Solid lipid nanoparticles (SLN) were developed at the beginning of the 1990s as an alternative to emulsions and liposomes for controlled drug delivery *(3)*. These particles are made from solid lipids (i.e., lipids that are solid at room temperature and also at body temperature) and are stabilized by surfactant(s). SLN can be formulated using highly purified triglycerides, complex glyceride mixtures, or even waxes. SLN have certain advantages over liposomes or emulsion for drug delivery applications, such as their good tolerability and biodegradation *(4)*, and their use in a wide-range of drug delivery applications, such as to improve ocular bioavailability of drugs *(5)*, for targeting of drugs to the brain *(6)*, and for drug delivery via parenteral *(7,8)*, pulmonary, and dermal routes *(9,10)*. These can be made "stealth" by incorporating polyoxyethylene glycol (PEG) or Pluronics into the formulation so that they are not recognized by the body's defense system and cleared rapidly by the reticuloendothelial system (RES) *(11)*. Stealth SLN have been shown to increase the tumor accumulation of antineoplastic agents, as well as can transport drugs to the brain which are otherwise incapable of crossing the blood-brain barrier (BBB) *(12)*.

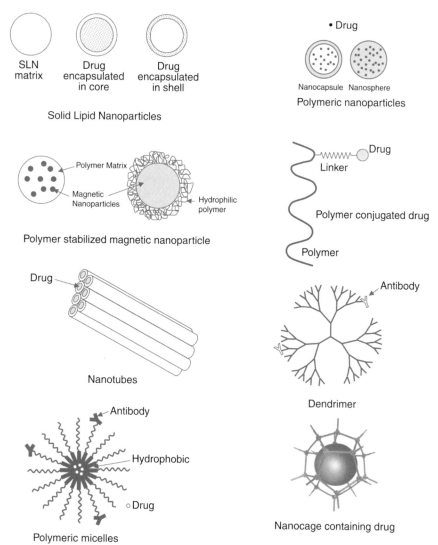

Fig. 1. Schematic drawing of different types of nano-sized carriers for drug delivery.

2.2. Polymeric Nanoparticles

Over the past few decades, there has been considerable interest in developing biodegradable nanoparticles as a drug delivery system *(13–15)*. Depending on the process used for their preparation, these can be nanoparticles, nanospheres, or nanocapsules. Nanospheres have a matrix-like structure, where active compounds can be either firmly adsorbed at their surface, entrapped, or dissolved

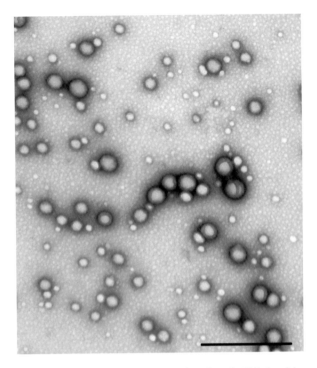

Fig. 2. Transmission electron micrograph of poly(DL-lactide-co-glycolide) nanoparticles (Bar = 500 nm).

in the matrix. Nanocapsules have a polymeric shell and an inner core. In this case, an active substance is usually dissolved in the core, but can be adsorbed at their surface *(13,14)*. The main advantages of nanoparticles for drug delivery applications is their small size, hence they can penetrate through smaller capillaries and are taken up by cells, which could allow efficient drug accumulation at the target sites *(16–19)*. Biodegradable materials used for the formulation of nanoparticles result in sustained drug release within the target site over a period of days or even weeks. Our laboratory has been investigating biodegradable nanoparticles formulated from poly dl-lactide co-glycolide (PLGA) and polylactide (PLA) polymers for sustained drug delivery (Fig. 2) *(14,20,21)*. Our interest also lies in studying their intracellular trafficking and determining the parameters that are critical to their efficient cellular uptake and retention. Recently, we have demonstrated rapid escape of nanoparticles from the endo-lysosomal compartment to the cytoplasmic compartment *(20)*. Greater and sustained antiproliferative activity was observed in vascular smooth muscle cells that were treated with dexamethasone-loaded nanoparticles whereas with drug in solution the effect was transient *(22)*. Nanoparticles were effective in sustaining intracellular dexamethasone levels, thus allowing a more efficient interaction of the drug with the glucocorticoid receptors, which are cytoplasmic *(22)*.

Multidrug resistance (MDR) is one of the major causes of treatment failure in cancer therapy, which is attributed to the reduced accumulation of drug in the tumor in addition to the possibility of membrane glycoprotein (P-gp)-dependent accelerated drug efflux *(23,24)*. To overcome the problem of efflux action of P-gp and to sustain the drug effect, various drug delivery systems have been developed. For example, doxorubicin formulated in nanoparticles following its chemical conjugation to the terminal end group of polymer by an ester linkage demonstrated sustained release over 1 mo *(25)*. Paclitaxel, a naturally occurring antineoplastic drug, despite its high potency, has posed challenges because of its poor solubility and low permeability associated with mucosal P-gp efflux action in cancer cells. Cremophor EL employed as a solubilizer caused serious side effects. Mu and Feng formulated a novel PLGA nanoparticle loaded with paclitaxel using vitamin E D-α-tocopheryl polyethylene glycol 1000 succinate (TPGS) as emulsifier. These nanoparticles demonstrated better drug encapsulation, uniform size distribution, morphological and physicochemical properties, and sustained drug release characteristics *(26)*.

Moghimi et al. recently reviewed applications of nanoparticles for targeted drug delivery *(27)*. Targeted delivery can be achieved by either active or passive targeting. Active targeting of a therapeutic agent is achieved by conjugating a therapeutic agent or the carrier system to a tissue- or cell-specific ligand. For example, if tumor-targeted drug delivery is desired, monoclonal antibodies such as antinuclear auto-antibodies (ANAs) can be attached to drug-loaded nanoparticles to promote drug delivery to tumor cells but not to normal cells *(28)*. ANAs have nucleosome-restricted specificity, which enables them to recognize the surface of tumor cells. Passive targeting is achieved as a result of enhanced permeation retention (EPR) effect *(29)*. This occurs because tumor vasculature is leaky; hence, circulating nanoparticles can accumulate more in the tumor tissue than in normal tissue. Another approach is the direct intratumoral delivery of anticancer agents using nanoparticles, an approach that can be used in the treatment of local cancers such as prostate or head and neck cancers. Recently, we have demonstrated that transferrin (Tf)-conjugated paclitaxel (Tx)-loaded biodegradable nanoparticles are more effective in demonstrating an antiproliferative effect of the drug than drug in solution or with unconjugated drug-loaded nanoparticles (Tx–NPs). In a human prostate cancer cell line (PC3), the IC_{50} (concentration of drug for 50% inhibition of cell growth) with Tf-conjugated nanoparticles was fivefold lower than that with Tx-NPs or drug in solution.

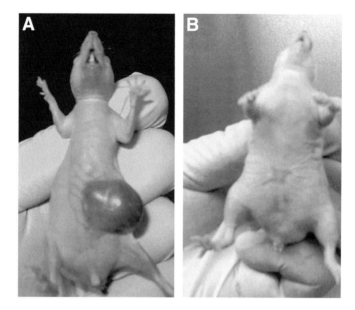

Fig. 3. Tumor regression with drug delivery in prostate tumor model. A representative animal from the group in which animals were treated with control nanoparticles (**A**) and transferrin-conjugated paclitaxel-loaded nanoparticles (**B**) at 45 d. PC3 cells (2×10^6 cells) were injected subcutaneously in athymic nude mice. Tumor nodules were allowed to grow to a diameter of about 5 to 6 mm prior to receiving a single intratumoral injection of nanoparticles. Animals that received transferrin-conjugated drug-loaded nanoparticles demonstrated complete tumor regression.

The better efficacy of conjugated nanoparticles was due to their greater cellular uptake and sustained intracellular retention in comparison to unconjugated nanoparticles or drug in solution. This characteristic of conjugated nanoparticles maintained higher intracellular drug levels than in cells treated with drug in solution or with unconjugated nanoparticles. Animals that received a single-dose intratumoral injection of Tf-conjugated drug-loaded nanoparticles (Tx dose = 24 mg/kg) demonstrated complete tumor regression (Fig. 3) and greater survival rate than those treated with equivalent doses of either unconjugated drug-loaded nanoparticles or drug in Cremophor EL formulation *(30)*.

Another characteristic function of nanoparticles is their ability to deliver drugs across biological barriers to the target site *(31,32)*. The brain delivery of a wide variety of drugs, such as antineoplastic and anti-HIV drugs, is markedly hindered because they have great difficulty in crossing the BBB *(33,34)*. The application of nanoparticles to brain delivery is a promising way of overcoming this barrier. Kreuter and colleagues demonstrated that poly(butylcyanoacrylate) nanoparticles coated with polysorbate-80 are effective in carrying different agents to the brain *(35)*. Although not fully elucidated, the most likely transport mechanism for these particles is via endocytosis across the endothelial cell lining of the BBB.

2.3. Ceramic Nanoparticles

The use of inorganic (ceramic) particles for drug delivery, especially bio-macromolecular therapeutics, is a new area *(36–38)*. Ceramic nanoparticles, because of their ultra-low size (less than 50 nm) and porous nature, are becoming an important drug delivery vehicle. Smaller-sized drug carrier systems are more effective in evading the uptake by the RES. In addition, ceramic nanoparticles do not show swelling or changes in porosity with pH. Therefore, these particles can effectively protect doped biomacromolecules such as enzymes against denaturation induced by changes in the external pH and temperature. In addition to silica, other materials, such as alumina, titania, etc., which are highly compatibile with biological systems because of their inert nature, are also being widely used for the formulation of nanoparticles. Further, their surfaces can be easily functionalized for conjugation to target-specific ligands such as monoclonal antibodies *(39,40)*. Roy et al. reported a novel nanoparticle-based drug carrier for photodynamic therapy *(38)*. The group synthesized ultra-fine organically modified silica-based nanoparticles (diameter ~30 nm), entrapping water-insoluble photosensitizing anticancer drug, 2-devinyl-2-(1-hexyloxyethyl) pyropheophorbide, in the nonpolar core of micelles formed by hydrolysis of triethoxyvinylsilane. The resulting drug-doped spherical monodispersed nanoparticles demonstrated uptake into the cytosol of the tumor cells which, when irradiated, generated singlet oxygen and damaged tumor cells. Silica nanoparticles have been also tested as a nonviral vector for gene delivery *(41)*.

2.4. Magnetic Nanoparticles

The most promising application of colloidal magnetic nanoparticles is for site-specific drug delivery. These nanoparticles can carry therapeutic agents on their surface or in their bulk when formulated using polymers, which can be driven to the target organ under an external magnetic field and then released there. For these applications, the size, charge, and surface chemistry of magnetic particles are particularly important, as these properties strongly affect both their blood circulation time and bioavailability of the particles within the body *(42)*. In addition, magnetic properties and internalization of particles in the target tissue depend strongly on the size of the magnetic particles *(43)*. For example, following systemic administration, larger particles with diameters greater than 200 nm are usually sequestered by the spleen as a result of mechanical filtration and are eventually removed by the cells of the phagocyte system, resulting in decreased blood circulation times. On the other hand, smaller particles with diameters of less than 10 nm are rapidly removed through extravasation and renal clearance. Particles ranging from ca. 10 to 100 nm are optimal for intravenous injection and demonstrate the most prolonged blood circulation times. Magnetic nanoparticles in the above size range are small enough to both evade the RES of the body and penetrate small capillaries within the body tissues, and therefore may offer the most effective distribution in certain tissues.

Several studies have demonstrated applications of magnetic nanoparticles for drug delivery. Widder et al. demonstrated the utility of magnetic albumin microparticles in animal tumor models in which they showed greater response in terms of tumor size regression and animal survival than with drug alone *(44)*. In another study, Gupta et al. demonstrated efficacy of magnetic microspheres in drug targeting, predominantly as a result of the magnetic effect and not to particle size *(45)*. Magnetic nanoparticles are usually injected into the arterial supply of the target organ to take advantage of first-pass organ extraction. Because these particles are smaller in diameter, they are able to pass through target capillaries, prior to systemic clearance. As magnetic particles traverse the target organ capillaries, an external magnetic field can retain them in small arterioles and capillaries, and they are ultimately taken up by the tumor cells. Gomez-Lopera et al. have described a method for preparing colloidal particles formed by a magnetite nucleus and a biodegradable poly(DL-lactide) polymer coating *(46)*. The method is based on the so-called double-emulsion technique, employed to obtain polymeric spheres loaded with therapeutic drugs, to be used as drug delivery vectors.

Recently, we developed a novel water-dispersible oleic acid (OA)-Pluronic-coated iron oxide magnetic nanoparticle formulation that can be loaded easily with high doses of water-insoluble anticancer agents (doxorubicin or paclitaxel) (Fig. 4) *(47)*. The magnetic property of the core iron oxide nanoparticles remains unaltered after drug loading. These nanoparticles further demonstrated sustained intracellular drug retention relative to drug in solution and a dose-dependent antiproliferative effect in breast and prostate cancer cell lines. This formulation can thus be used as an effective carrier system for systemic administration of water-insoluble drugs while simultaneously allowing magnetic targeting.

2.5. Polymer-Drug Conjugates as Drug Delivery System

Polymer chemistry for drug conjugation for drug targeting was conceptualized in 1975 *(48)*. Over the last decade, numerous polymer-drug conjugates have entered into phase I or II clinical trials as intravenous injectable anticancer drug therapy. These macromolecular prodrugs are comprised of a

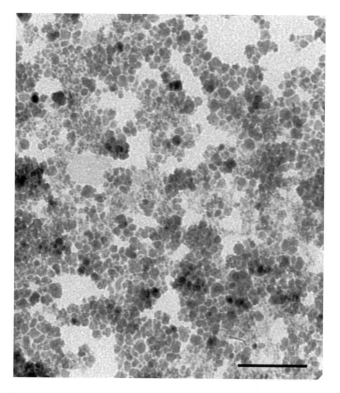

Fig. 4. Transmission electron micrograph of oleic acid-pluronic stabilized magnetic nanoparticles (Bar = 100 nm).

minimum of three components: a natural or synthetic water-soluble polymeric carrier (usually 10–100 kD); a biodegradable polymer-drug linkage, and a bioactive antitumor agent. In certain cases, ligands for receptor-mediated targeting have also been incorporated to enhance tumor-specific delivery of anticancer agents *(49)*. As some of the drug carriers often exert their effect via an intracellular pharmacological receptor, it is essential that drug carriers eventually access the correct intracellular compartment. Covalent attachment of drugs to a hydrophilic polymer enhances the aqueous solubility of poorly soluble drugs, and, with the aid of targeting moieties, tumor-selective delivery can be achieved. Following intravenous administration, most chemotherapeutic agents readily enter cells, crossing the plasma membrane by diffusion *(23)*. Conversely, a polymeric macromolecular prodrug enters cells by endocytosis *(50)*. This leads to a change in the pharmacokinetics of the drug once polymer-bound. In the 1980s, Maeda and colleagues recognized that long-circulating polymer conjugates demonstrate significant tumor uptake even in the absence of targeting moiety *(51)*. The phenomenon has been termed the EPR effect and it arises as a result of the hyperpermeability of the angiogenic tumor vasculature to circulating macromolecules and lack of tumor lymphatic drainage, which causes retention of the accumulating polymer conjugate *(52)*.

Most of the anticancer-drug conjugates that have been tested clinically used *N*-(2-hydroxypropyl)methacrylamide (HPMA) copolymers as a drug carrier. HPMA homopolymer was originally developed by Kopecek and colleagues as a plasma expander *(53)*. Collaborative research with Duncan and colleagues in the early 1980s produced two HPMA copolymer–doxorubicin conjugates that subsequently progressed into phase I/II evaluation under the co-sponsorship of the UK Cancer Research Campaign *(54)*.

Another important polymer–protein conjugate designed to provide localized treatment for hepatocellular carcinoma received market approval in Japan in 1990. It was demonstrated that chemical conjugation of the antitumor protein neocarzinostatin (NCS) with the synthetic copolymer of styrene maleic acid (SMA) significantly improves the physicochemical, biochemical, and pharmacological properties of anticancer drugs. This conjugate consists of two polymer chains of styrene maleic anhydride (SMA) covalently bound to the NCS *(55)*. The conjugate is called poly(styrene-co-maleic acid anhydride) neocarzinostatin (SMANCS), which demonstrated significant antitumor activity in a number of animal models *(55)*, and in clinical studies, 95% of the patients demonstrated a decrease in tumor size *(56)*. Several polymer–drug complexes are in clinical studies, such as HPMA copolymer conjugate of paclitaxel and camptothecin *(57)*, PGA-paclitaxel *(58)*, and PEG-camptothecin *(59)*. Phase II and Phase III clinical studies with these conjugates include combinations of PGA-paclitaxel with cisplatin and carboplatin to maximize the antitumor effect *(60)*.

2.6. Nanotubes, Nanowires, and Nanocages as Drug Delivery System

The design of new strategies for the delivery of drugs and molecular probes into cells has been necessitated by the poor cellular penetration of many small as well as macromolecules, including proteins and nucleic acids *(61)*. Strategies in which a poorly permeating drug or probe molecule is covalently attached to a transporter to produce a cell-penetrating conjugate offer a solution to this problem. Several classes of transporters have been investigated, including lipids, PEG, and, more recently, peptides *(62,63)*. The ability of new materials, such as nanotubes or nanocages, to serve as biocompatible transporters, has gained significance in recent years *(64–66)*. Nanotube structures that resemble tiny drinking straws are made from different materials such as organosilicon polymer, self-assembling lipid microtubes *(65)*, fullerene carbon nanotubes *(67)*, template-synthesized nanotubes, and peptide nanotubes *(68,69)*. Nanotubes offer some interesting advantages relative to spherical nanoparticles for biotechnological applications. For example, they have large inner volumes (relative to the dimensions of the tube), which can be filled with any desired chemical or biochemical species, ranging in size from small molecules to proteins. In addition, nanotubes have distinct inner and outer surfaces, which can be differentially modified for chemical or biochemical functionalization *(70,71)*. This creates the possibility, for example, of loading a drug inside of a nanotube with a particular biochemical payload, and at the same time imparting chemical features to the outer surface that render it biocompatible. The cylindrical geometry allows encapsulation of the drug followed by fabrication and modification of the vehicle. Drugs can be covalently bound on the surface, as has been demonstrated for testosterone *(72)*.

Thus far, the main activities with nanotubes include functionalization and immobilization of biomolecules on their surface for characterization, manipulation, separation, and for device applications such as biosensors *(73,74)*. Few reports exist on how carbon nanotubes interact with and affect living systems. Mattson et al. investigated the growth pattern of neurons on as-grown and functionalized multi-walled nanotubes *(75)*. Recently, Pantarotto et al. reported the internalization of fluorescently labeled nanotubes into cells with no apparent toxic effect *(76)*.

2.7. Dendrimers as Drug Delivery System

Dendrimers are polymeric macromolecular complexes that comprise a series of well-defined branches around an inner core *(77)* and have attracted significant attention in recent years for drug delivery applications because of their nanometer size range, ease of preparation and functionalization, and their ability to display multiple copies of surface groups for biological reorganization processes *(78,79)*. The higher generation dendrimers occupy a smaller hydrodynamic volume compared to the corresponding linear polymers, as a result of their globular structure *(80)*. Since their discovery in the early 1980s, dendrimers are the best examples of controlled hierarchical synthesis, allowing the generation of complex systems *(81)*. Initial studies with dendrimers as potential delivery systems were focused on their use as unimolecular micelles and "dendritic boxes" for the noncovalent encapsulation of drug molecules. For example, in early studies, DNA was complexed with polyamidoamine (PAMAM) dendrimers for gene delivery applications. These were then developed for the delivery of hydrophobic drugs and dye molecules for imaging applications *(82,83)*. In some cases, harsh conditions are required for dendrimer formulation *(83)*, whereas in others, the encapsulated drug is not well retained and the molecules are released relatively rapidly *(84)*. It has been shown that introduction of stabilizing polyethylene oxide (PEO) chains on the dendrimer periphery has expanded the scope of dendrimers, as it allows incorporation of anticancer drugs such as 5-fluorouracil, methotrexate, and doxorubicin and their release from the system at a sustained rate *(85)*.

An alternative approach to the development of dendrimers as anticancer drug carriers is to exploit their well defined multivalency for the covalent attachment of drug molecules to the dendrimer periphery. The drug loading can be tuned by varying the generation number of the dendrimer, and release of the drug can be controlled by incorporating degradable linkages between the drug and dendrimer. Some of these linkages can be peptide-based; these are broken down by specific enzymes that are overexpressed in disease conditions, such as overexpression of proteases in the arthritic condition. Duncan and co-workers *(86)* have prepared conjugates of PAMAM dendrimers with cisplatin, a potent anticancer drug with nonspecific toxicity and poor water solubility. The conjugates show increased solubility, decreased systemic toxicity, and selective accumulation in solid tumors. Zhou et al. have described the preparation of PAMAM dendrimers from a cyclic tetraamine core and the subsequent attachment of 5-fluorouracil to the dendrimer periphery *(87)*. Fréchet and co-workers prepared multivalent conjugates of folic acid and the drug methotrexate using polyaryl ether dendrimers *(88)*. Although the presence of several copies of folic acid or the hydrophilic drug molecule on the periphery of the dendrimer renders these conjugates water-soluble, the water solubility of the these dendrimers can be increased further by attaching PEO chains to the periphery *(84)*. By using a careful synthetic strategy with two different chain-end functionalities, it is also possible to attach both hydrophobic drugs and PEO moieties to the dendrimer periphery in a controlled manner *(89)*. The multivalent folic acid conjugates of dendrimers have important implications for targeting to tumor cells and also allow the attachment of various payloads, including targeting and diagnostic and therapeutic molecules, as well as combinations of these agents *(90,91)*. Recently, folate-modified PAMAM dendrimers have been successfully used as carriers of boron isotopes (^{10}B) in boron neutron-capture treatment of cancer tumors *(92)*. Boron neutron capture therapy is a cancer treatment based on a nuclear capture reaction *(93)*. When ^{10}B is irradiated with low energy or thermal neutrons, highly energetic α-particles and ^{7}Li ions are produced that are toxic to tumor cells. In addition, PAMAM dendrimer silver complexes that slowly release silver have shown antimicrobial activity against various Gram-positive bacteria *(94)*.

2.8. Polymeric Micelles as Drug Delivery System

Block copolymer micelles are promising drug carriers for various hydrophobic drugs including anticancer agents *(95)*. A polymeric micelle usually consists of several hundred block copolymers and has a diameter ranging from 20 to 50 nm *(13,95)*. There are two spherical concentric regions of a polymeric micelle, a densely packed core consisting of hydrophobic blocks and a shell consisting of a dense brush of hydrophilic block. PEO is frequently the hydrophilic block. Hydrophobic blocks include poly-alpha-hydroxy acids and poly-L-amino acids *(96,97)*. Because micelles can solubilize hydrophobic drugs, they can be used for intravenous administration of different water-insoluble drugs *(95,97)*. The hydrophilic corona of the micelles can prevent their interaction with blood components, and their small size can prevent their recognition by the immune system, and thus long circulation times in the blood stream may be achieved. Polymeric micelle-incorporated drugs may accumulate to a greater extent than free drug into tumors and demonstrate a reduced distribution in nontargeted areas. Accumulation of polymeric micelles in malignant tissue is due to increased vascular permeability and impaired lymphatic drainage *(98,99)*. Active targeting is also possible by modifying the peripheral chain ends of the polymers with conjugating ligands including antibodies to the micelle surface *(100)*. Recently, Torchilin et al. formulated antitumor antibody-conjugated polymeric micelles (immunomicelles) encapsulating the water-insoluble drug paclitaxel inside the hydrophobic core of the micelles *(101)*.

Pluronics are available in a wide range of molecular weights and ratios of PEG to poly(propylene glycol), thus varying in hydrophobicity *(102)*. Kabanov et al. have recently reviewed applications of Pluronics in drug delivery *(103)*. Besides drug solubilization, Pluronics exert interesting and important biological effects that have consequences on drug and gene delivery *(104)*. The group has demonstrated inhibition of drug efflux effect of P-gp with certain Pluronics *(105)*. They have also demonstrated increased drug uptake into the brain and drug absorption via oral administration using Pluronics *(106)*. Kataoka and colleagues have designed self-assembling polymeric micelles (NK911; 42 nm in diameter) using block copolymers of PEG (MW ~5000 g/mol)-poly(aspartic acid) where doxorubicin is covalently bound to the polymer (~45%). NK911 system accumulates preferentially in the tumor tissue as a result of the EPR effect, leading to a three- to fourfold improvement in targeting *(98)*.

3. CONCLUDING REMARKS

The successful development of nano-sized drug carriers for drug therapy requires an integrated approach, involving various disciplines such as polymer science, pharmaceutics, biology, engineering, etc. It requires a fundamental understanding of interactions of nano-sized systems with cell membranes and receptors, their intracellular pathways, and intracellular sorting mechanisms. Further, a better understanding of the biodistribution of nano-sized drug carriers and their interactions with blood proteins and tissues is necessary to explore their applications for drug targeting. In addition, the large-scale production of nano-sized systems requires expertise in engineering. The field of drug delivery thus has become a multidisciplinary enterprise.

ACKNOWLEDGMENTS

Grant support from the Nebraska Research Initiative and the National Institutes of Health (R01 EB003975 and R21 CA121751) is appreciated.

REFERENCES

1. http://nano.cancer.gov/
2. Roco MC. Nanotechnology: convergence with modern biology and medicine. Curr Opin Biotechnol 2003;14:337–346.
3. Kayser O, Lemke A, Hernandez-Trejo N. The impact of nanobiotechnology on the development of new drug delivery systems. Curr Pharm Biotechnol 2005;6:3–5.
4. Muller RH, Ruhl D, Runge S, Schulze-Forster K, Mehnert W. Cytotoxicity of solid lipid nanoparticles as a function of the lipid matrix and the surfactant. Pharm Res 1997;14:458–462.
5. Cavalli R, Gasco MR, Chetoni P, Burgalassi S, Saettone MF. Solid lipid nanoparticles (SLN) as ocular delivery system for tobramycin. Int J Pharm 2002;238:241–245.
6. Yang SC, Lu LF, Cai Y, Zhu JB, Liang BW, Yang CZ. Body distribution in mice of intravenously injected camptothecin solid lipid nanoparticles and targeting effect on brain. J Control Release 1999;59:299–307.
7. Wissing SA, Kayser O, Muller RH. Solid lipid nanoparticles for parenteral drug delivery. Adv Drug Deliv Rev 2004;56:1257–1272.
8. Kipp JE. The role of solid nanoparticle technology in the parenteral delivery of poorly water-soluble drugs. Int J Pharm 2004;284:109–122.
9. Fundaro A, Cavalli R, Bargoni A, Vighetto D, Zara GP, Gasco MR. Non-stealth and stealth solid lipid nanoparticles (SLN) carrying doxorubicin: pharmacokinetics and tissue distribution after i.v. administration to rats. Pharmacol Res 2000;42:337–343.
10. Maia CS, Mehnert W, Schafer-Korting M. Solid lipid nanoparticles as drug carriers for topical glucocorticoids. Int J Pharm 2000;196:165–167.
11. Uner M, Wissing SA, Yener G, Muller RH. Influence of surfactants on the physical stability of solid lipid nanoparticle (SLN) formulations. Pharmazie 2004;59:331–332.
12. Zara GP, Cavalli R, Bargoni A, Fundaro A, Vighetto D, Gasco MR. Intravenous administration to rabbits of non-stealth and stealth doxorubicin-loaded solid lipid nanoparticles at increasing concentrations of stealth agent: pharmacokinetics and distribution of doxorubicin in brain and other tissues. J Drug Target 2002;10:327–335.
13. Sahoo SK, Labhasetwar V. Nanotech approaches to drug delivery and imaging. Drug Discov Today 2003;8:1112–1120.

14. Panyam J, Labhasetwar V. Biodegradable nanoparticles for drug and gene delivery to cells and tissue. Adv Drug Deliv Rev 2003;55:329–347.

15. Brannon-Peppas L, Blanchette JO. Nanoparticle and targeted systems for cancer therapy. Adv Drug Deliv Rev 2004;56:1649–1659.

16. Desai MP, Labhasetwar V, Amidon GL, Levy RJ. Gastrointestinal uptake of biodegradable microparticles: effect of particle size. Pharm Res 1996;13: 1838–1845.

17. Desai MP, Labhasetwar V, Walter E, Levy RJ, Amidon GL. The mechanism of uptake of biodegradable microparticles in Caco–2 cells is size dependent. Pharm Res 1997;14:1568–1573.

18. Panyam J, Sahoo SK, Prabha S, Bargar T, Labhasetwar V. Fluorescence and electron microscopy probes for cellular and tissue uptake of poly(D,L-lactide-co-glycolide) nanoparticles. Int J Pharm 2003;262:1–11.

19. Thomas M, Klibanov AM. Conjugation to gold nanoparticles enhances polyethylenimine's transfer of plasmid DNA into mammalian cells. Proc Natl Acad Sci USA 2003;100:9138–9143.

20. Panyam J, Zhou WZ, Prabha S, Sahoo SK, Labhasetwar V. Rapid endo-lysosomal escape of poly(DL-lactide-co-glycolide) nanoparticles: implications for drug and gene delivery. FASEB J 2002;16:1217–1226.

21. Davda J, Labhasetwar V. Characterization of nanoparticle uptake by endothelial cells. Int J Pharm 2002;233:51–59.

22. Panyam J, Labhasetwar V. Sustained cytoplasmic delivery of drugs with intracellular receptors using biodegradable nanoparticles. Mol Pharm 2004;1:77–84.

23. Brigger I, Dubernet C, Couvreur P. Nanoparticles in cancer therapy and diagnosis. Adv Drug Deliv Rev 2002;54:631–651.

24. Vauthier C, Dubernet C, Chauvierre C, Brigger I, Couvreur P. Drug delivery to resistant tumors: the potential of poly(alkyl cyanoacrylate) nanoparticles. J Control Release 2003;93:151–160.

25. Yoo HS, Lee KH, Oh JE, Park TG. In vitro and in vivo anti-tumor activities of nanoparticles based on doxorubicin-PLGA conjugates. J Control Release 2000;68:419–431.

26. Mu L, Feng SS. Vitamin E TPGS used as emulsifier in the solvent evaporation/extraction technique for fabrication of polymeric nanospheres for controlled release of paclitaxel (Taxol). J Control Release 2002;80:129–144.

27. Moghimi SM, Hunter AC, Murray JC. Long-circulating and target-specific nanoparticles: theory to practice. Pharmacol Rev 2001;53:283–318.

28. Iakoubov L, Rokhlin O, Torchilin V. Anti-nuclear autoantibodies of the aged reactive against the surface of tumor but not normal cells. Immunol Lett 1995;47:147–149.

29. Maeda H, Wu J, Sawa T, Matsumura Y, Hori K. Tumor vascular permeability and the EPR effect in macromolecular therapeutics: a review. J Control Release 2000;65:271–284.

30. Sahoo SK, Ma W, Labhasetwar V. Efficacy of transferrin-conjugated paclitaxel-loaded nanoparticles in a murine model of prostate cancer. Int J Cancer 2004;112:335–340.

31. Fisher RS, Ho J. Potential new methods for antiepileptic drug delivery. CNS Drugs 2002;16:579–593.

32. Lockman PR, Mumper RJ, Khan MA, Allen DD. Nanoparticle technology for drug delivery across the blood-brain barrier. Drug Dev Ind Pharm 2002;28:1–13.

33. Kastin AJ, Akerstrom V, Pan W. Interleukin-10 as a CNS therapeutic: the obstacle of the blood-brain/blood-spinal cord barrier. Brain Res Mol Brain Res 2003;114:168–171.

34. Sun H, Dai H, Shaik N, Elmquist WF. Drug efflux transporters in the CNS. Adv Drug Deliv Rev 2003;55:83–105.

35. Kreuter J, Ramge P, Petrov V, et al. Direct evidence that polysorbate-80-coated poly(butylcyanoacrylate) nanoparticles deliver drugs to the CNS via specific mechanisms requiring prior binding of drug to the nanoparticles. Pharm Res 2003;20:409–416.

36. Cherian AK, Rana AC, Jain SK. Self–assembled carbohydrate-stabilized ceramic nanoparticles for the parenteral delivery of insulin. Drug Dev Ind Pharm 2000;26:459–463.

37. Jain TK, Roy I, De TK, Maitra AN. Nanometer silica particles encapsulating active compounds: a novel ceramic drug carrier. J Am Chem Soc 1998;120:11,092–11,095.

38. Roy I, Ohulchanskyy TY, Pudavar HE, et al. Ceramic-based nanoparticles entrapping water-insoluble photosensitizing anticancer drugs: a novel drug-carrier system for photodynamic therapy. J Am Chem Soc 2003;125:7860–7865.

39. Lal ML, Kim KS, He GS, et al. Silica nanobubbles containing an organic dye in a multilayered organic/inorganic heterostructure with enhanced luminescence. Chem Mater 2000;19:2632–2639.

40. Badley RD, Ford WT, McEnroe FJ, Assink RA. Surface modification of colloidal silica. Langmuir 1990;6:792–801.

41. Roy I, Ohulchanskyy TY, Bharali DJ, et al. Optical tracking of organically modified silica nanoparticles as DNA carriers: a nonviral, nanomedicine approach for gene delivery. Proc Natl Acad Sci USA 2005;102:279–284.

42. Chouly C, Pouliquen D, Lucet I, Jeune JJ, Jallet P. Development of superparamagnetic nanoparticles for MRI: effect of particle size, charge and surface nature on biodistribution. J Microencapsul 1996;13:245–255.

43. Chatterjee J, Haik Y, Chen CJ. Modification and characterization of polystyrene-based magnetic microspheres and comparison with albumin-based magnetic microspheres. J Mag Mag Mat 2001;225:21–29.

44. Widder KJ, Senyei AE, Ranney DF. In vitro release of biologically active adriamycin by magnetically responsive albumin microspheres. Cancer Res 1980;40:3512–3517.

45. Gupta PK, Hung CT. Magnetically controlled targeted micro-carrier systems. Life Sci 1989;44:175–186.

46. Gomez-Lopera SA, Plaza RC, Delgado AV. Synthesis and characterization of spherical magnetite/biodegradable polymer composite particles. J Colloid Interface Sci 2001;240:40–47.

47. Jain TK, Morales MA, Sahoo SK, Leslie–Pelecky DL, Labhasetwar V. Iron-oxide nanoparticles for sustained delivery of anticancer agents. Mol Pharm 2005;2:194–205.

48. Ringsdorf H. Structure and properties of pharmacologically active polymers. J Polym Sci Symposium 1975;51:135–153.

49. Greish K, Fang J, Inutsuka T, Nagamitsu A, Maeda H. Macromolecular therapeutics: advantages and prospects with special emphasis on solid tumour targeting. Clin Pharmacokinet 2003;42:1089–1105.

50. Luo Y, Prestwich GD. Cancer-targeted polymeric drugs. Curr Cancer Drug Targets 2002;2:209–226.

51. Matsumura Y, Maeda H. A new concept for macromolecular therapeutics in cancer chemotherapy: mechanism of tumoritropic accumulation of proteins and the antitumor agent smancs. Cancer Res 1986;46:6387–6392.

52. Maeda H, Sawa T, Konno T. Mechanism of tumor-targeted delivery of macro-molecular drugs, including the EPR effect in solid tumor and clinical overview of the prototype polymeric drug SMANCS. J Control Release 2001;74:47–61.

53. Sprincl L, Exner J, Sterba O, Kopecek J. New types of synthetic infusion solutions. III. Elimination and retention of poly-[N-(2-hydroxypropyl) methacrylamide] in a test organism. J Biomed Mater Res 1976;10:953–963.

54. Vasey PA, Kaye SB, Morrison R, et al. Phase I clinical and pharmacokinetic study of PK1 [N-(2-hydroxypropyl)methacrylamide copolymer doxoru-bicin]: first member of a new class of chemotherapeutic agents-drug-polymer conjugates. Cancer Research Campaign Phase I/II Committee. Clin Cancer Res 1999;5:83–94.

55. Maeda H. SMANCS and polymer-conjugated macromolecular drugs: advan-tages in cancer chemotherapy. Adv Drug Deliv Rev 2001;46:169–185.

56. Fang J, Sawa T, Maeda H. Factors and mechanism of "EPR" effect and the enhanced antitumor effects of macromolecular drugs including SMANCS. Adv Exp Med Biol 2003;519:29–49.

57. Schoemaker NE, van Kesteren C, Rosing H, et al. A phase I and pharmaco-kinetic study of MAG-CPT, a water-soluble polymer conjugate of camp-tothecin. Br J Cancer 2002;87:608–614.

58. Li C, Yu DF, Newman RA, et al. Complete regression of well-established tumors using a novel water-soluble poly(L-glutamic acid)-paclitaxel conju-gate. Cancer Res 1998;58:2404–2409.

59. Greenwald RB, Choe YH, McGuire J, Conover CD. Effective drug delivery by PEGylated drug conjugates. Adv Drug Deliv Rev 2003;55:217–250.

60. Singer JW, Baker B, De Vries P, et al. Poly-(L)-glutamic acid-paclitaxel (CT-2103) [XYOTAX], a biodegradable polymeric drug conjugate: characteriza-tion, preclinical pharmacology, and preliminary clinical data. Adv Exp Med Biol 2003;519:81–99.

61. Smith DA, van de Waterbeemd H. Pharmacokinetics and metabolism in early drug discovery. Curr Opin Chem Biol 1999;3:373–378.

62. Bendas G. Immunoliposomes: a promising approach to targeting cancer therapy. BioDrugs 2001;15:215–224.

63. Wright LR, Rothbard JB, Wender PA. Guanidinium rich peptide transporters and drug delivery. Curr Protein Pept Sci 2003;4:105–124.

64. Dai H. Carbon nanotubes: synthesis, integration, and properties. Acc Chem Res 2002;35:1035–1044.

65. Selinger JV, Schnur JM. Theory of chiral lipid tubules. Phys Rev Lett 1993;71:4091–4094.

66. Spector MS, Schnur JM. DNA ordering on a lipid membrane. Science 1997;275:791–792.

67. Zheng LX, O'Connell MJ, Doorn SK, et al. Ultralong single-wall carbon nanotubes. Nat Mater 2004;3:673–676.

68. Ghadiri MR, Granja JR, Milligan RA, McRee DE, Khazanovich N. Self-assembling organic nanotubes based on a cyclic peptide architecture. Nature 1993;366:324–327.

69. Ghadiri MR, Granja JR, Buehler LK. Artificial transmembrane ion channels from self-assembling peptide nanotubes. Nature 1994;369:301–304.

70. Lee SB, Mitchell DT, Trofin L, Nevanen TK, Soderlund H, Martin CR. Antibody-based bio-nanotube membranes for enantiomeric drug separations. Science 2002;296:2198–2200.

71. Mitchell DT, Lee SB, Trofin L, et al. Smart nanotubes for bioseparations and biocatalysis. J Am Chem Soc 2002;124:11,864–11,865.

72. Goldstein AS, Amory JK, Martin SM, Vernon C, Matsumoto A, Yager P. Testosterone delivery using glutamide-based complex high axial ratio microstructures. Bioorg Med Chem 2001;9:2819–2825.

73. Liu J, Rinzler AG, Dai H, et al. Fullerene pipes. Science 1998;280: 1253–1256.

74. Chen RJ, Zhang Y, Wang D, Dai H. Noncovalent sidewall functionalization of single-walled carbon nanotubes for protein immobilization. J Am Chem Soc 2001;123:3838–3839.

75. Mattson MP, Haddon RC, Rao AM. Molecular functionalization of carbon nanotubes and use as substrates for neuronal growth. J Mol Neurosci 2000;14:175–182.

76. Pantarotto D, Tagmatarchis N, Bianco A, Prato M. Synthesis and biological properties of fullerene-containing amino acids and peptides. Mini Rev Med Chem 2004;4:805–814.

77. Boas U, Heegaard PM. Dendrimers in drug research. Chem Soc Rev 2004;33:43–63.

78. Padilla De Jesus OL, Ihre HR, Gagne L, Frechet JM, Szoka FC, Jr. Polyester dendritic systems for drug delivery applications: in vitro and in vivo evaluation. Bioconjug Chem 2002;13:453–461.

79. Quintana A, Raczka E, Piehler L, et al. Design and function of a dendrimer-based therapeutic nanodevice targeted to tumor cells through the folate receptor. Pharm Res 2002;19:1310–1316.

80. Nierengarten JF, Eckert JF, Rio Y, Carreon MP, Gallani JL, Guillon D. Amphiphilic diblock dendrimers: synthesis and incorporation in Langmuir and Langmuir-Blodgett films. J Am Chem Soc 2001;123:9743–9748.

81. Morgan JR, Cloninger MJ. Heterogeneously functionalized dendrimers. Curr Opin Drug Discov Devel 2002;5:966–973.

82. Newkome GR, Moorefield CN, Baker GR, Saunders MJ, Grossman SH. Chemistry of micelles. 13. Monomolecular micelles. Angew Chem 1991;103:1207–1209.

83. Jansen JFGA, Meijer EW, de Brabander-van den Berg EMM. The dendritic box: shape-selective liberation of encapsulated guests. J Am Chem Soc 1995;117:4417–4418.

84. Liu M, Kono K, Frechet JM. Water-soluble dendritic unimolecular micelles: their potential as drug delivery agents. J Control Release 2000;65:121–131.

85. Kojima C, Kono K, Maruyama K, Takagishi T. Synthesis of polyamidoamine dendrimers having poly(ethylene glycol) grafts and their ability to encapsulate anticancer drugs. Bioconjug Chem 2000;11:910–917.

86. Malik N, Evagorou EG, Duncan R. Dendrimer-platinate: a novel approach to cancer chemotherapy. Anticancer Drugs 1999;10:767–776.

87. Zhuo RX, Du B, Lu ZR. In vitro release of 5-fluorouracil with cyclic core dendritic polymer. J Control Release 1999;57:249–257.

88. Kono K, Liu M, Frechet JM. Design of dendritic macromolecules containing folate or methotrexate residues. Bioconjug Chem 1999;10:1115–1121.

89. Liu M, Frechet JMJ. Designing dendrimers for drug delivery. Pharm Sci Technol Today 1999;2:393–401.

90. Esfand R, Tomalia DA. Poly(amidoamine) (PAMAM) dendrimers: from biomimicry to drug delivery and biomedical applications. Drug Discov Today 2001;6:427–436.

91. Sudimack J, Lee RJ. Targeted drug delivery via the folate receptor. Adv Drug Deliv Rev 2000;41:147–162.

92. Barth RF, Soloway AH, Fairchild RG, Brugger RM. Boron neutron capture therapy for cancer. Realities and prospects. Cancer 1992;70:2995–3007.

93. Barth RF, Soloway AH. Boron neutron capture therapy of primary and metastatic brain tumors. Mol Chem Neuropathol 1994;21:139–154.

94. Balogh L, Swanson DR, Tomalia DA, Hagnauer GL, McManus AT. Dendrimer-silver complexes and nanocomposites as antimicrobial agents. Nano Lett 2001;1:18–21.

95. Kakizawa Y, Kataoka K. Block copolymer micelles for delivery of gene and related compounds. Adv Drug Deliv Rev 2002;54:203–222.

96. Kwon GS, Okano T. Soluble self-assembled block copolymers for drug delivery. Pharm Res 1999;16:597–600.

97. Lavasanifar A, Samuel J, Kwon GS. Poly(ethylene oxide)-block-poly(L-amino acid) micelles for drug delivery. Adv Drug Deliv Rev 2002;54:169–190.

98. Nakanishi T, Fukushima S, Okamoto K, et al. Development of the polymer micelle carrier system for doxorubicin. J Control Release 2001;74:295–302.

99. Yokoyama M, Okano T, Sakurai Y, Fukushima S, Okamoto K, Kataoka K. Selective delivery of adriamycin to a solid tumor using a polymeric micelle carrier system. J Drug Target 1999;7:171–186.

100. Torchilin VP. Structure and design of polymeric surfactant-based drug delivery systems. J Control Release 2001;73:137–172.

101. Torchilin VP, Lukyanov AN, Gao Z, Papahadjopoulos-Sternberg B. Immunomicelles: targeted pharmaceutical carriers for poorly soluble drugs. Proc Natl Acad Sci USA 2003;100:6039–6044.

102. Kabanov AV, Alakhov VY. Pluronic block copolymers in drug delivery: from micellar nanocontainers to biological response modifiers. Crit Rev Ther Drug Carrier Syst 2002;19:1–72.
103. Kabanov AV, Batrakova EV, Alakhov VY. Pluronic block copolymers as novel polymer therapeutics for drug and gene delivery. J Control Release 2002;82:189–212.
104. Kabanov AV, Batrakova EV, Miller DW. Pluronic block copolymers as modulators of drug efflux transporter activity in the blood-brain barrier. Adv Drug Deliv Rev 2003;55:151–164.
105. Kabanov AV, Batrakova EV, Alakhov VY. Pluronic block copolymers for overcoming drug resistance in cancer. Adv Drug Deliv Rev 2002;54:759–779.
106. Kreuter J. Influence of the surface properties on nanoparticle-mediated transport of drugs to the brain. J Nanosci Nanotechnol 2004;4:484–488.

14
Gene and Drug Delivery System with Soluble Inorganic Carriers

Jin-Ho Choy, Man Park, and Jae-Min Oh

Summary

Inorganic-based delivery systems are attracting increased attention partially because their inertness gives rise to safety and stability in biosystems and partially because their frameworks can be readily and exactly manipulated. Among the diverse inorganic candidates, such as nanoparticles and clays, we have focused our attention on layered double hydroxides (LDH) with anion-exchange capacity. Diverse anionic molecules can be loaded into the interlayer space of LDH by various routes, such as coprecipitation, anion exchange, and reconstruction. Furthermore, the loaded molecules can be completely discharged by both anion exchange and framework disintegration. When DNA is hybridized with LDH, the intercalated DNA is safely protected against harsh conditions, such as strong alkaline to weakly acidic environments, as well as against DNase attack, which enabled us to develop a novel gene delivery system. HL-60 cells treated with As-myc/LDH hybrids exhibited time-dependent inhibition of cell proliferation, indicating nearly 65% inhibition of growth compared to the untreated cells, after 4 d. On the other hand, LDH itself was noncytotoxic towards HL-60, indicating its biocompatibility. We also demostrated the successful application of a drug–LDH hybrid to in vitro cancer treatment in which LDH played an essential role in the protected delivery of methotrexate (MTX). The initial proliferation of Saos-2 cells was more strongly suppressed by treatment with the MTX-LDH hybrid than with MTX alone. A series of genetic and efficacy assays indicated that LDH does not exert appreciably harmful effects on either normal or cancer cells, and that the action mechanism of MTX is not affected by hybridization. Furthermore, LDH could be vein-injected without any significant effects on tissues or organs below a dose rate of 100 mg/kg. Therefore, we believe that, in the not-so-distant future, LDH will serve as next-generation drug carriers for a broad spectrum of drugs.

Key Words: Antisense; bio-LDH hybrids; inoganic-organic-hybrids; layered double hydroxides (LDH); methotrexate.

From: *NanoBioTechonology: BioInspired Devices and Materials of the Future*
Edited by: Oded Shoseyov and Ilan Levy © Humana Press Inc., Totowa, NJ

1. INTRODUCTION

Protected and controlled delivery of the desired amount of drug to a specific site promises immeasurable benefits for the treatment of various severe diseases in humans, animals, and plants. Not only can it minimize the side effects resulting from frequent and excessive dosages, but it can also allow administration of many potential drugs that cannot be applied today as a result of their inherent problems, such as liability, high toxicity, low availability, etc. Recently, inorganic-based delivery systems have attracted attention, partially because their inertness gives rise to safety and stability in biosystems and partially because their frameworks could be readily and exactly manipulated *(1–19)*. Among the diverse inorganic candidates, such as nanoparticles and clays, particular attention has been paid to layered double hydroxides (LDH) with anion-exchange capacity (AEC) *(5–20)*. Their fascinating features for drug delivery stem from their unique structure and physicochemical properties.

The fundamental structure of LDH is based on hydrotalcite, which can be easily synthesized, typically by coprecipitation. Depending on the arrangement of an octahedral layer and an interlayer, two polytypes of layer structure are well recognized. The one (hydrotalcite) has a rhombohedral unit cell containing three stacked repeat units, whereas the other (manasseite) has a hexagonal unit cell containing two stacked repeat units *(21,22)*. These two polytypes exhibit the same local topology of layer–interlayer bonding, but differ in the long-distance layer–layer interactions. The sheet in these layer structures can be constructed in two distinct modes: hydrotalcite-like and hydrocalumite-like. The hydrotalcite-like sheet is constructed with a stacking brucite structure of $Mg(OH)_2$, in which $Mg(OH)_6$ octahedra are connected through edge sharing into two-dimensional sheets with a layer thickness of 4.8 Å. Some divalent cations in the brucite layer are substituted by trivalent cations such as Al^{3+}, which develop a permanent positive layer charge. On the other hand, Ca-containing LDH have hydrocalumite-like sheets with corrugated brucite-like main layers *(22)*. Ca atoms are hepta-coordinated, with six hydroxides and one interlayered water, which are edge-shared with octahedral trivalent metal cations. The general formula of hydrotalcite-type LDH is $[M^{2+}_{1-x}M^{3+}_x(OH)_2]^-(A^{m-})_{x/m}, nH_2O$, where the M^{n+} are metal cations (M^{2+} = $Mg^{2+}, Zn^{2+}, Ni^{2+}, Cu^{2+},..., M^{3+} = Al^{3+}, Fe^{3+},...$) and A^{m-} are interlayer anions ($A^{m-} = CO_3^{2-}, NO_3^-, SO_4^{2-}$, and other anionic species). Whereas LDH with hydrocalumite-like sheets are formed in an almost fixed molar ratio of $2 Ca^{2+}/M^{3+}$, which results in the general formula of $[Ca_2M(OH)_6]^+ A^- \cdot mH_2O$, the kind of M^{3+} in the hydroxide layer is very limited, typically Fe^{3+} and Al^{3+}. Another unique group of LDH are layered hydroxide salts with AEC *(22–25)*. The layered hydroxide salts can be classified into two structural types based on the structure of either zinc or copper hydroxide nitrate with the typical compositions, $Zn_5(OH)_8(NO_3)_2 \cdot 2H_2O$ and $Cu_2(OH)_3NO_3$, in which the interlayered nitrate is exchangeable. A fraction of the Zn or Cu cations can be isomorphically substituted for other divalent cations. These hydroxide salts are fundamentally built of brucite-like layers in which one-fourth of the octahedral sites are vacant. In particular, vacant zinc sites are occupied with tetrahedral zinc cations.

Their exceptional properties, which are very attractive for drug delivery, include high AEC (2–5 meq/g), great flexibility of chemical composition, swelling properties, extreme affinity to carbonate ion, pH-dependent solubility, and availability *(9–11,20,21)*. In particular, their preference for carbonate ion as interlayer anion and acid liability along with exact control of framework composition make LDH compatible with a variety of biosystems, such as humans. These properties allow a variety of anionic substances to be intercalated into the interlayer space through electrostatic interaction. Fortunately, there are many important biomolecules to be negatively charged within the biosystem. They include oligomers and polymers like nucleotides and double-stranded DNA, as well as simple molecules like nucleic acids and vitamins. Hybridization of these biosubstances with LDH can offer unique features such as stabilization against physicochemical and biological degradation, controlled release, enhanced cellular uptake, and, presumably, targeted delivery. In fact, there have been an increasing number of attempts to apply LDH as a drug delivery matrix for therapeutic and pharmaceutical purposes *(5–19)*. These studies showed that LDH can successfully function as delivery matrices to strikingly enhance drug efficiency without noticeably harmful effects. Here, we review the feasibility of LDH for advanced delivery of biologically active substances to develop their therapeutic and pharmaceutical applications.

2. STRUCTURAL ASPECTS OF BIO-LDH HYBRIDS

Bio-LDH can be prepared by three distinct routes: anion exchange, coprecipitation, and reconstruction. The choice of method depends mainly on the inherent property of the guest biosubstance. Except in the coprecipitation route, pristine LDH are required as the host matrix for the guests. In this review, we focus on the preparation of bio-LDH hybrids by the anion-exchange route *(4–9)*. LDH can be easily prepared by the typical coprecipitation method under N_2 atmosphere following the conventional route *(5,23,24)*. A mixed aqueous solution containing Mg^{2+} [0.024 *M*, from $Mg(NO_3)_2$] and Al^{3+} [0.012 *M*, from $Al(NO_3)_3$] is titrated dropwise into a 0.1 *M* NaOH solution with vigorous stirring at room temperature to give rise to $Mg_2Al(NO_3)$-LDH. For the pristine $Zn_2Al(NO_3)$-LDH, a mixed aqueous solution containing Zn^{2+} [0.02 mol, from

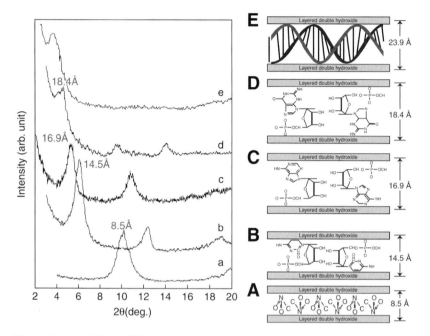

Fig. 1. Powder X-ray diffraction patterns and structural diagrams of **(A)** the pristine MgAl- layered double hydroxides (LDH), **(B)** CMP-LDH hybrid, **(C)** AMP-LDH hybrid, **(D)** GMP-LDH hybrid, and **(E)** DNA-LDH hybrid.

$Zn(NO_3)_2 \cdot 6H_2O$] and Al^{3+} [0.01 mol, from $Al(NO_3)_3 \cdot 9H_2O$] is titrated drop-wise with NaOH (0.1 *M*) solution. During the titration, the solution pH is adjusted to a value of 10 ± 0.2 and 7 ± 0.2 for MgAl-LDH and ZnAl-LDH, respectively.

The powder X-ray diffraction (PXRD) patterns and the schematic diagrams for each hybrid are displayed in Figs. 1–5. According to the PXRD results, the interlayer distance of LDH increases from 0.85 nm for the pristine LDH to 1.45 nm for cytidine-5′-monophosphate (CMP), 1.69 nm for adenosine-5′-monophosphate (AMP), 1.84 nm for guanosine-5′-monophosphate (GMP), 2.39 nm for DNA, 1.94 nm for ATP, 1.88 nm for fluorescein 5-isothiocyanate (FITC), and 1.71 nm for As-myc, 2.19 nm for methotrexate (MTX), 1.05 nm for vitamin C, and 5.38 nm for vitamin E, upon intercalation of biologically functionalized molecules into the interlayer space of the hydroxide layers *(5–10)*. Considering the thickness of the LDH layer (~0.48 nm) and the size of each molecule, we can deduce that the biologically functionalized molecules are intercalated in the interlayer space of LDH by electrostatic interaction. The lack of changes in structure or chemical identity were further confirmed by

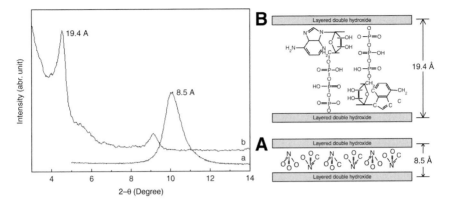

Fig. 2. Powder X-ray diffraction patterns and structural diagrams of **(A)** the pristine MgAl- layered double hydroxides (LDH), and **(B)** ATP-LDH hybrid.

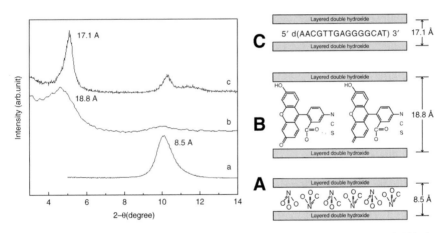

Fig. 3. Powder X-ray diffraction patterns and structural diagrams of **(A)** the pristine MgAl- layered double hydroxides (LDH), **(B)** FITC-LDH hybrid, and **(C)** As-myc-LDH hybrid.

various spectroscopic analyses, such as ionizing radiation (IR) and ultraviolet (UV)-vis spectrophometry.

3. PROPERTIES AND CELLULAR BEHAVIORS OF BIO-LDH HYBRIDS

3.1. Monophosphate- and DNA-LDH Hybrids

Monophosphate- and DNA-LDH hybrids were prepared by ion-exchanging the interlayer nitrate ions in the pristine MgAl-LDH with nucleoside monophosphates such as AMP, GMP, CMP, and herring testis DNA, at pH 7.0.

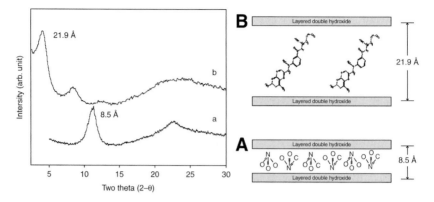

Fig. 4. Powder X-ray diffraction patterns and structural diagrams of **(A)** the pristine MgAl-layered double hydroxides (LDH), and **(B)** MTX-LDH hybrid.

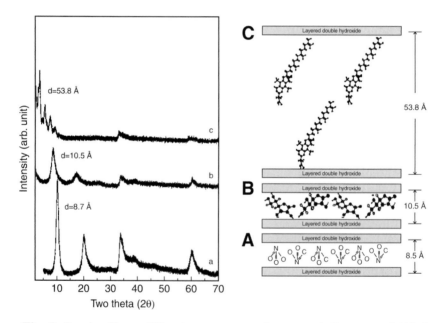

Fig. 5. Powder X-ray diffraction patterns and structural diagrams of **(A)** the pristine ZnAl-layered double hydroxides (LDH), **(B)** vitamin C-LDH hybrid, and **(C)** vitamin E-LDH hybrid.

For the DNA-LDH hybrid, the protein-free DNA was extracted from the crude materials and then sheared off to the size of 500–1000 bp prior to intercalation. The pristine LDH was dispersed in a decarbonated aqueous solution containing an excess of dissolved AMP, GMP, CMP, and DNA, and

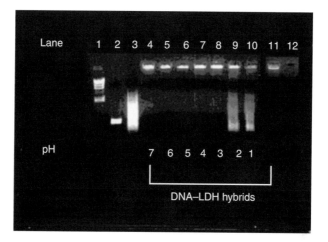

Fig. 6. Electrophoretic analyses of the DNA-layered double hydroxides (LDH) hybrids with respect to pH. The pH of the solution dispersed with hybrid was adjusted to 7.5, 6.0, 5.0, 4.0, 3.0, 2.0, and 1.0, respectively, by adding 1 *M* HCl. Lane 1, λ/Hind III-cut DNA marker (descent to 23.1, 9.4, 6.5, 4.3, 2.3, 2.0 kbp); lane 2, 500-bp DNA marker; lane 3, DNA; lanes 4–10, DNA-LDH hybrids at pH 7.5, 6.0, 5.0, 4.0, 3.0, 2.0, and 1.0, respectively; lane 11, DNA-LDH hybrid treated with DNase I and DNA recovered by acid treatment; lane 12, DNA only treated with DNase I.

reacted for 48 h with steady stirring. The reaction products were then isolated and washed by centrifugation.

Electrophoretic analysis was carried out to prove that DNA-LDH could play a role as a gene reservoir. To test the resistance of the hybrid against DNase I, 96 U of DNase I was added to the nanohybrids (8 μg) and the native DNA (15 μg) directly and treated for 0.5, 1, and 24 h at 37°C. DNA was recovered from the hybrid after the reaction was quenched with DNase I stop solution (0.2 *M* NaCl, 40 m*M* EDTA, and 1% sodium dodecyl sulfate [SDS]) and then the medium was acidified to pH 2.0 to dissolve the host lattice. Then DNase I-treated samples and acid-treated samples were analyzed through gel eletrophoresis.

The DNA/LDH hybrid has a gallery height of 19.1 Å, which is consistent with the thickness of a DNA molecule (20 Å) in a double-helical conformation. It is clear that the interlayer DNA molecules are arranged parallel to the basal plane of the hydroxide layers. From the circular dichroism (CD) analysis, it was determined that the intercalated DNA was stable between the hydroxide layers because the CD band of the DNA/LDH hybrid was observed at the same wavelength as the band of ordinary B-form DNA. Figure 6 represents the electrophoretic analysis of DNA/LDH hybrid, which shows that the DNA/LDH hybrid has pH-dependent properties *(5)*. There are no DNA bands beyond pH ≈ 3.0, indicating that the DNA molecules in the hybrid system were quite stable, even in a weakly acidic atmosphere. However, the DNA bands appeared when the hybrids were treated in a strong acidic medium below pH ≈ 2.0. This result indicated that the hydroxide layers are dissolved under such acidic conditions (Fig. 6, lanes 1–10). From the DNA elution, as shown in Fig. 6, lanes 11 and 12, it can be deduced that the DNA/LDH hybrid can protect DNA from DNase I enzyme. Consequently, the electrophoretic analysis revealed that the DNA/LDH hybrid plays a role as a gene reservoir.

3.2. ATP-LDH Hybrid

The ATP-LDH hybrid was prepared by the ion-exchange route using pristine MgAl-LDH and ATP (containing 40 mCi of [γ-^{32}P]ATP). An excess amount of ATP was added to the aqueous dispersion of pristine MgAl-LDH and reacted at pH 7.0 for 48 h with constant stirring. The reaction products were then isolated and washed as described above. The isotope-labeled ATP-LDH hybrid and [γ-^{32}P]ATP only were added to the HL-60 cells in 20 mL of RPMI-1640 with 10% heat-inactivated fetal bovine serum, and then incubated in a 5% CO_2 incubator at 37°C for 1, 2, 4, 6, 20, or 24 h. For each reaction time, 1 mL of sample was taken and centrifuged, then the separated supernatant was collected and the cell pellet washed once with 1 mL of phosphate buffer (10 mM Na_2HPO_4, pH 7.4, 150 mL NaCl) followed by sedimentation. The supernatant was again separated and collected, and the cell pellet was lysed in 200 mL of lysis buffer (10 mM Tris/Cl, pH 7.4, 150 mL NaCl, 1% SDS) and then extracted with 200 mL of phenol. After separating the aqueous phase, the phenol phase was extracted again with 200 mL of water. Aliquots of the combined aqueous extracts, cell walls, and culture-medium supernatant were analyzed by liquid scintillation counting. The percentage of hybrid and [γ-^{32}P]ATP taken up by the cells was calculated by dividing the counts in the combined aqueous phases of the cell pellet extract by the total counts in the cell pellet, cell wash, and culture-medium supernatant. All of the procedures were repeated three times to check the reproducibility.

Figure 2 clearly demonstrates that LDH also successfully encapsulates ATP via the intercalation route. Fourier transform infrared (FT-IR) spectroscopy proved that the ATP molecules stabilized in the interlayer space of LDH retained their chemical and biological integrity. The biomolecules were well stabilized in the LDH lattice. However, they could be, if necessary, deintercalated by ion-exchange reaction with other anions and carbonate from CO_2 atmosphere. These features allow LDH to be applied as new drug or gene carriers if the transfer efficiency of the bio-hybrid to target organs or cells is proved to be enhanced *(6)*.

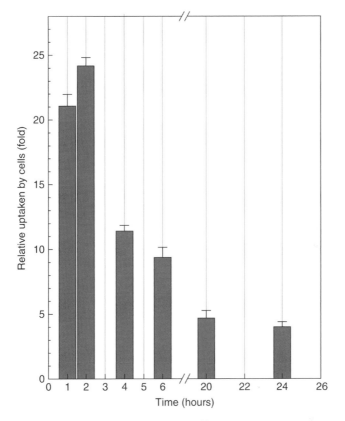

Fig. 7. Histogram for uptake efficiency of [γ-^{32}P] radioactive isotope labeled ATP-layered double hydroxides (LDH) hybrids into HL-60 cells. The uptake efficiency of ATP-LDH hybrids was normalized to that of ATP only. After it was carried out three times, the error bar was plotted from their standard deviation.

To elucidate the transfer efficiency, isotope-labeled [γ-^{32}P]ATP-LDH hybrid was applied for the cellular uptake test with eukaryotic cells, and its cellular behavior was monitored with respect to incubation time. Figure 7 clearly demonstrates that the exogenously introduced ATP-LDH hybrid can enter HL-60 cells effectively within a relatively short time *(8)*. The transfer efficiency was found to be much higher, up to about 25-fold after 2 h incubation, than that of ATP only, whereas after 4 h incubation, the uptake amount of the hybrids decreases, below 12-fold. The triphosphate group of [γ-^{32}P]ATP has a negative charge, which inhibits [γ-^{32}P]ATP's internalization into the cell through the negatively charged cell walls. The hybridization between ATP and LDH neutralizes the surface charge of anionic phosphate groups in ATP as a result of the cationic charge of LDH, which becomes favorable for endocytosis of cells, and eventually results in enhanced transfer efficiency. The longer the incubation time in a CO_2 atmosphere, the more ATP will be released from the interlayer space of the hydroxide lattice. In spite of this kind of ATP release, the transfer efficiency of the hybrid remains higher than that of ATP alone, up to about fourfold after 24 h incubation. This result indicates that the hybridization between cationic layers and anionic biomolecules would greatly enhance the transfer efficiency of biomolecules to mammalian cells or organs.

3.3. Antisense and FITC-LDH Hybrids

The c-myc antisense oligonucleotide (As-myc) with sequence 5'-AACG TTGAGGGGCAT-3' and FITC were intercalated into pristine MgAl-LDH through ion-exchange reaction at pH 7.0 for 48 h.

To verify the cellular uptake behavior of FITC-LDH hybrid, laser-scanning confocal microscopy (LSCM) analysis was carried out. NIH3T3 cells were grown on round cover-slips in a 12-well culture plate and cultured for a day. The 1 μM and 3 μM of FITC-LDH hybrids were added to the cells and incubated for 1, 2, 4, 6, and 8 h, respectively. For comparison, a control experiment was performed with only 5 μM of FITC itself under the same conditions. All of the samples were washed with phosphate-buffered saline (PBS) buffer three times and fixed with 3.7% formaldehyde in PBS. After washing again with PBS, the samples were observed with LSCM (Carl Zeiss LSM 410).

The cellular-uptake experiments were carried out for As-myc-LDH hybrids. HL-60 cells were used to prove that the LDH could act as a drug-delivery vector in gene therapy. HL-60 cells were exposed to As-myc or As-myc-LDH hybrid at a final concentration of 5, 10, and 20 M, respectively. Cell viability was estimated by spectrophotometry measurement of the samples treated with 3-(4,5-dimethylthiazol-2-yl)-2,5-diphenyl tetrazolium bromide (MTT) assay. MTT assay is a colorimetric method that measures the reduction of MTT reagent by mitochondrial succinate dehydrogenase. Because the reduction of MTT can only occur in metabolically active cells, the level of activity is a measure of the viability of the cells.

We also performed cellular uptake experiments using FITC as a probe to verify the delivery potential. The evidence of cellular uptake of the FITC-LDH hybrid was obtained directly from LSCM experiments. FITC-LDH hybrids (1 μM and 3 μM) were added to NIH3T3 cells and then incubated for 1, 4, and 8 h. Then the cells were washed with PBS, and fixed with 3.7% formaldehyde prior to measurements. Figure 8 shows the cellular localization of the fluorophore obtained after a fixed incubation time. The fluorophores were detected in cells within an hour of incubation, and the fluorescence

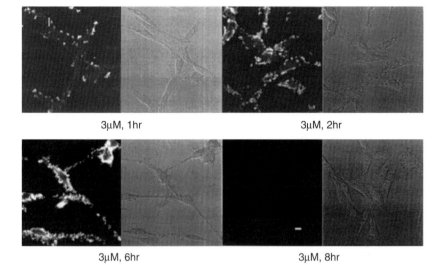

3µM, 1hr   3µM, 2hr

3µM, 6hr   3µM, 8hr

Fig. 8. Laser confocal fluorescence microscopy of fluorophore in NIH3T3 cells. Cells were incubated with 1 µ*M* FITC-layered double hydroxides (LDH) for 1, 4, and 8 h, respectively. The other fluorescence micrograph was obtained with 5 µ*M* FITC only. The bar is 10 µm.

intensities were increased continuously up to 8 h. The fluorophores in the cells were distributed mainly in peripheral and cytosol parts, although some were found in the nucleus. Moreover, the cells treated with 3 m*M* FITC-LDH hybrids showed more intense fluorescence than those with the 1 µ*M* (Fig. 8). In contrast, the cells treated with 5 µ*M* FITC remained dark regardless of incubation time (Fig. 9), because cells could not take up FITC itself, even at high concentrations. It is obvious that LDH plays an important role in mediating the cellular uptake of FITC. All of the cells can engulf the neutralized nanoparticles through phagocytosis or endocytosis. Moreover, it has been confirmed that FITC in the hybrid can be partially released gradually by ion-exchange reaction in a physiological salt condition over 8 h. It has also been found that LDH as encapsulating materials can be gradually dissolved in a weakly acidic solution. Therefore, we conclude that the intracellular fluorescence was created by deintercalated FITC and some by FITC-LDH hybrids in the cell.

It is necessary to test the cytotoxicity of LDH themselves in the cells for use as antisense oligonucleotide delivery carriers. In fact, one critical factor for the overall transfection efficacy of an oligonucleotide delivery system is cytotoxicity. Cell damage resulting from a cytotoxic delivery system is detrimental because the subsequent delivery in the cell must be capable of supporting

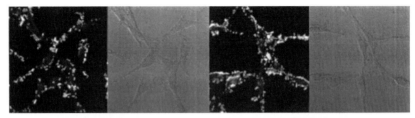

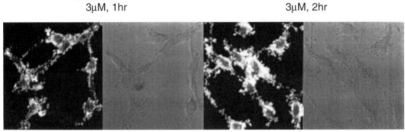

3µM, 1hr   3µM, 2hr

3µM, 6hr   3µM, 8hr

Fig. 9. Laser confocal fluorescence microscopy of fluorophore in NIH3T3 cells. Cells were incubated with 3 µ*M* FITC-layered double hydroxides for 1, 2, 6, and 8 h, respectively.

translation and transcription. As shown in Fig. 10, LDH themselves had no effect on the viability of HL-60 cells, the human promyelocytic leukemia cell, when administered at levels below 1000 mg/mL for up to 4 d. However, many previously examined cationic lipid complexes have been found to be toxic to cells at concentrations near their effective doses if exposure times are extended to several hours *(26–32)*. This suggests that the molecules cannot easily be metabolized. For example, polylysine was the first polymer used to mediate the transfection of cells *(27)*, and polyethyleneimine (PEI) was of the new "proton sponge" category hypothesized to mediate the escape of plasmid DNA from the endosomal pathway. However, although these poly- mers were demonstrated to be the optimal polycations to mediate transfec- tion, they were associated with a considerable degree of cytotoxicity *(29,30)*. When COS-7 cells (immortalized African green monkey kidney fibroblasts) were incubated with polylysine and PEI, their cell viability was approx 50% at 20 µg/mL and approx 2% at 10 µg/mL after 48 h, respec- tively *(30)*. In addition, when Vero cells (African green monkey kidney) were exposed to the various concentrations of DOTMA/DOPE (Lipofectin™, a 1:1 [w/w] formulation of the cationic lipid, DOTMA), at 100 mg/mL, there were no viable cells left after 4 d of continuous treatment *(31)*. Thus, less toxic and more efficient delivery vehicles are needed for oligonucleotide-based therapeutics. In contrast, LDH themselves show no

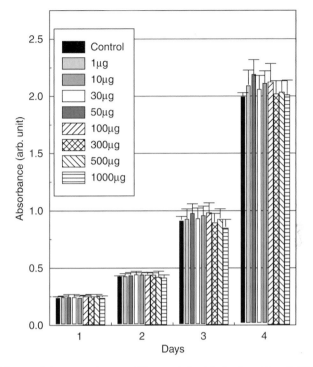

Fig. 10. Cytotoxicity experiment of layered double hydroxides (LDH) in HL-60 cells. Various amounts of LDH were added to cells at final concentrations from 1 µg/mL to 1000 µg/mL.

discernible cytotoxic effects on HL-60 cells when administered at levels below 1000 mg/mL for up to 4 d.

3.4. Methotrexate-LDH Hybrid

MTX, a folic acid antagonist, is also well known for the treatment of certain human cancers, such as bone cancer and leukemia, among others *(33–37)*, and it has been used in clinical diagnostics for more than 50 yr. MTX acts as an inhibitor of dihydrofolate reductase (DHFR) during the folate cycle, in which the thymidylate is synthesized for DNA *(35,36)*. Upon DHFR inhibition, the folate cycle stops, and consequently, the cell dies as a result of a lack of newly synthesized DNA. Because DNA synthesis is most essential for tumor cell proliferation, MTX can effectively suppress tumors. However, affinity of MTX to DHFR is fairly low compared with dihydrofolate. Therefore, the maintenance of high cellular MTX levels is essential for drug activity. Unfortunately, the very short plasma half-life and high efflux vs influx rate of MTX necessitates administering a high dose of drugs to maintain

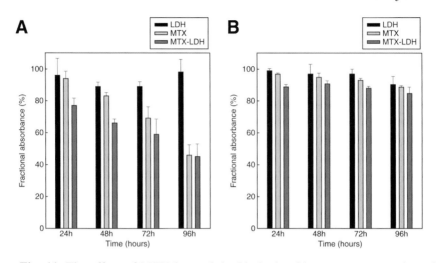

Fig. 11. The effect of MTX-layered double hydroxides at a concentration of 5.0 µg/mL on (**A**) tumor cell growth and (**B**) normal cell growth.

the cellular MTX level, and this could lead to drug resistance and nonspecific toxicities in normal proliferating cells.

MTX-LDH hybrids were prepared by the ion-exchange route. The pristine MgAl-LDH was dispersed in aqueous solution containing a threefold excess amount of MTX for AEC, and stirred vigorously at 60°C for 3 d under N_2 atmosphere. The resulting solids were washed, vacuum-dried, and used for subsequent investigations.

To evaluate the tumor-suppression effects of MTX-LDH hybrids, the human osteosarcoma cell line, Saos-2, was used. Four columns of 96-well plates were prepared, each plate containing 1.2×10^4 Saos-2 cells. Each column was treated with MTX and MTX-LDH hybrids. The MTX treatment concentrations were 500, 50, 5, 5×10^{-1}, 5×10^{-2}, 5×10^{-3}, 5×10^{-4}, 5×10^{-5}, 5×10^{-6} µg/mL. The tumor-suppression effects were evaluated every 24 h from 24 h to 96 h by MTT assay *(9)*.

We verified the drug-delivery efficacy of MTX-LDH hybrid with the cancer cell line Saos-2 (osteosarcoma, human), and also proved the cancer specificity of the hybrid with a normal fibroblast cell line (tendon, human). Figures 11 and 12 show the results of cell-line test with Saos-2 and fibroblast cell lines, respectively. In the fibroblast cell-line, MTX itself, MTX-LDH, and LDH had no significant effect on the proliferation of normal cells. However, MTX and MTX-LDH suppressed the proliferation of tumor cells, whereas LDH itself exerted no significant effect. During the initial period, MTX-LDH was more efficient than MTX in the suppression of tumor cells. The clear difference in

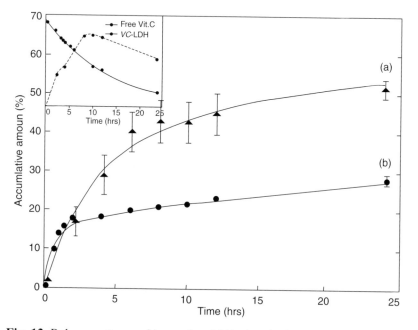

Fig. 12. Release patterns of intercalated **(A)** vitamin C and **(B)** vitamin E. Inset shows free vitamin C (solid line) and deintercalated vitamin C (dotted line).

drug efficiency between MTX and MTX-LDH lasted for 3 d after administration. This result indicates that MTX delivery to the tumor cell is noticeably enhanced by hybridization with LDH. It seems that MTX itself needs a lag time to be effective because of its very short plasma half-life, which results in a decrease in its concentration during the initial period. In the hybrid system, however, MTX can reach the tumor cell membrane without any early decomposition, because the MTX molecules are stabilized and protected in the interlayer space of the LDH lattice.

This result clearly confirms that the intercalation reaction causes no drug denaturation and also enhances both the permeability of the drug into the target cell and cancer cell specificity.

3.5. Vitamin-LDH Hybrids

Vitamin-LDH hybrids with various kinds of vitamin molecules, such as vitamins C and E, were prepared for pharmaceutical and cosmetic uses. The vitamin C (VC)-LDH hybrid was synthesized by anion-exchange reaction of the pristine ZnAl-LDH with ascorbic acid at pH 7.0. The pristine LDH (0.001 mol) was dispersed in decarbonated aqueous solution containing an excess amount of ascorbic acid (0.003 mol), and reacted for 48 h in an ice-bath (at 0°C) with constant stirring. The reaction product was isolated, washed, and freeze-dried. The vitamin E (VE)-LDH hybrid was directly synthesized by coprecipitation reaction under N_2 atmosphere. A mixed aqueous solution containing Zn^{2+} (6.657×10^{-4} mol) and Al^{3+} (2.219×10^{-4} mol) was titrated dropwise with NaOH (0.1 M) solution in the presence of vitamin [(+)-α-tocopherol acid succinate; 3.768×10^{-4} mol) with vigorous stirring. The solution pH was adjusted to 7.5 ± 0.2 at 25°C. The resulting precipitates were washed thoroughly with deionized water and ethanol.

Deintercalation of vitamin C from the hybrid was carried out in CO_3^{2-}-saturated aqueous solution, whereas that of vitamin E was performed in 50% ethanol solution saturated with CO_3^{2-}, respectively. Each sample was stirred in an incubating shaker for 0, 2, 4, 6, 8, 10, 12, and 24 h, respectively. The concentrations of deintercalated vitamin were estimated by measuring the UV absorbance of vitamins C and E at 265 and 277 nm, respectively.

From the PXRD pattern shown in Fig. 5, we can deduce that vitamins C and E are stabilized between the LDH layers. The FT-IR, thermogravimetry, and UV-vis spectroscopic results also confirmed that the vitamin molecules were well encapsulated and well oriented by hydroxide layers. To utilize vitamin-LDH hybrids for various applications, such as pharmaceuticals or cosmetics, it is important to be able to recover the intercalated vitamin molecules in various media. We examined the deintercalation kinetics of vitamin molecules from vitamin-LDH hybrids *(10)*.

Deintercalation of the VC-LDH hybrid was carried out in CO_3^{2-}-saturated aqueous solution, whereas that of the VE-LDH hybrid was performed in 50% ethanol. The concentrations of deintercalated vitamins were estimated by the characteristic absorption peaks at 265 and 277 nm in the UV/vis spectra of vitamins C and E, respectively. The amounts of deintercalated vitamins were plotted with reaction times. Figure 12 clearly shows that vitamin release occurs relatively slowly. Both hybrids exhibit the L-type release pattern that typically results from ion-exchange reaction. VC-LDH showed a considerable deviation from the other. This deviation could be explained by its rapid degradation into oxalic aicd, L-threonic acid, CO_2, and L-xylonic acid in the presence of O_2. Because water-dissolved O_2, even at extremely low concentrations, results in quick oxidation of deintercalated vitamin C, it was nearly impossible to obtain the exact cumulative amount for the long time in CO_3^{2-}-saturated aqueous condition. In fact, Fig. 12 (inset) shows that free vitamin C (solid line) and deintercalated vitamin C (dotted line) contents decrease exponentially after 8 h. To revise the cumulative amount of vitamin C, the deintercalated amount was added to the decomposed amount of free vitamin C. However, the overall release patterns of the vitamins clearly indicate that intercalated vitamins are mainly anion-exchanged by dissolved carbonates.

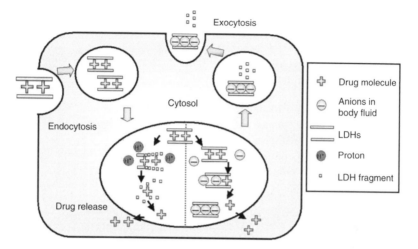

Fig. 13. Schematic illustration of the delivery of biomolecules by layered double hydroxides and its behavior within cells.

The release rates of vitamins from vitamin-LDH hybrids are likely to be affected by factors such as solvent, kind of anion, and charge density, which can be easily controlled. Therefore, it is highly feasible that vitamins can be dispensed in a controlled manner even in under ambient conditions.

4. CONCLUDING REMARKS

LDH exhibit unique features, such as high anion-exchange capacity (2–5 meq/g), exceptionally high affinity for carbonate ion, and pH-dependent solubility, which make them one of the most interesting biocompatible carriers for diverse anionic biosubstances. The results summarized in this review indicate that a positively charged framework enables LDH not only to accommodate a considerable amount of biosubstances, but also to decrease the cell-wall barrier for enhanced cellular uptake of intercalated biosubstances. In addition, acid liability and carbonate affinity can lead to targeted delivery as well as controlled release. The suggested mechanism governing the delivery of biomolecules by LDH is illustrated in Fig. 13. Although the in vivo evidence is still not fully validated, hybridization of biosubstances with LDH could open a new area in advanced drug-delivery systems.

REFERENCES

1. Carretero MI. Clay minerals and their beneficial effects upon human health. A review. Appl Clay Sci 2002;21:155–163.
2. Lin FH, Lee YH, Jian CH, Wong JM, Shieh MJ, Wang CY. A study of purified montmorillonite intercalated with 5-fluorouracil as drug carrier. Biomaterials 2002;23:1981–1987.

3. Ito T, Sugafuji T, Maruyama M, Ohwac Y, Takahashic T. Skin penetration by indomethacin is enhanced by use of an indomethacin/smectite complex. J Supramol Chem 2001;1:217–219.
4. Lee WF, Fu YT. Effect of montmorillonite on the swelling behavior and drug-release behavior of nanocomposite hydrogels. J Appl Polym Sci 2003;89: 3652–3660.
5. Choy JH, Kwak SY, Park JS, Jeong YJ, Portier J. Intercalative nanohybrids of nucleoside monophosphates and DNA in layered metal hydroxide. J Am Chem Soc 1999;121:1399–1400.
6. Choy JH, Kwak SY, Jeong YJ, Park JS. Inorganic layered double hydroxides as nonviral vectors. Angew Chem Int Ed Engl 2000;39:4041–4045.
7. Choy JH, Park JS, Kwak SY, Jeong YJ, Han YS. Layered double hydroxide as gene reservoir. Mol Cryst Liq Cryst Sci Technol, Sect A 2000;341:1229–1233.
8. Choy JH, Kwak SY, Park JS, Jeong YJ. Cellular uptake behavior of [gamma-P-32] labeled ATP-LDH nanohybrids. J Mater Chem 2001;11:1671–1674.
9. Choy JH, Jung JS, Oh JM, et al. Layered double hydroxide as an efficient drug reservoir for folate derivatives. Biomaterials 2004;25:3059–3064.
10. Choy JH, Son YH. Intercalation of vitamer into LDH and their controlled release properties. Bull Kor Chem Soc 2004;25:122–126.
11. Choy JH, Oh JM, Choi SJ. Nanoceramics-biomolecular conjugates for gene and drug delivery. Adv Sci Tech 2006;45:769–778.
12. Choy JH, Park M, Oh JM. Bio-nanohybrids based on layered double hydroxide. Curr Nanosci 2006;2:275–281.
13. Oh JM, Choi SJ, Kim ST, Choy JH. Cellular uptake mechanism of an inorganic nanovehicle and its drug conjugates: enhanced efficacy due to clathrin-mediated endocytosis. Bioconjugate Chem 2006;17:1411–1417.
14. Oh JM, Kwak SY, Choy JH. Intracrystalline structure of DNA molecules stabilized in the layered double hydroxide. J Phys Chem Solids 2006;67: 1028–1031.
15. Oh JM, Park M, Kim ST, Jung JY, Kang YG, Choy JH. Efficient delivery of anticancer drug MTX through MTX-LDH nanohybrid system. J Phys Chem Solids 2006;67:1024–1027.
16. Choy JH, Choi SJ, Oh JM, Park T. Clay minerals and layered double hydroxides for novel biological applications. Appl Clay Sci 2007;36:122–132.
17. Kwak SY, Jeong YJ, Park JS, Choy JH. Bio-LDH nanohybrid for gene therapy. Solid State Ionics 2002;151:229–234.
18. Yang JH, Lee SY, Han YS, Park KC, Choy JH. Efficient transdermal penetration and improved stability of L-ascorbic acid encapsulated in an inorganic nanocapsule. Bull Kor Chem Soc 2003;24:499–503.
19. Hwang SH, Han YS, Choy JH. Intercalation of functional organic molecules with pharmaceutical, cosmeceutical and nutraceutical functions into layered double hydroxides and zinc basic salts. Bull Kor Chem Soc 2001;22:1019–1022.
20. Khan AI, O'Hare D. Intercalation chemistry of layered double hydroxides: recent developments and applications. J Mater Chem 2002;12:3191–3198.
21. Cavani F, Trifirò F, Vaccari A. Hydrotalcite-type anionic clays: Preparation, properties and applications. Catal Today 1991;11:173–301.

22. Rousselot I, Taviot-Gueho C, Leroux F, Leone P, Palvadeau P, Besse JP. Insights on the structural chemistry of hydrocalumite and hydrotalcite-like materials: Investigation of the series $Ca_2M_3+(OH)(6)Cl \cdot 2H(_2)O$ (M3+: Al3+, Ga3+, Fe3+, and Sc3+) by X-ray powder diffraction. J Solid State Chem 2002;167:137–144.
23. Stahlin W, Oswald HR. The crystal structure of zinc hydroxide nitrate, $Zn_5(OH)_8(NO_3)_2.2H_2O$ Acta Cryst B 1970;26:860–863.
24. Morioka H, Tagaya H, Karasu M, Kadokawa JI, Chiba K. Preparation of hydroxy double salts exchanged by organic compounds. J Mater Res 1998;13:848–851.
25. Meyn M, Beneke K, Lagaly G. Anion-exchange reactions of hydroxy double salts. Inorg Chem 1993;32:1209–1215.
26. Bennet CF, Chiang MY, Chan H, Shoemaker JEE, Mirabelli CK. Cationic lipids enhance cellular uptake and activity of phosphorothioate antisense oligonucleotides. Mol Pharmacol 1992;41:1023–1033.
27. Wu G Y, Wu CH. Receptor-mediated in vitro gene transformation by a soluble DNA carrier system. J Biol Chem 1987;262:4429–4432.
28. Zanta MA, Boussif O, Adib A, Behr JP. In vitro gene delivery to hepatocytes with galactosylated polyethylenimine. Bioconjug Chem 1997;8:839–844.
29. Choksakulnimitr S, Masuda S, Tokuda H, Takakura Y, Hashida M. In-vitro cytotoxicity of macromolecules in different cell-culture systems. J Controlled Release 1995;34:233–241.
30. Brazeau GA, Attia S, Poxon S, Hughes JA. In vitro myotoxicity of selected cationic macromolecules used in non-viral gene delivery. Pharm Res 1998;15:680–684.
31. Putnam D, Langer R. Poly(4-hydroxy-L-proline ester): low-temperature poly-condensation and plasmid DNA complexation. Macromolecules 1999;32:3658–3662.
32. Guy-Caffey JK, Bodepudi V, Bishop JS, Jayaraman K, Chaudhary N. Novel polyaminolipids enhance the cellular uptake of oligonucleotides. J Biol Chem 1995;270:31,391–31,396.
33. Hitchings GH, Smith SL. Dihydrofolate reductases as targets for inhibitors. Adv Enz Regul 1980;18:349–371.
34. Bertino JR. Karnofsky memorial lecture. Ode to methotrexate. J Clin Oncol 1993;11:5–14.
35. Bielack SS, Kempf-Bielack B, Heise U, Schwenzer D, Winkler K. Combined modality treatment for osteosarcoma occurring as a second malignant disease. J Clin Oncol 1999;17:1164–1174.
36. Eksborg S, Alberioni F, Rask C, Beck O, Palm C, Schroeder H, Peterson C. Methotrexate plasma pharmacokinetics: importance of assay method Cancer Lett 1996;108:163–169.
37. Genestier L, Paillot R, Quemeneur L, Izeradjene K, Revillard JP. Mechanisms of action of methotrexate. Immunopharmacology 2000;47:247–257.

15

Molecules, Cells, Materials, and Systems Design Based on Nanobiotechnology Use in Bioanalytical Technology

Tetsuya Haruyama

Summary

Bioanalytical chemistry has undergone remarkable development in the past 30 yr, along with the progress of related fields. Connected technologies belong to many different fields, because bioanalytical technology has expanded its scope of application and its principles. And because bioanalysis requires the most sophisticated tools, bioanalytical chemistry has adopted novel concepts and techniques. Clinical applications, such as the design of novel drugs, are the most important kind of application in bioanalytical chemistry. Usually, analytical chemistry is interested in the accurate quantification of specific molecules. However, novel concepts have now been developed that enable scientists to obtain biological information from cultured cells and tissues. In cellular/tissular biosensing, the goal is not to determine the existence and amount of specific molecules, but to survey the effects of chemical/physical stimuli on cells/tissues or organs. In order to develop such novel analytical technology, an interdisciplinary approach is required. By using the nanobiotechnology methods focused on molecular design, model-cell fabrication, and instrumental system fabrication, a novel bio-inspired technology has been developed for novel biosensing. The author has developed the concept of cellular/tissular biosensing, and has created practical techniques. Therefore, here I review recent progress in the area of bioanalytical technology, which is based on nanobiotechnology and is focused on the biosensing of biological information, e.g., cellular/tissular biosensing.

Key Words: Artificial enzyme; biosensor; cellular biosensing; engineered cell; nanobiotechnology; synapse model; tissular biosensing.

1. INTRODUCTION

In the life sciences, much valuable information has been obtained and many viable technologies have been developed through the great progress in biotechnology. In particular, molecular biology and cellular/tissular

From: *NanoBioTechonology: BioInspired Devices and Materials of the Future*
Edited by: Oded Shoseyov and Ilan Levy © Humana Press Inc., Totowa, NJ

engineering have progressed, and new techniques have enabled the creation of new functional proteins and cells/tissues. Moreover, computer/digital technology has progressed rapidly in the last 20 yr. Simulation technology is one of the great outcomes of this progress. The advent of three-dimensional (3D) simulation software for molecular designing has made it possible to visualize the relationship between molecular structures and their function. These new techniques are of great help in creating a new paradigm for biotechnology.

Another new area of knowledge, concerned with the novel phenomenon of nanospace, has been keenly pursued. Nanostructures may reveal hitherto-unknown functions of known materials. In this paper, the author presents his own and others' recent achievements in the area of designed molecules, cells, materials, and systems. These achievements have been obtained through molecular biology, cellular/tissular technology, simulation technology, and nanotechnology.

2. DESIGNED MOLECULES FOR USE IN BIOANALYTICAL TECHNOLOGY

Many functional molecules have already been designed and synthesized. Some of them are out on the market as functional materials. These are not based on mimetic molecular design but on the synergy effect of functional groups. On the other hand, biomolecules, especially enzyme molecules, have been the target of biomolecular mimetic design. Enzymes are catalytic functional molecules, and are frequently employed as bioreactors, in assay kits, in biosensors, and for other industrial purposes. Therefore, artificial enzymes are very much in demand. The molecular design of artificial enzymes is studied from the perspective of mimetic molecular design and minute organic synthesis. Many types of artificial enzyme molecules have been designed and synthesized. For instance, cyclodextrin is a circular compound often employed as the base skeletal structure of an active center. Because the active center of natural enzymes is always surrounded by functional residues, the hole at the center of the cyclodextrin molecule provides a suitable frame for the design of artificial enzymes. Tsutsumi et al. have obtained artificial protease by attaching several functional residues to the interior of the cyclodextrin molecule *(1)*. Most enzymes are metalloenzymes, and are mimicked through minute organic synthesis. Komiyama has designed artificial nucleases based on a minute-synthesized metallocomplex molecule *(2)*. Most of these molecular designs are aimed at either high catalytic activity or strict molecular selectivity such as that found in natural enzyme molecules. These excellent studies have developed biomimetic engineering techniques in the course of producing novel molecules.

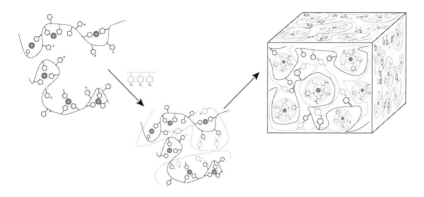

Fig. 1. The bottom-up synthesis of integrated metal complex sites as active centers (nanocavities) in an artificial enzyme, the polymer–metal–polymer complex.

This means, however, that the technological achievements of these studies are not wholly practical. Nevertheless, they have provided important information for further the development of practical, usable artificial functional molecules. Artificial enzymes are probably the solution to the problem. Because natural enzymes are digested and degraded by proteases that exist under biological conditions, enzyme sensors are difficult to employ in cellular/tissular biosensing. However, artificial enzymes can be designed at the molecular level so that their characteristics can make them invulnerable to digestion.

The author has designed an artificial enzyme (a polymer–metal–polymer, or PMP, complex) as a molecular transducer for biosensors that are able to detect adenosine triphosphate (ATP). ATP and its derivatives (e.g., ADP, PPi) are key molecules in living systems, providing a crucial parameter for the determination of both cellular/tissular viability and genetic information. However, there is as yet no sophisticated method of measuring ATP *in situ*.

As shown in Fig. 1, the PMP complex created by the author consists of one or more metal ions and two or more polymers, and is synthesized by coordinative self-assembly. First, the metal-ion coordinative polymer is formed and shrinks through the formation of a complex with the metal ion(s). In the case of ATP-sensing, copper ions and poly-L-histidine are employed. At the same time, functional polymers, such as polystyrene sulfonate, form a polyion–polyion complex with each other, leading to the strengthening of the structure. Through a bottom-up process, the metal complex sites of the active center (nanocavities) are integrated into the matrix, as shown in Fig. 1. The PMP complex has thus been created with ATP biosensing in mind. As illustrated in Fig. 2, the PMP complex is laid out as a thin membrane on an electrode surface of the sensor device. The PMP complex membrane allows for three modes of molecular selectivity: (1) molecular sieving enabled by

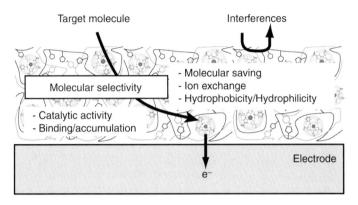

Fig. 2. Hypothetical illustration of molecular selection by the polymer–metal–polymer complex layer on the surface of an electrode.

the stitched structure of the polyion-polyion complex, (2) hydropathical selectivity through the mixing ratio of the polymers, and (3) electrostatic selectivity through the charged polymers *(3,4)*. The PMP complex also acts as a molecular transducer. In the case of the PMP complex designed for the hydrolysis of ATP, the PMP complex hydrolyzes the ATP in the nanocavity and then the PO_4^{3-} broken away by hydrolysis is accumulated between the PMP complex layer and the electrode. The accumulated PO_4^{3-} is reduced electrochemically, and this represents the sensory response (Fig. 3) *(4)*. The response current is proportional to the bulk ATP concentration. In the ordinary system, phosphate cannot be measured electrochemically. However, in the present system, phosphate ions (PO_4^{3-}) are produced through the hydrolysis of ATP, a reaction catalyzed by the PMP complex, an artificial enzyme. The phosphate ion is immediately reduced electrochemically, before it is protonated and is stabilized by H_2O.

The sensor device can determine ATP concentrations from 1 n*M* to 10 m*M*. This high sensitivity is practical; this is the first instance of an artificial enzyme successfully applied in biosensor fabrication.

We have also designed and synthesized a PMP complex for the detection of another biological molecule. Cellular nitric oxide (NO) detection has been performed in cell/tissue culture media *(5)*. NO is a well known biological substance which acts as a regulation factor in various organs, e.g., blood vessels, the myocardium, macrophages, and the cerebellum. Regarding the PMP complex design for NO detection, the NO diffuses smoothly from the bulk solution into the interior of the PMP complex. In the PMP complex layer, NO binds to the metal complex in the nanocavities, and is accumulated in NO-metal form in the nanocavities. Then, the NO-metal is oxidized by pulsed potential in order to obtain a sensory response.

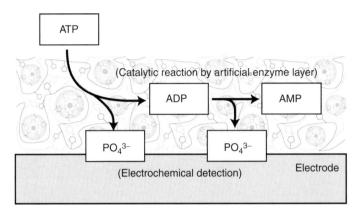

Fig. 3. Schematic reaction of ATP in the polymer–metal–polymer complex layer on the surface of an electrode.

The PMP complex has been designed and synthesized based on a unique bottom-up assembly. The molecular design makes it possible to detect molecules that cannot be detected by conventional biosensors.

3. ENGINEERED CELLS DESIGNED FOR USE IN BIOANALYTICAL TECHNOLOGY

Bioassay is a major technique for obtaining the titer of chemicals (clinical drugs) in the living body. However, animal experimentation makes the research problematic in terms of both ethics and expense. On the other hand, animal cells and tissues can be cultured easily as a result of recent technical developments. As cells represent the minimum functional and integrated communicable units of a living system, cellular biosensing can be performed to determine the presence of extracellular chemical/physical stimuli. Cultured cells transduce and transmit a variety of chemical and physical signals, in the form of specific substances or proteins produced by them. Such signals can be employed as parameters for clinical drug screening through a high-throughput process. Many types of cultured-cell- or tissue-based bioassay methods have been developed. However, neural systems have very peculiar functions with regards to signal transduction. Of course, the central nervous system has been keenly studied with the aim of developing drugs. Most bioassays for the nervous system have been performed in animal experiments. For the primary screening of drugs targeting the nervous system, the high-throughput bioassay method is appropriate *(6)*.

The author's research group has developed a unique strategy to both demonstrate and monitor the channel-gated receptor function on engineered cells *(7)*. We employed a glutamate receptor (GluR; channel-gated receptor)

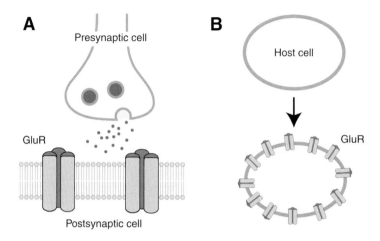

Fig. 4. Glutamate channel-gated receptor (GluR) expressed on the surface of an engineered cell. (**A**) Hypothetical illustration of a native nerve postsynaptic nerve membrane with GluR. (**B**) Hypothetical illustration of engineered cell (a cell expressing GluR on its surface).

which was cloned from a postsynaptic nerve membrane. The cloned gene was transfected into insect cells that can be easily handled. The engineered cell expressed the GluR on the cell surface, in a similar way to the postsynaptic membrane, as illustrated in Fig. 4. As shown in the illustration, GluR was overexpressed and accrued on the cell surface. This type of engineered cell can be employed as a synaptic model cell for the evaluation of the chemical effectiveness of the channel-gate receptor function. In the process of the evaluation, the synaptic model cell was connected to a microelectrode, which in turn was connected with a reference electrode through an amplifier and a potentiometer, as illustrated in Fig. 5.

This system can both perform and monitor the postsynaptic function. The principle on which this system is based is the ionic flux intermediated by GluR between the outside and the inside of the cell. When the agonistic ligand (e.g., glutamic acid) binds to the GluR, a channel opens and a Na^+ ion enters the cell selectively through the GluR channel. This entrains the exit of a K^+ ion from the cell, as illustrated in Fig. 5. The selective ion flux leads to the localization of Na^+ ions between the vicinity of the cell and the bulk solution. This concentration of ions creates a difference in the electrode potential when the electrode system is set up as illustrated in Fig. 5. In the case of the engineered cell, GluR was overexpressed and accrued on the cell surface. In this system, Na^+ ion flux is highly intensified in comparison to the native postsynaptic membrane. Therefore, the dynamic behavior of the potential profile of the GluR channel-gated receptor can be monitored based

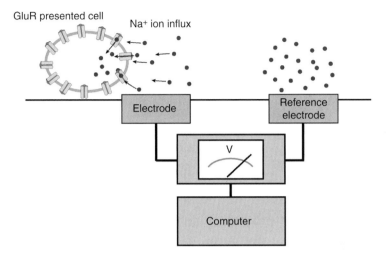

Fig. 5. The dynamic function of Na$^+$-ion influx through the GluR channel of the engineered cell in an OCP system.

on the influx of Na$^+$ ions through the GluR *(7)*. In short, bioassay of the postsynaptic function can be performed using engineered GluR cells and an outer cell potential (OCP) recording system, as illustrated in Fig. 5.

Engineered cells can be genetically designed to transmit specific signals in response to specific effects on cells. Such cellular signals are not very easy to detect. However, with a sophisticated system employing engineered cells and sensor devices, such as the present one, cellular biosensing can be achieved as a practical and feasible bioassay for drug screenings, pharmaceutical studies, and chemical safety proofing.

4. DESIGNED MATERIAL SURFACE FOR BIOANALYTICAL TECHNOLOGY

The modification of existing materials is an important way of obtaining functional surfaces. In the field of bioanalytical technology, the molecular modification of metal is a key technology for creating biosensor devices. In the case of enzyme biosensors, the detection principle is based on the redox reaction of an enzyme located near an electrode. In other words, the enzyme biosensor can detect analytes by the electron transfer between the enzyme molecule (coenzyme or catalytic product) and an electrode. The design of the interface between the molecule and the electrode (or other transuding device) is crucial for smooth electron transfer from the molecule.

Many kinds of molecular immobilization techniques have been studied. Most of these immobilization methods are based on either chemical covalent

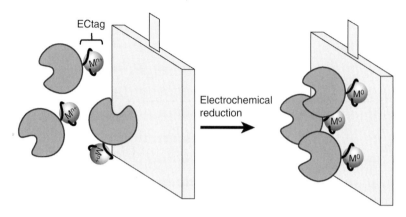

Fig. 6. Molecular immobilization through the ECtag method.

bonding *(8–11)* or specific/nonspecific adsorption *(12–14)*. However, chemical modification might lead to the unwanted loss of the functional characteristics of the proteins used, because the functional residues are annihilated indiscriminately by the chemical coupling reaction. In the case of protein immobilization on a metal substrate, thiol-Au self-assembly has been virtually the only available method. Thiol residues (–SH) can form a type of coordinate bond (mercaptide bond) with Au or Pt *(15)*. The mercaptide bond has already been employed in immobilizing protein on a metal surface. However, because the thiol mercaptide bond is very much dependent on the flatness (atomic epitaxy) of the solid metal surface, the method does not have wide application. More sophisticated types of molecular immobilization on metal substrates are necessary for more advanced protein/metal applications. In order to develop such new molecular immobilization methods, a novel basis as for the interaction between molecules and substrate materials is probably required.

We discovered the novel phenomenon of the formation of a neutral metal complex on a redox surface *(16)*. This finding suggests that the neutral metal complex can be employed as an anchor to immobilize molecules on a redox surface, such as that of an electrode. The author developed the ECtag method for molecular immobilization based on the formation of a neutral metal complex on a redox surface. The ECtag method can be employed to achieve molecular immobilization through a simple process. As hypothetically shown in Fig. 6, the ECtag method is based on the electrochemical deposition on the redox surface of the metal, which forms a neutral metal complex with the ECtag ligand. The ECtag ligand (metal coordinative ligand with a flexible structure) is tagged onto a protein molecule or any other molecule which is to be immobilized. The structure of the ECtag ligand can be designed in

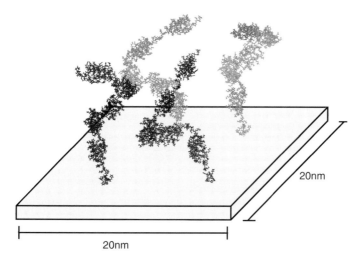

20nm

20nm

Fig. 7. Simulated surface of protein A immobilized on the surface of an electrode by the ECtag method. The value of the molecular density is based on the experimental data.

many ways, and the ECtag can be tagged onto the molecule either genetically or chemically. The ECtag-tagged part coordinates the metal ion. The coordinated metal ion is reduced and deposited on the redox surface as a result of the electrode potential. However, the reduced metal (M^0) is kept coordinated by the ECtag ligand. Through an electrochemical reaction, the tagged molecule is immobilized on the electrode surface because the molecule is tagged with the ECtag as illustrated in Fig. 6. The ECtag method can be applied to most molecules and most metals and conductive materials.

In the case of protein immobilization using the ECtag method, the molecular orientation can be controlled genetically. Figure 7 shows a simulated image of immobilized protein A (a bio-affinity protein from a *Staphylococcus* strain) immobilized on a platinum electrode by the ECtag method. The image simulation is based on both the actual amount of immobilized protein A-ECtag and its actual structure. As clearly shown in the simulation image, the protein A molecules are immobilized at substantial density in perpendicular orientation.

The ECtag method can be used to immobilize molecules other than genetically constructed proteins. As described above, the ECtag ligand can be selected from a variety of metal-coordinative residues and can be structured in many ways, and the ECtag can be tagged onto molecules either genetically or chemically. In other words, the ECtag method can immobilize various molecules on various conductive materials. Moreover, because

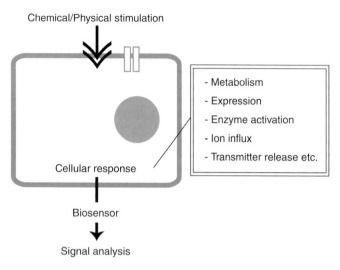

Chemical/Physical stimulation

- Metabolism
- Expression
- Enzyme activation
- Ion influx
- Transmitter release etc.

Cellular response

Biosensor

Signal analysis

Fig. 8. The concept of cellular biosensing through a cascade of cellular responses.

tagging and insertion of the ECtag can be designed through either genetic or chemical techniques, the interfacial interaction between the molecule and the electrode can be optimized for the smooth transmission of molecular information. The interface design can potentially be sophisticated enough to enable both catalytic and affinity biosensing. Furthermore, this ECtag technology will contribute to the creation of biocompatible metallic materials for clinical use.

5. CELLULAR/TISSULAR SENSING SYSTEMS FOR USE IN BIOANALYTICAL TECHNOLOGY

Cells represent the minimum units of living systems, particularly in terms of molecular response to extracellular stimuli (chemical and physical). Cultured cells transduce and transmit both chemical and physical signals in response to extracellular stimuli. Such cellular responses are probably useful when surveying chemical information about living systems. This type of surveillance can be employed as a screening tool to ensure both pharmaceutical and chemical safety, as schematically illustrated in Fig. 8. However, such cellular signals are mostly very weak and not easily detected with conventional analytical methods. To enable the significant assessment of cellular responses, signal profiles are needed, and to monitor the cellular-response profiles, *in situ* analysis is required.

Electrochemical assay, including measurement with biosensors, is a superior method when performing *in situ* analysis. Both electrochemical assay

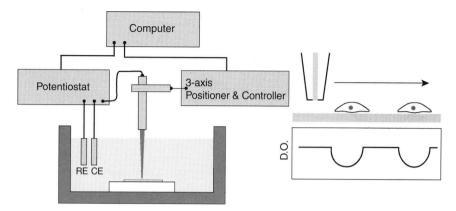

Fig. 9. Scanning electrochemical microscopy (SECM).

devices and biosensor devices can be miniaturized to size scale of cells. They can be positioned near the cells, and this allows them to detect cellular signals before they are attenuated.

Scanning electrochemical microscopy (SECM) is an example of biological activity imaging. As illustrated in Fig. 9, SECM is a type of scanning probe microscope similar to atomic force microscopy (AFM) and scanning tunneling microscopy (STM). If a micro dissolved-oxygen (DO) electrode is employed as the scanning probe, the dimensional localization of oxygen concentrations can be obtained. If the cell undergoes lively respiration, the oxygen concentration in its vicinity decreases, leading to a decreased current response from the micro oxygen electrode. Using this system, the localization of cellular viability can be monitored.

The microarray electrode, shown in Fig. 10, is another example of a cellular biosensing device. It has been used to monitor active potential propagation across neural networks.

Both examples of biosensing systems have been developed within semiconductor engineering. It is a fundamental technology for constructing the miniaturized systems described here.

These are also examples of systems which can monitor the functions of native cultured cells. As described under Subheading 3, engineered cells can be fitted with additional features through a genetic design that allows them to transmit signals in response to extracellular influences on "specific parts of living systems." It is of the utmost importance to obtain significant information from cells/tissues for an evaluation of the effects of chemicals on living systems.

If the specific cellular responses can be obtained, nonminiaturized biosensor devices can gather sophisticated information through cellular biosensing.

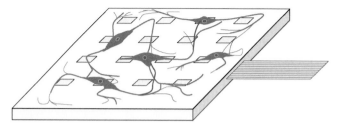

Fig. 10. The application of a microarray electrode for cellular biosensing.

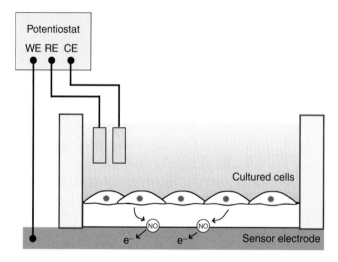

Fig. 11. Cellular biosensing based on a cellular-nitric oxide sensor device equipped with a cell-culture bottle.

Figure 11 shows an instrumental setup for the cellular biosensing and monitoring of cellular NO. The biological roles of NO have been well documented *(17–19)*. NO regulates typical biological functions in various organs, as well as in various cultured cells and tissues. On the other hand, the production profile of cellular and tissue NO is a significant marker of vital information both in vitro and in vivo. A number of techniques have been developed for the direct measurement of biological NO. However, *in situ* measurement of NO is difficult with conventional analytical methods, because NO as a biological signal is only present in low concentrations. Furthermore, NO has a very short half-life under physiological conditions. We previously developed several types of principles for the detection of NO *(20,21)*. In our most current achievement in the field of NO sensing, the NO sensor material not only monitors the cellular NO profile but also provides cell-adhesive activity in order to culture mammalian cells on their own *(5)*. Because the sensor material is designed and synthesized through coordinative self-assembled structure (*see* Subheading 2), it is easy to co-materialize with cell-adhesive polymer through the process of coordinative self-assembly. In the PMP complex, there are metal-coordinating nanocavities. Cellular NO can smoothly diffuse into these nanocavities, because the cells adhere to the PMP complex layer tightly. Therefore, NO diffuses into the PMP complex quickly and accumulates in the nanocavities. The accumulation of NO enables its high-sensitivity electrochemical detection.

6. CONCLUSION

Nanostructured materials display new characteristics which enable *in situ* detection of biological information. These functions also allow the development of new concepts in bioanalysis. For example, a synapse model cell and its instrumental system can be used in a high-throughput bioassay to survey the effect of chemicals on synapse function. In the case of nanostructured artificial enzyme, the PMP complex can perform *in situ* sensing of biological substances that cannot be monitored with conventional assay methods. These are successful achievements in the field of bioanalysis based on bionanotechnology. As described above, nanotechnology and bionanotechnology yield both novel concepts and novel technologies for the development of bioanalytical techniques. A novel field of analytical technology has been opened.

REFERENCES

1. Tsutsumi H, Hamasaki K, Mihara H, Ueno A. Cyclodextrin-peptide hybrid as a hydrolytic catalyst having multiple functional groups. Bioorg Med Chem Lett 2000;10:741–744.
2. Kitamura Y, Sumaoka J, Komiyama M. Hydrolysis of DNA by cerium(IV)/EDTA complex. Tetrahedron 2003;59:10,403–10,408.
3. Haruyama T, Asakawa H, Migita S, Ikeno S. Bio-, nano-technology for cellular biosensing. Curr Appl Phys 2005;5:108–111.
4. Ikeno S, Haruyama T. Biological phosphate ester sensing using an artificial enzyme PMP complex. Sensors and Actuator B 2005;108:646–650.
5. Asakawa H, Ikeno S, Haruyama T. Molecular design of a PMP complex and its application for molecular transducer on cellular NO sensing. Sensors and Actuator B 2005;108:608–612.
6. Drews J. Drug discovery: a historical perspective. Science 2000;287:1960–1964.
7. Haruyama T, Bongsebandhuphubhade S, Nakamura I, et al. A biosensing system based on extracellular potential recording of ligand-gated ion channel function overexpressed in insect cells. Anal Chem 2003;75(4):918–921.
8. Afrin R, Haruyama T, Yanagida Y, Kobatake E, Aizawa M. Catalytic activity of teflon particle immobilized protease in aqueous solution. J Mol Catal B 2000;9:259–267.

9. Korecka L, Bilkova Z, Holeapek M, et al. Utilization of newly developed immobilized enzyme reactors for preparation and study of immunoglobulin G fragments. J Chromatography B 2004;808(1):15–24.

10. Ma ZW, Gao CY, Ji JA, Shen JC. Protein immobilization on the surface of poly-L-lactic acid films for improvement of cellular interactions. Eur Polymer J 2002;38(11):2279–2284.

11. Kang IK, Kwon BK, Lee JH, Lee HB. Immobilization of proteins on poly (methyl methacrylate) films. Biomaterials 1993;14(10):787–792.

12. Kanno S, Yanagida Y, Haruyama T, Kobatake E, Aizawa M. Assembling of engineered IgG-binding protein on gold surface for highly oriented antibody immobilization. J Biotechnol 2000;76:207–214.

13. Bucur B, Andreescu S, Marty JL. Affinity methods to immobilize acetyl cholinesterases for manufacturing biosensors. Anal Lett 2004;37(8): 1571–1588.

14. Kroger D, Liley M, Schiweck W, Skerra A, Vogel H. Immobilization of histidine-tagged proteins on gold surfaces using chelator thioalkanes. Biosensors and Bioelectronics 1999;14:155–161.

15. Jung HH, Do WY, Shin S, Kim K. Molecular dynamics simulation of ben-zenethiolate and benzylthiolate on Au(111). Langmuir 1999;15(4):1147–1154.

16. Haruyama T, Sakai T, Matsuno K. Protein layer coating on metal surface by reversible electrochemical process through genetical introduced Tag. Biomaterials 2005;26/24:4944–4947.

17. Chiavegatto S, Dawson VL, Mamounas LA, Koliatsos VE, Dawson TM, Nelson RJ. Brain serotonin dysfunction accounts for aggression in male mice lacking neuronal nitric oxide synthase. Proc Natl Acad Sci 2001;98:1277–1281.

18. Hisamoto K, Ohmichi M, Kurachi H, et al. Estrogen induces the Akt-dependent activation of endothelial nitric-oxide synthase in vascular endothelial cells. J Biol Chem 2001;276:3459–3467.

19. Savchenko, Barnes S, Kramer RH. Cyclic nucleotide-gated channels in synaptic terminals mediate feedback modulation by nitric oxide. Nature 1997;390:694–698.

20. Haruyama T, Shiino S, Yanagida Y, Kobatake E, Aizawa M. Two types of electro-chemical nitric oxide (NO) sensing system with heat-denatured Cyt C and radical scavenger PTIO. Biosensors and Bioelectronics 1998;13:763–769.

21. Kamei K, Haruyama T, Mie M, Yanagida Y, Aizawa M, Kobatake E. Construction of endothelial cellular biosensing system for blood pressure control drugs with electrochemical monitoring of nitric oxide. Biosensors and Bioelectronics 2004;19:1121–1124.

V
DE NOVO DESIGNED STRUCTURES

16
Self-Assembly of Short Peptides
for Nanotechnological Applications

Ehud Gazit

Summary

The self-association of molecules to form nanoscale assemblies is a key element in "bottom-up" nanotechnological design. Biomolecules represent a unique case of self-assembling modules because of their inherent biological specificity. Such specificity can mediate precise molecular recognition processes that lead to the formation of well ordered nanoscale structures from very simple building blocks. Furthermore, the biochemical nature of the biomolecules facilitates a variety of chemical and biological modifications that allow the formation of highly functional self-assembled material. In this review, we will focus on the properties of natural and designed self-associating short peptide fragments. This class of biomolecules is of special interest as a result of its large chemical diversity, small size, biocompatibility, and simple synthesis in large amounts. The peptides may contain any of the natural amino acids but also hundreds of nonnatural ones, which results in matchless chemical diversity and remarkable molecular properties. The rationale of the selection and design of the peptides, as well as the mechanism of molecular recognition and self-assembly that includes hydrophobic, electrostatic, and aromatic interactions, are described. Moreover, specific applications that include biomaterial fabrication, cell-support scaffold preparation, biomineralization, and bio-inorganic patterning and composite formation are reviewed. Finally, future prospects for the use of peptide-based nanoscale assemblies are described.

Key Words: Nanotechnology; peptide building blocks; peptide nanotubes; peptide nanospheres; bio-inspired assembly; molecular recognition; self-assembly.

1. BOTTOM-UP ASSEMBLY OF NANOSCALE DEVICES

The process of miniaturization revolutionized the world of electronics in the 20th century. In less than half a century, the number of components per electronic circuit rose from tens to hundreds of millions. This remarkable

From: *NanoBioTechonology: BioInspired Devices and Materials of the Future*
Edited by: Oded Shoseyov and Ilan Levy © Humana Press Inc., Totowa, NJ

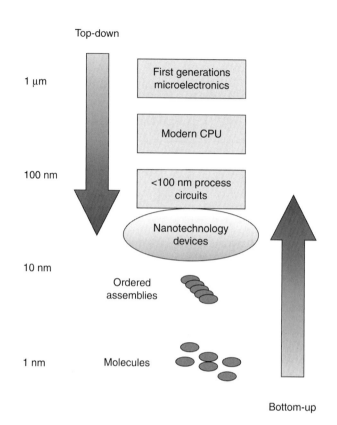

Fig. 1. Top-down vs bottom-up approaches. Modern microelectronics utilized miniaturization and optimization of lithography and etching techniques to achieve increasingly complex integrated circuits (as well as electromechanical devices; *see* text). This allows the organization of hundreds of millions of transistors on a single chip. Currently, a 90-nm process is used for the production of modern electronic devices. The bottom-up scheme, on the other hand, is based on the molecular recognition and self-assembly of molecules in the range of nanometers or even smaller. The molecules self-assemble to structures of the size of several tens of nanometers or larger. These structures can be then modified, functionalized, and integrated into functional devices.

achievement was mainly the result of very sophisticated and inspired top-down attempts that allowed continuous improvement of the lithography process (Fig. 1). Current industrial procedures have already reached a standard process at the 90-nm scale using advanced lithography and etching techniques *(1,2)*.

In spite of the impressive success in the advancement of top-down miniaturization, it appears that these techniques are about to reach a critical limit that is due to the basic physical properties of matter. To address this limitation, various approaches have been devised, the most important being those that are "bottom-up" (Fig. 1). These directions of fabrication at the nanoscale are not based on the continuous improvement of successful existing techniques but on the use of much simpler building blocks for the fabrication of complex nanoscale systems. The bottom-up process is mainly derived from the ability of very small and simple elements to self-associate into much more complex structures (Fig. 1). The process of self-association or self-assembly is based on molecular recognition principles. The essence of molecular recognition is the preferred interaction between two or more molecules that is, on the one hand, energetically favorable, but on the other, specific. To achieve such strong yet specific interactions, the molecular associations that mediate the process are comprised of a set of different noncovalent interactions. These include hydrogen bonding, electrostatic interactions, hydrophobic interactions, and aromatic interactions, among many others. The combination of the entire set of interactions vs the entropic tendency for dissociation defines the level of interaction.

A limitation of the top-down approach stems from its predominant restriction to silicon-based technology. This technique is very useful for the fabrication of electronic elements such as transistors. Moreover, the technique has been used not only in microelectronics but also for the fabrication of micro-mechanical and micro-electromechanical (MEMS) systems. Microscale accelerometers, inductors, and tunable capacitors can be fabricated using well optimized silicon-based processes *(3)*. Yet, the predominant silicon-based technology limits many biological-related applications that occur in aqueous solution near biocompatible organic surfaces.

2. CARBON NANOTUBES

Carbon nanotubes (CNTs) are considered to be major building blocks for future bottom-up applications *(4)*. From their discovery in 1991, the superb mechanical, electrical, and thermal properties of CNTs were pinpointed to suggest their use as key features for the fabrication of nanoscale devices. The CNTs, ranging in diameter from less than a nanometer (for single-walled CNTs [SWNTs]) to tens of nanometers (in the case of multi-walled nanotubes [MWNTs]) is indeed appropriate for the fabrication of nanoscale electronic, mechanical, and electromechanical devices.

In spite of the superb properties of the CNTs, they also present many limitations, especially for biological-related applications. One major limitation of the CNTs is their remarkable hydrophobicity. The sp2 graphite-like aromatic carbon organization results in extremely hydrophobic material that tends to aggregate in aqueous solution. Another limitation of the tubular carbon material is its lack of chemical functionality. The simple carbon structure is completely nonreactive. This results in a significant restriction of the orientation and integration of the carbon structures into functional elements. Only end-derivation or coating with functional molecules (especially organic molecules) allow further medication. Furthermore, some of the CNTs are remarkable conductors with metal-like properties. Yet, CNTs are synthesized as a random mixture of conductive and semi-conductive species with no way to sort between the two classes of tubes. Finally, CNTs are especially expensive materials. The price of CNT samples is in the range of hundreds of dollars per gram and is not expected to go any lower than $10 per gram in the near future. Therefore, many applications that have been suggested for CNTs, including field emitters for flat panel displays and hardening elements for composite materials, are currently not commercially viable.

3. BIOLOGICAL APPROACH TO SELF-ASSEMBLY

The self-assembly of complex structures from simple building blocks is very common in biology *(5)*. The essential boundary of cellular entities, the cell membrane, is formed by the self-assembly of phospholipids into closed-cage micro-scale structures. This assembly, which is based on simple hydrophobic interactions, results in the formation of two-dimensional (2D) bilayers that are closed into spherical structures. The phospholipid membranes also possess some further chemical information due the various phospholipid head groups that may be acidic, zwitterionic, or, on rare occasion, basic. Biological phospholipid bilayer membranes may, in and of themselves, serve as ordered structures at the nanoscale level (such as the well studied self-assembled monolayers [SAM] *(6)*). Furthermore, phospholipids may form nanoscale spherical assemblies such as small unilamellar vesicles (SUVs), which may range in diameter from tens to hundreds of nanometers. Such SUVs may have ample uses as nanocontainers in drug delivery, biosensing, nanofluidics, and other nanotechnology applications.

Biological systems may of course assemble into much more complex entities. Elaborate biological machines, such as the DNA polymerase assembly, the photosynthetic system, the protein-synthesizing ribosome machine, or molecular motors, are based on the precise assembly of tens or hundreds of proteins with or without the addition of nucleic acids or small organic and inorganic elements. These structures are practically nanoscale machines that are built entirely on the basis of molecular self-assembly *(7)*.

The basis for the formation of such complex structures from biological building blocks stems from the enormous molecular information and structural diversity that is available for biological molecules. Protein molecules are especially rich in molecular information. A simple protein of 100 amino acids represents a structural diversity of 20^{100} conformations, as any of the 20 naturally occurring amino acids can be present in the polypeptide chain.

Naturally amino acids represent a variety of chemistries with functional groups that include amines, carboxyls, thiols, alcohols, aliphatic moieties of various lengths and branching, and more. Furthermore, synthetic peptide chemistry allows further incorporation of hundreds of nonnatural amino-acid derivatives. Peptide synthesis by solid-phase chemistry is a well established procedure and the synthesis of 30-amino-acid peptides is easily performed. Nucleic acids, which have four building blocks, represent less diverse structural species, but the structural diversity of a 100-mer nucleic acid is 4^{100}, i.e., more than 10^{60}.

The assembly process is then based on a molecular recognition process. The summation of the noncovalent interactions results in the fabrication of well ordered assemblies that execute complicated biochemical activities.

4. MODEL SYSTEMS FOR THE FORMATION OF NANOSTRUCTURES

As mentioned above, synthetic peptides represent a favorable molecular system for molecular self-assembly. This is based on the remarkable and diverse molecular recognition abilities that may be achieved in relatively simple molecules. Furthermore, the synthesis of peptides is a straightforward procedure and peptides may be easily synthesized in gram or even kilogram quantities. For various application, peptides can be synthesized in a D-isomer conformation. The use of these steroisomers not only increases the conformational space but is also highly useful for attaining biostable derivatives that are not sensitive to degradation by natural biological enzymes.

Peptide nanostructures have been in use for more than a decade in various nanotechnological applications. The first major example of self-assembled peptide nanostructures was based on the use of cyclic peptide building blocks *(8–10)*. These alternating D- and L-isomers of cyclic peptides self-assemble into ordered arrays of peptide nanotubes, of which the internal diameter can be controlled by the size of the peptide building blocks (Fig. 2). The cyclic peptides transform into a beta-sheet conformation upon assembly, which results in an alternation of 90° in the direction of the composing residues. The molecular organization of the tubular structures resembles the organization of CNTs with nearly 1D characteristic. The application of the peptide nanotube arrays was demonstrated in various systems. The most interesting was the utilization of the peptide nanotubes as artificial ion channels *(9)*. Upon their application to cells, the peptide structures form transmembrane pores of nanometric diameter that cause the uncontrolled release of solutes from the cell, which finally leads to their death. Thus, these agents can serve as novel antibacterial and cytotoxic agents *(10)*. Another recent application of the cyclic peptide nanotubes is for molecular electronics *(10)*. In those applications,

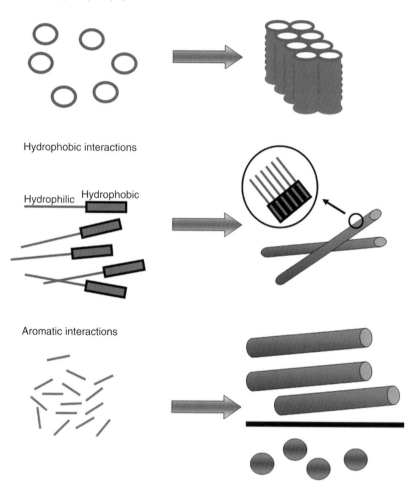

Fig. 2. Molecular schemes to achieve peptide-based self-assembly. Peptide structures are highly useful building blocks for nanoscale assemblies. The first demonstration of peptide-based nanotechnological assemblies is that of the cyclic peptide
nanotubes. These tubes are based on the stacking of closed cyclic structures to form an
array of tubes *(8–10)*. Another well studied approach is the use of peptide amphiphile
(11–16). The assembly in this case is based on hydrophobic core formation. A recent
approach is based on very short aromatic building blocks *(18–21)*. These short peptides
(as short as dipeptides) can form either tubular or spherical structures.

aromatic moieties of small band gap are conjugated into the side chain of the
peptide to form well ordered arrays of electron-transferring moieties that are
adjacent in space. Thus, these organic materials allow the directional transfer of
electrons in space to achieve a conducting biomaterial at the nanoscale.

5. HYDROPHOBIC LINEAR PEPTIDES AND PEPTIDE CONJUGATES

Other studies using peptide systems are based on the self-assembly of linear building blocks via hydrophobic interactions (Fig. 2). This mechanism of assembly is quite similar to the above-described formation of biological membranes.

The first examples of hydrophobic peptide assembly are based on the conjugation of aliphatic hydrophobic moieties into polar peptide building blocks *(11–15)*. The conjugated structures therefore have an amphiphilic structure, with a polar peptide moiety and a hydrophobic aliphatic one. Therefore, upon transfer of the organic structures into aqueous solution, a micellization process occurs (Fig. 2). The hydrophobic parts of the peptides form a water-protected core whereas the polar parts are directed to the aqueous phase. Appropriate design of the architecture of the molecules results in a well ordered structure on the nanoscale. In most cases, the desired organization is that of fibrillar or tubular structures that resemble to some extent the hexagonal-phase organization of phospholipid molecules.

Another approach that is based solely on peptide building blocks is the use of hydrophobic (aliphatic) and hydrophilic amino-acid moieties to attain a molecule with only amphiphilic peptides *(16)*. As with the conjugated peptide structure, the peptide-only amphiphilic molecules self-assemble into micelle-like structures on the nanoscale. These peptides are designated "surfactant-like peptides" to reflect their amphiphilic nature. It was recently demonstrated that such peptide structures, as short as octapeptides, can form well ordered arrays of peptide nanotubes and nanovesicles.

A variation on these types of peptide-only building blocks is based on peptides that have not only hydrophobic and hydrophilic parts, but also complementary charges *(17)*. The electrostatic interaction between positively charged moieties such as lysines and arginines and negatively charged ones such as aspartic and glutamic acids results in the rapid assembly of the building blocks into supramolecular structures on the nanoscale.

6. AROMATIC PEPTIDE NANOTUBES AND OTHER PEPTIDE NANOSTRUCTURES

Another direction for the assembly of peptide structures on the nanoscale is based on the interaction of short aromatic peptides *(18–21)*. These peptides are, in most cases, significantly shorter than the aforedescribed peptide systems. The interaction of this peptide is also based on hydrophobic, entropy-driven interaction, like the amphiphilic peptide. However, in this case the geometrically restricted interactions between aromatic moieties allow the achievement of highly ordered structures using very small building blocks (Fig. 2).

The most significant case of the self-assembly of such aromatic peptides is the formation of nanotubes by the diphenylalanine peptide. This peptide is derived from the core recognition motif of the β-amyloid polypeptide *(18–20)*. The notion for the use of this peptide is derived from insights into the formation of nanofiber amyloid structures by various polypeptides. Amyloid structures are ordered semi-crystalline fibers with a diameter of 7 to 10 nm that are associated with major human diseases *(21,22)*. It has been shown that the assembly of many of the amyloid-forming peptides is facilitated by aromatic interactions *(23–25)*. The search for minimal peptide fragments that contain the molecular information needed for the assembly of amyloid fibers resulted in the discovery of the nanotube-forming dipeptides. These peptides self-assemble into hollow tubular structures at the nanoscale. The ability of the tubes to serve as a casting mold for inorganic nanomaterials was also demonstrated. In these experiments, metal ions, such as silver ions, are reduced to solid metal within the lumen of the tubes. The peptide structure is then enzymatically degraded to end up with metal nanowires with a diameter of about 20 nm *(18)*. For other applications in which stable nanotubes are needed, the tubes can be formed by D-amino-acid isomers. These tubes are identical to the L-isomer nanotubes but are completely resistant to proteolytic degradation *(18)*. The use of the peptide nanotubes for electrochemical-based biosensing was recently demonstrated *(26)*.

The search for even simpler aromatic peptide building blocks led to the study of the nonnatural diphenylglycine building blocks *(27)*. These dipeptides self-assemble into closed-cage nanospheres with a diameter of several tens of nanometers (Fig. 2). These spherical structures may be used for a variety of applications, such as the nanocontainers described above for the SUVs. But beyond that, the peptide vesicles offer a variety of chemical modifications and structural diversity that is far greater than that of the phospholipid vesicles.

7. THE USE OF SELF-ASSEMBLED PEPTIDE STRUCTURES FOR TISSUE-ENGINEERING

The development of novel biomaterials is another key area for advancing medical technology. Biologically compatible scaffolds are an emerging material for controlled drug release, tissue repair, and tissue engineering *(28)*. This is due to the fact that many diseases cannot be treated solely by small-molecule drugs, and thus cell-based therapy is emerging as an alternative approach. Several types of collagen-based biological scaffolds, their derivatives, and other biocompatible polymers have shown great promise in this area.

Self-assembling peptide systems are especially attractive for this purpose because of their ability to serve as biologically compatible scaffolds. Previous studies have described the use of ionic self-complementary oligopeptides for this purpose *(28)*. These biological-material scaffolds are hydrogels that consist of greater than 99% water content (peptide content 1–10 mg/mL). Fabricated biomaterials made from these peptides have been shown to support cell attachment and differentiation, neurite outgrowth, and the formation of functional synapses of primary and cultured neuronal cells *(28)*.

8. FUTURE PROSPECTS

The fabrication of well ordered nanostructures by simple biological building blocks has been clearly demonstrated. The next step in the utilization of these assemblies is the move from ordered structures to functional ones. One of the greatest advantages of the bio-based building blocks is the ability to modify them with chemical and biochemical moieties. This should allow the incorporation of bio-recognition as well as enzymatic activities into the ordered bio-nanoassemblies.

Another key challenge is the ability to form artificial 3D assemblies using bio-inspired building blocks. Although the biological machines are mostly 3D, the self-assembled structures describeD here are mostly 1D or 2D. Much of the current effort in the field is being directed toward this goal. The combination of DNA recognition elements together with peptide building blocks may be highly useful in achieving this goal.

9. CONCLUSIONS

Bottom-up fabrication of nanomaterials by the self-assembly of small building blocks represents a key alternative to top-down fabrication schemes. Biological building blocks serve as especially attractive components because of the elaborate and diverse recognition between biological moieties. Among the biological building blocks, short peptides are the most studied. Assembly of peptide fragments based on hydrophobic, electrostatic and/or aromatic interactions has been demonstrated as a useful route for the fabrication of nanoscale materials. The use of these nanostructures has already been demonstrated in various applications.

ACKNOWLEDGMENTS

Support from the TAU Nanoscience and Nanotechnology Project and the Israel Science Foundation (F.I.R.S.T program) is gratefully acknowledged.

REFERENCES

1. Wang KL. Issues of nanoelectronics: a possible roadmap. J Nanosci Nanotechnol 2002;2:235–266.
2. Whitesides GM, Christopher Love J. The art of building small. Sci Am 2001;285:38–47.
3. Maboudian R, Carraro C. Surface chemistry and tribology of MEMS. Annu Rev Phys Chem 2004;55:35–54.

4. Endo M, Hayashi T, Kim YA, Terrones M, Dresselhaus MS. Applications of carbon nanotubes in the twenty-first century. Philos Transact A Math Phys Eng Sci 2004;362:2223–2238.
5. Zhang S. Fabrication of novel biomaterials through molecular self-assembly. Nat Biotechnol 2003;21:1171–1178.
6. Chaki NK, Vijayamohanan K. Self-assembled monolayers as a tunable platform for biosensor applications. Biosens Bioelectron 2002;17:1–12.
7. Drexler KE. Molecular nanomachines: physical principles and implementation strategies. Annu Rev Biophys Biomol Struct. 1994;23:377–405.
8. Ghadiri MR, Granja JR, Milligan RA, McRee DE, Khazanovich N. Molecular nanomachines: physical principles and implementation strategies. Nature 1993 Nov 25;366(6453):324–327.
9. Ghadiri MR, Granja JR, Buehler LK. Artificial transmembrane ion channels from self-assembling peptide nanotubes. Nature 1994;369:301–304.
10. Horne WS, Ashkenasy N, Ghadiri MR. Modulating charge transfer through cyclic D,L-alpha-peptide self-assembly. Chemistry 2005;11:1137–1144.
11. Hartgerink JD, Beniash E, Stupp SI. Self-assembly and mineralization of peptide-amphiphile nanofibers. Science 2001;294:1684–1688.
12. Djalali R, Chen YF, Matsui H. Au nanocrystal growth on nanotubes controlled by conformations and charges of sequenced peptide templates. J Am Chem Soc 2003;125:5873–5879.
13. Silva GA, Czeisler C, Niece KL, Beniash E, Harrington DA, Kessler JA, Stupp SI. Selective differentiation of neural progenitor cells by high-epitope density nanofibers. Science 2004;303:1352–1355.
14. Djalali R, Samson J, Matsui H. Doughnut-shaped peptide nano-assemblies and their applications as nanoreactors. J Am Chem Soc 2004;126:7929–7935.
15. Guler MO, Soukasene S, Hulvat JF, Stupp SI. Presentation and recognition of biotin on nanofibers formed by branched peptide amphiphiles. Nano Lett 2005;5:249–252.
16. Vauthey S, Santoso S, Gong H, Watson N, Zhang S. Molecular self-assembly of surfactant-like peptides to form nanotubes and nanovesicles. Proc Natl Acad Sci USA 2002;99:5355–5360.
17. Zhang S, Holmes T, Lockshin C, Rich A. Spontaneous assembly of a self-complementary oligopeptide to form a stable macroscopic membrane. Proc Natl Acad Sci USA 1993;90:3334–3338.
18. Reches M, Gazit E. Casting metal nanowires within discrete self-assembled peptide nanotubes. Science 2003;300:625–627.
19. Lu K, Jacob J, Thiyagarajan P, Conticello VP, Lynn DG. Exploiting amyloid fibril lamination for nanotube self-assembly. J Am Chem Soc 2003;125:6391–6393.
20. Song Y, Challa SR, Medforth CJ, et al. Synthesis of peptide-nanotube platinum-nanoparticle composites. Chem Commun 2004;9:1044–1045.
21. Dobson CM. Protein folding and misfolding. Nature 2003;426:884–890.
22. Rochet JC, Lansbury PT Jr. Amyloid fibrillogenesis: themes and variations. Curr Opin Struct Biol 2000;10:60–68.
23. Gazit E. A possible role for pi-stacking in the self-assembly of amyloid fibrils. FASEB J 2002;16:77–83.

24. Tartaglia GG, Cavalli A, Pellarin R, Caflisch A. The role of aromaticity, exposed surface, and dipole moment in determining protein aggregation rates. Protein Sci. 2004;13:1939–1941.
25. Makin OS, Atkins E, Sikorski P, Johansson J, Serpell LC. Molecular basis for amyloid fibril formation and stability. Proc Natl Acad Sci USA 2005;102:315–320.
26. Yemini M, Reches M, Rishpon J, Gazit E. Novel electrochemical biosensing platform using self-assembled peptide nanotubes. Nano Lett 2005;5:183–186.
27. Reches M, Gazit E. Formation of closed-cage nanostructures by self-assembly of aromatic dipeptides. Nano Lett 2004;4:581–585.
28. Narmoneva DA, Oni O, Sieminski AL, Zhang S, Gertler JP, Kamm RD, Lee RT. Self-assembling short oligopeptides and the promotion of angiogenesis. Biomaterials 2005;26:4837–4846.

17
Nanotube Membranes for Biotechnology

Lane A. Baker and Charles R. Martin

Summary

In this chapter, we discuss nanopore membranes developed for applications in the field of bio/nanotechnology. These nanopore membranes are suitable for template synthesis, bioseparations, and chemical and biochemical sensor development. We will briefly review the materials and methods of nanotube membrane technology and then discuss research and applications of these membranes with respect to template synthesis, separations, and sensing.

Key Words: Sensor; separations; template; bioanalytical; analytical; materials; electrochemistry.

1. INTRODUCTION

Nanomaterials are transforming bioanalytical chemistry. At the nanometer scale, the physical properties of the macroscale world no longer hold. Therefore, the physical and chemical properties of a nanoscale material may be altogether different from an analogous macroscale material. Synthesizing and manipulating materials at the nanometer scale is a fundamental requirement for the application of nanotechnology across a wide variety of disciplines. We have been investigating membranes with pores of controllable size, geometry, and surface chemistry *(1–3)*. The pores of these membranes can be modified, creating nanometer-scale tubes that retain the geometry of the original pores and imparting additional functionality to the membrane. These nanotube membranes show promise for use in the fields of bioanalysis and biotechnology. Specifically, these membranes and materials prepared from these membranes can be used for chemical and biochemical separations, as platforms for biochemical sensing, and for the template synthesis of biofunctionalized materials. In this chapter, we will review the materials

From: *NanoBioTechonology: BioInspired Devices and Materials of the Future*
Edited by: Oded Shoseyov and Ilan Levy © Humana Press Inc., Totowa, NJ

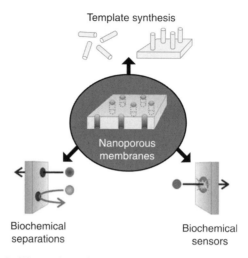

Fig. 1. Schematic illustration of the uses of nanotube-based membrane systems in biotechnology. Clockwise from top: template synthesis, biochemical sensors, and biochemical separations.

and techniques used to create, manipulate, and interrogate nanotube membrane systems.

Nanotube membranes are an attractive platform for nanotechnology in large part because of the simple, yet effective, ways in which they can be used. Membrane approaches to nanotechnology offer a facile way to handle and manipulate nanomaterials without the use of highly specialized equipment. Further, homogenous pores ensure homogenous nanomaterials, a characteristic that is often not easily achieved at these small scales. Appropriate membranes can be purchased commercially, or can be fabricated, and relatively simple techniques can be used to chemically or physically modify the membrane properties.

There are three general membrane-based strategies that we have used to prepare nanomaterials. These strategies are illustrated in Fig. 1. In the first, template synthesis, nanometer-scale pores are used to synthesize and modify materials with at least one dimension that is nanometer in scale. In a second strategy, nanometer-scale pores are used to separate species that translocate a nanotube membrane. In a third approach, we describe steps toward the use of nanotube membrane-based sensors. In this review, we will discuss these uses of nanotube membranes in the context of biotechnology. We will briefly review the materials and methods of nanotube membrane technology and then discuss biochemically oriented research and applications of these membranes with respect to separations, sensing, and template synthesis.

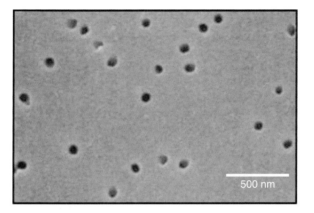

Fig. 2. Scanning electron micrograph of a polycarbonate membrane.

2. MATERIALS AND MEASUREMENTS

2.1. Track-Etch Membranes

Membranes prepared by the track-etch procedure are created by bombarding (or "tracking") a thin film of the material of interest with high-energy particles, creating damage tracks. The damage tracks are then chemically developed (or "etched") to produce pores. A variety of membrane materials are compatible with this technique; however, polymer films have shown the greatest utility. Porous poly(carbonate), poly(ethylene terephthatlate), and poly(imide) membranes are all commonly produced with this method. Track-etch membranes are available from commercial sources or can be fabricated using tracked material. Pore dimensions can be controlled by development conditions, including pH, temperature, and time. The density of pores can range from a single pore to millions of pores, and is controlled by the fluence of impinging particles in the tracking process. An example of a poly(carbonate) membrane produced using this method is shown in Fig. 2 *(4)*.

2.2. Porous Alumina Membranes

Alumina membranes are obtained through the electrochemical growth of a thin, porous layer of aluminum oxide from aluminum metal in acidic media. Membranes of this type may be obtained commercially with a variety of pore sizes, or can be grown using well established procedures. Pores with dimensions from 200 to 5 nm can be obtained in millimeter-thick membranes. The pores created are nominally arranged in a hexagonally packed array. Pore densities can be as high as 10^{11} pores cm^{-2}. An example of such a membrane is shown in Fig. 3 *(5)*.

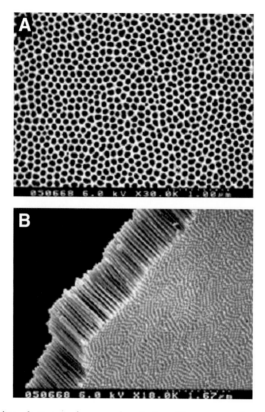

Fig. 3. Scanning electron micrographs: A, top, B, cross section, of an anodically etched alumina membrane.

2.3. Membrane Measurements

Measuring the transport or separation of an analyte with a nanotube membrane typically requires the use of a U-tube permeation cell or a conductivity cell. Membranes are initially mounted in a holder to ensure a tight seal, with the membrane separating the two halves of the cell. A schematic diagram of a typical configuration for mounting a membrane using parafilm spacers and glass slides is shown in Fig. 4 *(6)*. The membrane is then placed between two half-cells of a U-tube cell (schematically illustrated in Fig. 5), and the entire assembly is held together with a clamp. By placing a solution with species to be separated in the half-cell on one side of the membrane (feed side) and monitoring the concentration of species present in the opposite half-cell (permeate side) as a function of time, the flux of a species across the membrane may be determined. Typically, the flux is monitored using ultraviolet (UV)/vis spectroscopy or chromatographic methods. Flux of species across the membrane may be modulated by applying a voltage

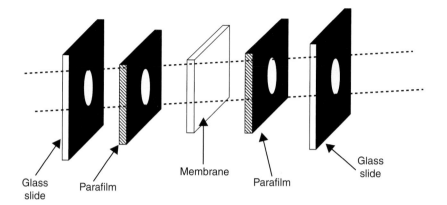

Glass slide   Parafilm   Membrane   Parafilm   Glass slide

Fig. 4. Schematic illustration of the procedure for mounting a nanoporous membrane in a support structure.

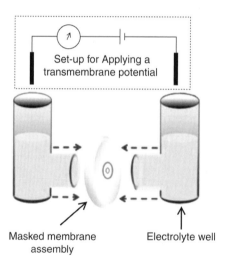

Set-up for Applying a transmembrane potential

Masked membrane assembly   Electrolyte well

Fig. 5. Schematic illustration of a U-tube cell used for preparing and measuring the properties of nanotube membranes.

between the two half-cells, resulting in electrophoretic movement, or by applying an anisotropic pressure between the two half-cells, resulting in pressure-driven flow.

2.4. Electroless Deposition of Gold

We have found electroless plating of materials in and on these nanoporous membranes to be a powerful method for both synthesizing materials and providing a membrane that is amenable to surface modification. We have described our method for electroless plating in detail previously *(7)*. Briefly, template membranes are "sensitized" by soaking in a solution of Sn(II)Cl, which results in adsorption of Sn(II) to the surface of the membrane. The Sn(II) coated membrane is then soaked in a solution of $Ag(NO_3)$. The Sn(II) adsorbed to the surface of the membrane is oxidized by the Ag(I) present in solution (Eq. 1), resulting in the deposition of Ag(0) nanoparticles on the membrane surfaces.

$$Sn(II)_{surf} + 2Ag(I)_{aq} \rightarrow Sn(IV)_{aq} + 2Ag(0)_{surf} \qquad (1)$$

The membrane with Ag particles adsorbed to the surface is then soaked in a commercial Au-plating solution. The silver particles at the surface reduce Au(I) present in plating solution, resulting in the deposition of Au(0) nanoparticles on the membrane surfaces (Eq. 2). The deposited Au particles serve as

$$Ag(0)_{surf} + Au(I)_{aq} \rightarrow Ag(I)_{aq} + Au(0)_{surf} \qquad (2)$$

autocatalysts for further Au deposition using formaldehyde as a reducing agent. The Au films deposited on the membranes cover both the pore walls and the surface of the membrane, but do not close the pore mouths. Further, the gold surface layer can be selectively removed, leaving Au nanotubes present in the pores. Electroless plating affords additional control over two critical parameters in nanotube membrane systems, pore size, and surface chemistry. By controlling the plating time and conditions, the amount of Au deposited can be controlled. This translates into more precise control over the pore diameter. The use of Au allows facile Au-thiol chemistry to be utilized to control the surface chemistry of the pores. Adsorption of charged thiols, thiolated DNAs, or functional thiols (which can undergo further chemical modification) allows the incorporation of appropriate chemistries for separations and sensing into the pores.

2.5. Sol-Gel Deposition

Another method we have found useful for materials preparation using membranes is sol-gel chemistry *(8)*. This method has been extremely versatile and can be used to produce nanotubes or nanowires, as desired. In the sol-gel method, a sol of tetraethyl orthosilicate is formed in an acidic ethanol solution. A template membrane is then sonicated in the sol solution, is removed, and is then dried and cured overnight at 150°C. Adsorption of the sol on the membrane produces a thin film of silica on the pore walls and membrane surface. Control of time and sol concentration during deposition allows control of thickness at the nanometer level. Deposited silica can be easily modified further using silane chemistry, allowing the incorporation of almost any functionality. Further, by mechanical polishing, the silica present at either or both faces of the membrane can be removed. When the membrane is dissolved, a silica negative of the original template membrane is obtained. Additionally, other inorganic materials, such as TiO_2, ZnO, and WO_3 can be templated using this sol-gel method.

3. SEPARATIONS WITH NANOTUBE MEMBRANES

Nanotube membranes may be used in biochemical separations by placing a mixture of molecules on one side of a membrane (feed side); selected molecules then translocate the membrane to the opposite side (permeate side) either by passive diffusion or a diffusion-assisted process, such as electrophoresis. Selectivity of nanotube membranes for species in the feed solution can be achieved through several means, including the size of the nanotubes in the membrane, the surface charge of the nanotubes in the membrane, and the addition of selective complexing agents to the nanotubes of the membrane. We have used these methods, and others, to create membranes for separating species of biochemical interest, including ions, small molecules, proteins, and nucleic acids. We have used both alumina and polymer membranes in a variety of configurations. Experiments related to biochemical separations will be discussed in this section.

3.1. Separation of Proteins by Size

The simplest and most straightforward use of nanotube membranes for separations uses the size of the inner diameter of the nanotube to physically select for a molecule based on its hydrodynamic radius. In this case, the molecules being separated are proteins *(9)*. Six-micrometer thick polycarbonate membranes with pores either 30 or 50 nm in diameter are plated with gold using electroless deposition. After deposition, the inner diameter (i.d.) of the gold-plated 50 nm pore membrane is 45 nm. Membranes with smaller diameter pores are obtained after plating the 30-nm pore membranes. The gold plated membranes are soaked 6 d in a 1 mM solution of a thiol-terminated poly (ethylene glycol) (PEG-thiol; molecular weight [MW] = 5000 Da). Previous measurements have shown the thin film formed from this polyethylene glycol (PEG)-thiol is approx 2.4 nm. The PEG-thiol film is used to prevent nonspecific adsorption (and resulting pore blockage) to the membrane surfaces.

Gold nanotube membranes with PEG-thiol monolayers are placed in a U-tube permeation cell and solution from the feed side of the membrane is forced through the cell by applying 20 psi pressure. The concentration of protein is monitored by periodically sampling the permeate side of the U-tube using UV/vis spectroscopy. The results of permeation experiments with single

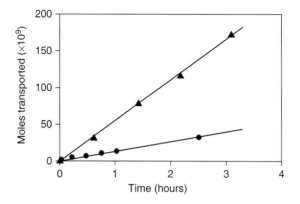

Fig. 6. Plots of moles transported vs time for lysosome (triangles) and bovine serum albumin (circles) across a 40-nm Au inner diameter nanotube membrane.

protein solutions of lysozyme (Lys; MW = 14 kDa) and bovine serum albumin (BSA; MW = 67 kDa) through a 40-nm-i.d. nanotube membrane are shown in Fig. 6. The Stokes radii for BSA and Lys are 3.6 and 2 nm, respectively. Using the Stokes-Einstein equation, the calculated diffusion coefficient for Lys should be 1.8 times higher than BSA. From the data in Fig. 6, a much higher flux for Lys relative to BSA is observed. The higher flux is a consequence of the smaller size of Lys relative to BSA, as BSA is physically hindered from translocating the membrane.

Experiments with two proteins in the feed solution were also performed. In this case, the experiments are carried out in a similar fashion, except the feed side of the membrane contained an equimolar ratio of Lys and BSA, and the results were monitored using high-performance liquid chromatography (HPLC). The results for two protein Lys/BSA permeation experiments as a function of nanotube inner diameter are shown in Fig. 7. In Fig. 7A, HPLC data of the initial feed solution are shown. Figure 7B shows HPLC data of the permeate solution after transport through a 45-nm-i.d. nanotube membrane. In analogy to the single-protein permeation experiments shown in Fig. 6, the permeation of BSA is hindered relative to Lys, as observed by the diminished BSA peak. Figure 7C shows the HPLC of the permeate solution of a 30-nm-i.d. nanotube membrane: in this case, the BSA peak is highly attenuated, indicating a small relative amount of BSA has permeated. In the case of a 20-nm-i.d. nanotube membrane (Fig. 7D), there is no detectable permeation of BSA. (Although it is unlikely there is no BSA present in the permeate solution, the amount of BSA present is below the detection limit.) The selectivity coefficient, defined as the ratio of the concentration

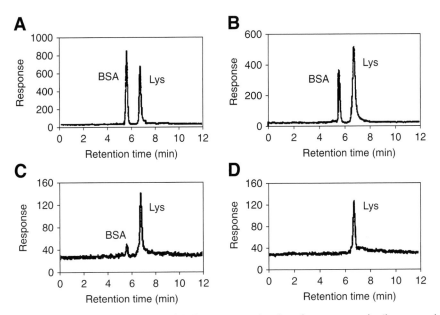

Fig. 7. High-performance liquid chromatography data for two-protein (lysosome/ bovine serum albumin) permeation experiments. **(A)** Feed solution. Permeate solutions after transport through (inner diameter) 45-nm **(B)**, 30-nm **(C)**, and 20-nm **(D)** nanotube membranes.

of Lys to the concentration of BSA present in the permeate solution, is a quantifiable measurement of nanotube membrane performance. The selectivity coefficient decreases with increasing nanotube diameter, from ≥20 to 13 to 2.2 for nanotube i.d.s of 20, 30, and 45 nm, respectively. There is a tradeoff for increased selectivity, however, as flux decreases with decreasing nanotube i.d., resulting higher selectivity but lower productivity for the smaller-i.d. nanotube membranes.

3.2. Charge-Based Separation of Ions

In the case of the experiments just discussed, we have used a chemisorbed thiol to prevent nonspecific adsorption to the membrane and nanotube walls. We have also demonstrated that this same strategy, adsorption of a functional thiol, can induce chemical selectivity to gold plated nanotube membranes. Early experiments showed the adsorption of a hydrophobic or hydrophilic thiol could promote transport of a chemical species based on the hydrophobicity of the permeate molecule *(7,10)*. In a similar manner, we have demonstrated that the chemisorbtion of L-cysteine (through the sulfur-bearing side chain) to the interior of the nanotube walls affords a

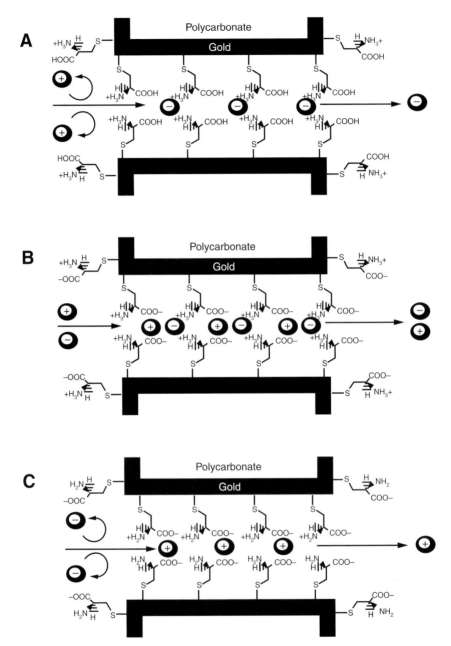

Fig. 8. Schematic representation showing the three states of protonation and the resulting ion-permselectivity of the chemisorbed cysteine. (**A**) Low pH, cation-rejecting/anion-transporting state. (**B**) pH 6.0, non-ion-permselective state. (**C**) High pH, anion-rejecting/cation-transporting state. Note: ion transport in one direction (e.g., anions from left to right in A) is balanced by an equal flux of the same charge in the opposite direction, so that electroneutrality is not violated in the two solution phases.

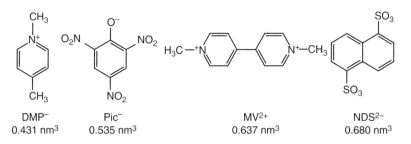

DMP⁻	Pic⁻	MV²⁺	NDS²⁻
0.431 nm³	0.535 nm³	0.637 nm³	0.680 nm³

Fig. 9. Chemical structures and molecular volumes for the permeate ions that were investigated.

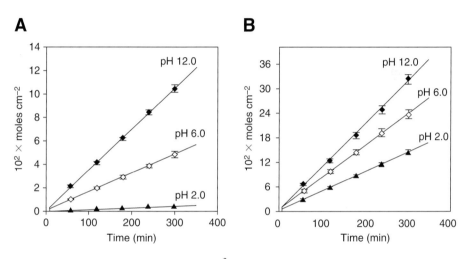

Fig. 10. Permeation data for **(A)** MV^{2+} and **(B)** DMP^+ at various pH values for the 1.4-nm-inner diameter Au nanotubule membrane that was modified with L-cysteine. The error bars represent the maximum and minimum values obtained from three replicate measurements.

method of separating ionic species based on charge, creating a pH-switchable ion-transport selective membrane *(11)*. A schematic representation of this is shown in Fig. 8. Polycarbonate membranes 6 mm thick with 30-nm-diameter pores are plated with gold to varying inner diameters. L-cysteine is chemisorbed to the gold nanotube walls by soaking the plated membranes in a 2 m*M* solution overnight. The chemical structures and molecular volumes of species separated, methylviologen (MV^{2+}), 1,5-naphthalene disulfonate (NDS^{2-}), 1,4-dimethylpyridinium iodide (DMP^+), and picric acid (Pic^-) are shown in Fig. 9. Cysteine-modified gold nanotube membranes are mounted in a U-tube for permeation experiments. The transport of species across the membrane as a function of time is monitored using UV/vis spectroscopy at molecule-appropriate wavelengths.

Table 1
Flux Data

Permeate ion	Nanotubule i.d.	Flux ($10^7 \times$ mol cm^{-2} h^{-1})		
		pH 2.0	pH 6.0	pH 12.0
DMP$^+$	0.9	0.25	0.66	1.6
Pic$^-$	0.9	0.89	0.44	0.22
MV^{2+}	0.9	0.018	0.18	0.40
NDS^{2-}	0.9	0.12	0.042	0.030
DMP$^+$	1.4	2.9	4.8	6.5
Pic$^-$	1.4	4.7	3.6	2.6
MV^{2+}	1.4	0.086	0.98	2.1
NDS^{2-}	1.4	1.7	0.75	0.14
MV^{2+}	1.9	1.5	6.8	15
NDS^{2-}	1.9	1.5	6.3	2.0
MV^{2+}	3.0	3.1	11	21
NDS^{2-}	3.0	20	11	3.4

The effect of pH on the transport of cations, MV^{2+} and DMP$^+$, through the L-cysteine-modified gold nanotubes is shown in Fig. 10. At pH 2.0, the nanotube walls are positively charged, resulting in low cation flux due to electrostatic repulsion of the like-charged cations. In the case of pH 12.0, the nanotube walls are negatively charged and a high cation flux is observed. At pH 6.0, close to the isoelectric point of cysteine, an intermediate flux is observed. In the case of transport of anions, NDS^{2-} and Pic$^-$, the opposite effects are observed (not shown). At pH 12.0, the nanotube walls are negatively charged, resulting in low anion flux due to electrostatic repulsion of the like-charged anions. In the case of pH 2.0, the nanotube walls are positively charged and a high anion flux is observed. Again, at pH 6.0, close to the isoelectric point of cysteine, an intermediate flux is observed. The results of these experiments in terms of nanotube i.d. and flux (Table 1) and selectivity coefficient, as defined by the ratio permeate transported as a function of pH (Table 2), are shown. A detailed investigation of the mechanism of the observed pH transport properties determined two electrostatic effects responsible for the selectivity observed. One electrostatic effect, an electrostatic accumulation effect, occurs when the permeate ion has a charge opposite the charge on the nanotube wall. A second electrostatic effect, an electrostatic rejection effect, occurs when the permeate ion has the same charge as the charge on the nanotube walls. These experiments clearly demonstrate the ability to design membranes selective for ionic species, in this case, using an chemisorbed amino acid.

Table 2
αpH12/pH2 and αpH2/pH12 Values

| Permeate ion | Nanotubule i.d. (nm) | Selectivity coefficient | |
		αpH12/pH2	αpH2/pH12
DMP$^+$	0.9	6.4	
Pic$^-$	0.9		4.0
MV^{2+}	0.9	22	
NDS^{2-}	0.9		4.0
DMP$^+$	1.4	2.2	
Pic$^-$	1.4		1.8
MV^{2+}	1.4	24	
NDS^{2-}	1.4		12
MV^{2+}	1.9	10	
NDS^{2-}	1.9		6.8
MV^{2+}	3.0	6.9	
NDS^{2-}	3.0		5.9

3.3. Separations Using Molecular Recognition

3.3.1. Enzymatic Molecular Recognition

One of the earliest examples of biochemical separations with a nanotube membrane uses enzymes immobilized in a polymeric membrane as a selective molecular recognition agent *(12)*. The membrane used for this separation is a 10-mm thick polycarbonate membrane with 400-nm-diameter pores. A cartoon of the final modified membrane is shown in Fig. 11. To modify the membrane for biochemical separations, a thin gold film is then sputtered across one face of the membrane. This sputtered film is too thin to close the membrane pores, but is thick enough to provide a conductive electrode layer. This electrode is then used to electropolymerize a thin (~100 nm) polypyrrole layer, forming plugs of polypyrrole that were porous enough for solvent molecules to permeate, but were not porous enough for larger enzymes to permeate. A thin gold film is then sputtered on the other side of the membrane and a solution of an apoenzyme is vacuum filtered through the membrane from the open to closed end. An apoenzyme is chosen as a molecular recognition agent, because without the addition of a cofactor, substrate molecules would not be catalyzed by the apoenzyme, allowing the substrate to be selected for without chemical conversion. After the membrane is loaded with an apoenzyme, a layer of

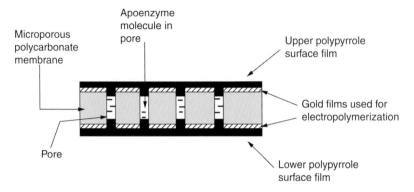

Fig. 11. Schematic cross-section of the polypyrrole/polycarbonate/polypyrrole sandwich membrane with the apoenzyme entrapped in the pores. The membrane is drawn as coming out of the plane of the paper. The various components are not drawn to scale.

polypyrrole is electropolymerized across the top layer of the membrane, encapsulating the apoenzyme in a porous matrix permeable to solvent and substrate molecules.

Membranes modified with alcohol dehydrogenase apoenzyme (apo-ADH) are mounted in a U-tube permeation cell. The membranes are then subjected to pure and mixed solutions of ethanol and phenol. (Ethanol is a substrate for apo-ADH, but phenol is not.) The results of transport experiments with this membrane and a control membrane with no apoenzyme loaded are shown in Fig. 12. In Fig. 12A, which shows the control vs the apo-modified membrane, it is clear the amount of ethanol transported by the apo-ADH-modified membrane is higher than the unmodified membrane. In Fig. 12B, the transport of ethanol and phenol with the apo-ADH-modified membrane is shown. The ratio of the slopes of the flux of ethanol and phenol yields a selectivity coefficient of 9.2 for ethanol. The selectivity from a mixed solution (shown in Fig. 12B) is analogous to the selectivity obtained when transport experiments of the individual molecules were performed. Membranes are also modified with a variety of other apoenzymes, including apo aldehyde dehydrogenase, and apo D-amino acid oxidase (apo-D-AAO). Apo-D-AAO binds only to D-amino acids, allowing us to interrogate the ability of this type of membrane to separate enantiomers. In these enantioselective membranes, a selectivity coefficient of 3.3 is obtained for D-phenylalanine vs L-phenylalanine. By using smaller pores, this selectivity coefficient could be increased to 4.9, because of an increase in the amount of permeate transported through

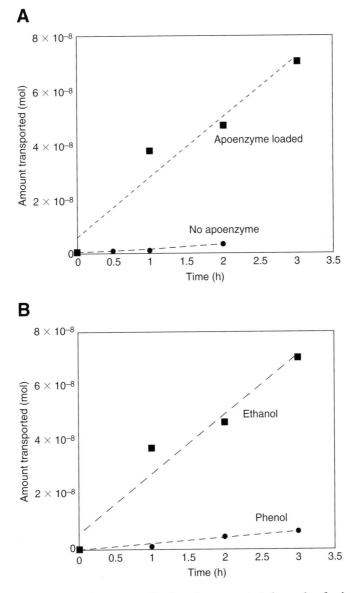

Fig. 12. (A) Plots of amount of ethanol transported from the feed solution through the membrane and into the permeant solution vs time for a membrane loaded with apo-ADH and for an apo-ADH-freemembrane. The feed solution was 0.5 m*M* in both ethanol and phenol. The slopes of these lines provide the ethanol flux across the membrane. **(B)** Plots of amount of ethanol (substrate) and phenol (nonsubstrate) transported vs time for the apo-ADH-loaded membrane. Feed solution as in A. The ratio of the slopes provides the selectivity coefficient for ethanol vs phenol transport.

facilitated mechanisms as compared to the amount of permeate transported through passive diffusion.

3.3.2. Chiral Separation

We have also used antibody modified alumina membranes to perform enantiomeric separations of a drug molecule *(13)*. In these experiments, alumina membranes with initial pore diameters of 20 and 35 nm are used. A sol-gel method (similar to that described previously) is used to deposit silica nanotubes in the pores of the membrane. The silica nanotubes are then modified with an aldehyde silane. Biochemical recognition is incorporated into this membrane by coupling the aldehyde groups of the silanes with primary amines present in antibody Fab fragments. Antibodies are selected that bind the drug 4-[3-(4-fluorophenyl)-2-hydroxy-1-[1,2,4]triazol-1-yl-propyl]-benzonitrile, an inhibitor of aromatase enzyme activity. This molecule has two chiral centers, yielding four possible isomers, RR, SS, SR, and RS. Fab fragments (anti-RS) of this antibody that selectively bind the RS relative to the SR form of the drug are used to modify membranes.

A racemic mixture of the drug molecule is placed on one side of a U-tube permeation cell, and the flux of each species is monitored as a function of time by periodically monitoring the concentration of each enantiomer present in the permeate solution with a chiral chromatographic method. A selectivity of 2 was obtained for the RS relative to the SR enantiomer, indicating that the membranes transport the RS form twice as fast as the SR form. A facilitated transport mechanism was determined to be responsible for transport in these membranes. As in the case of the apoenzyme modified membranes, by decreasing the pore diameter the selectivity coefficient is increased to 4.5 (at the expense of lower total flux). It was also found that by adding dimethyl sulfoxide (DMSO) to the feed and permeate solutions in concentrations from 10 to 30%, the rate of transport for the RS form of the drug could be regulated. This occurs because DMSO weakens the affinity of the anti-RS Fab fragment for the RS enantiomer. Thus, at 30% DMSO content, the relative transport rates for the RS and SR enantiomers were essentially equal. Because antibodies can be developed for a wide variety of species of biochemical interest, this method should be highly adaptable to a wide variety of targets.

3.3.3. Separation of Nucleic Acids

We have also used nanotube membranes to perform separations of DNA with single-base mismatch selectivity *(14)*. In these experiments, 6-mm thick polycarbonate membranes with 30-nm-diameter pores are coated with gold using electroless deposition. The diameter of the pores after gold deposition is determined to be 12 ± 2 nm. Linear DNA or hairpin DNA are used as the molecular recognition agent in these experiments. DNA hairpins contain complementary sequences at the each end of the molecule, and under appropriate conditions, form a stem–loop structure. As a result of this structure, hybridization of complementary DNA is very selective: in optimal cases a single-base mismatch will not hybridize. A 30-base DNA hairpin with a thiol modification at the 5′ end allowed facile chemisorption of the molecular recognition agent to the gold-coated nanotubes. The six bases at each end of the DNA strand were complementary, forming the stem, with the loop comprised of the remaining 18 bases in the middle of the DNA strand. The thiol-modified linear DNA molecular recognition modifiers used the same 18 bases in the middle of the molecule, but the 6 bases at each end were not complementary; thus, these linear sequences do not form the stem–loop structure. DNA molecules to transport are 18 bases long and are either perfect complements to the bases in the loop or contained one or more mismatches.

DNA-modified membranes are mounted in a U-tube permeation cell and molecules to transport are added to the feed side of the membrane. Transport is monitored by measuring the UV/vis absorbance of the permeate solution as a function of time. These systems also demonstrated a facilitated transport mechanism for complementary sequences of DNA. In the case of linear DNA, the selectivity coefficient for perfect-complement DNA (PC-DNA) vs single-base mismatch DNA is 1; that is to say, there was no selectivity. PC-DNA vs a seven-base mismatch showed a selectivity coefficient of 5. In the case of hairpin DNA-modified membranes, transport plots of PC-DNA through a modified and unmodified membrane are shown in Fig. 13. In Fig. 13A, the flux of DNA through an unmodified membrane is significantly lower than transport through the membrane modified with a perfectly complementary hairpin DNA. In Fig. 13B, the "Langmurian" shape characteristic of facilitated transport is observed for the PC-DNA, whereas diffusive flux is observed for the membrane with no DNA modification. In the case of hairpin DNA molecular recognition elements, a selectivity coefficient of 3 is obtained for a PC-DNA sequence vs a single-base mismatch sequence. A selectivity coefficient of 7 is obtained for a PC-DNA sequence vs a seven-base mismatch.

Nanotube membranes have shown the ability to separate an amazingly diverse field of biochemical species, from ions to DNA to proteins to drug molecules. The selectivity in each of these separations is governed by the inherent selectivity in the immobilized biochemical species used to effect recognition or through physical properties of the nanotubes themselves.

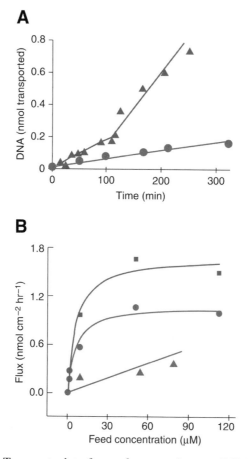

Fig. 13. (A) Transport plots for perfect-complement (PC)-DNA through gold nanotube membranes with (triangles) and without (circles) the immobilized hairpin-DNA transporter. The feed solution concentration was 9 μ*M*. **(B)** Flux vs feed concentration for PC-DNA. The data in red and blue were obtained for a gold nanotube membrane containing the hairpin-DNA transporter. At feed concentrations of 9 μ*M* and above, the transport plot shows two linear regions. Squares, data obtained from the high slope region at longer times; circles, data obtained from the low slope region at shorter times; triangles, data obtained for an analogous nanotube membrane with no DNA transporter.

4. TOWARDS NANOTUBE MEMBRANES FOR BIOCHEMICAL SENSORS

Many of the principles of biochemical sensing with nanotube membranes are inspired by results obtained with separations using such membranes. The small, often molecular, sizes of the nanotubes prepared offer new approaches to bioanalytical chemistry at the nanometer scale. We have previously described composite membranes with thin polymer skins that function as chemical sensors. In this review, we will discuss our results with nanotube membranes that function as ion channel mimics. These experiments are our first steps towards constructing nanotube based biochemical sensors that function in a manner analogous to biological channels.

4.1. Ligand-Gated Membranes

Ligand-gated ion channels in biochemical systems respond to an external chemical stimulus by switching between an off (no current or low current) and on (high current) state *(6)*. We have created synthetic nanotube membranes that can mimic the function of natural ligand-gated ion channels. Our ion channel mimics start in the low- or no-current state and convert to the high-current state in the presence of the appropriate analyte, in analogy to the functioning of acetylcholine-gated channels found in nature. Sixty-micrometer thick alumina membranes with 200-nm-diameter pores were modified with an octadecyl silane or gold-coated alumina membranes were modified with an octadecyl thiol. This creates a highly hydrophobic membrane that does not become wet when placed in water. When the membrane is mounted in a U-tube permeation cell and a transmembrane potential is applied, the hydrophobicity of the pores results in the passage of zero or very low currents, effectively an off state. Initial experiments using an ionic surfactant, dodecylbenzene sulfonate, showed that when $10^{-6}–10^{-5}$ M were added to one side of the membrane, it partitions into the pore. This creates a more hydrophilic environment inside the pore, allowing the pores to become wet. This wetting results in a dramatic drop in resistance and the passage of a measurable current, effectively an on state.

These ligand-gated ion channel mimics can also be used to detect drug molecules. In these experiments, the effects of the hydrophobicity of three drug molecules, bupivacaine, amiodarone, and amitriptyline, on the observed transmembrane resistances were investigated. The hydrophobicity of these molecules, a function of molecular weight, and polarity, increases in the order: bupivacaine < amitriptyline < amiodarone. If the hydrophobic nature of these molecules is responsible for the partitioning of these molecules into the membrane, and thus turning on the current, then transitions from the off to on state of the membrane would occur at the lowest concentrations of amiodarone. This is what is observed experimentally (Fig. 14). Bupivacaine is the least hydrophobic of these compounds, and it is also

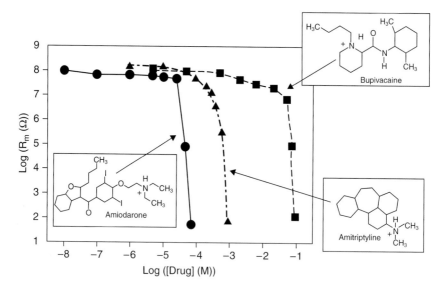

Fig. 14. Plots of log membrane resistance vs log[drug] for the indicated drugs and a C18-modified alumina membrane.

observed experimentally that bupivacaine requires the highest concentration to effect gating from off to on.

4.2. Voltage-Gated Conical Nanotube Membranes

In addition to ligand-gated ion channels, we have also mimicked the properties of voltage-gated ion channels *(15)*. In these studies, we have used polymer membranes with a single pore. The single-pore membranes are prepared either by isolating individual pores in low-density tracked films, or using films with a single damage track. This approach allows us to investigate the properties of a single nanopore rather than the ensemble of pores present in conventional membranes. By applying a transmembrane current, we are able to monitor the flow of ionic currents through the pore analogous to ion channels in lipid bilayers using traditional patch clamp techniques. Current can be monitored as a function of time or as a function of applied voltage. Pores used for these studies are anisotropically etched to create conical, rather than cylindrical, pores. The use of conical pores lowers the total resistance of the pore, allowing higher currents to flow, while retaining the nanometer dimension at the tip of the conical pore. Single-conical-pore membranes used in these studies have been plated with gold through electroless deposition to permit the

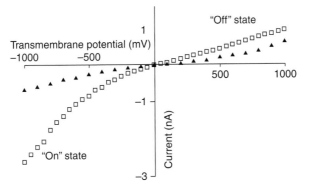

Fig. 15. *I-V* curves in 0.1 *M* KCl (squares) and 0.1 *M* KF (triangles).

chemisorption of functionalized thiols that enable us to control the surface chemistry of the nanotube walls.

In the first study, 12-mm thick poly(ethylene terephthalate) membranes with a single damage track were obtained from GSI (Darmstadt, Germany). The track was anisotropically etched using a basic solution on one side and an acidic stopping medium on the other. This results in the formation of a conical pore. By controlling the etching time and concentrations of base and acid, pores with nominal cone tips 20 nm in diameter and cone bases 600 nm in diameter can be obtained. Conical pores are then plated with gold, forming conical gold nanotubes. After plating, the small diameter (cone tip) of the pore was nominally 10 nm. The membrane was then mounted in a conductivity cell with a solution of 0.1 *M* KCl on both sides of the half-cell. In the case of a bare gold membrane (Fig. 15), ionic currents are rectified, creating a two-state system. At negative potentials, the pore is "on" whereas at positive potentials, the pore is "off." This phenomenon is observed as a result of the adsorption of Cl⁻ to the walls of the gold nanotube, creating a high negative charge at the nanotube surface. When the solution is changed from 0.1 *M* KCl to 0.1 *M* KF rectification is not observed (Fig. 15). This is due to the fact that F⁻ does not adsorb to gold, as Cl⁻ does.

The effect of charge on the nanotube walls was further investigated by measuring current-voltage curves of nanotubes with chemisorbed 2-mercaptopropionic acid (Fig. 16) or mercaptoethyl ammonium (not shown) to respective nanotube membranes. In the case of 2-mercaptopropionic acid, the carboxylate group can be protonated or deprotonated by varying the solution pH. At pH 6.6, the carboxylic acids are deprotonated resulting in a negatively charged nanotube surface. Current-voltage curves at this pH

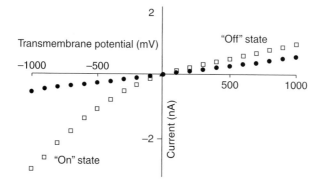

Fig. 16. *I-V* curves in 0.1 *M* KF for gold nanotubes modified with 2-mercaptopropionic acid; pH = 6.6 (squares) and pH = 3.5 (circles).

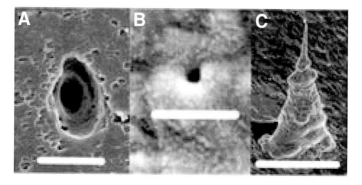

Fig. 17. Electron micrographs showing **(A)** large-diameter (scale bar = 5.0 μm) and **(B)** small-diameter (scale bar = 333 nm) opening of a conical nanopore, and **(C)** a liberated conical Au nanotube (scale bar = 5.0 μm).

showed rectification, similar to that observed in the case of Cl⁻ adsorbed to bare gold nanotubes. When the pH was lowered to 3.5, the carboxylic acid groups were protonated, removing the negative charge at the surface. Current-voltage curves at this pH showed no rectification, as observed when KF was used as the electrolyte. By using mercaptoethyl ammonium, a positively charged cation, current rectification can be reversed, meaning that at positive potentials higher current is passed and the nanotube is "on," and at negative potentials, low current is passed and the nanotube is "off." A detailed model of the mechanism of rectification based on the formation of an electrostatic trap that arises as a result of the inherent asymmetry in charged conical pores was developed to explain the observed current-voltage curves and rectification.

Table 3
Nanotube Mouth Diameter (*d*), DNA Attached, r_{max}, Radius of Gyration of DNA (*r_g*), and Extended Chain Length (*l*)

d(nm)	DNA attached	r_{max}	r_g	*l* (nm)[a]
41	12-mer	1.5	1.4	5.7
46	15-mer	2.2	1.6	6.9
42	30-mer	3.9	2.9	12.9
38	45-mer	7.1	4.0	18.9
98	30-mer	1.1	2.9	12.9
59	30-mer	2.1	2.9	12.9
39	30-mer	3.9	2.9	12.9
27	30-mer	11.5	2.9	12.9
13	30-mer	4.7	2.9	12.9
39	30-mer hairpin	1.4	n/a	6.9

[a]Includes the $(CH_2)_6$ spacer.

4.3. Electromechanically-Gated Conical Nanotube Membranes

In an effort to design more sophisticated biomimetic conical nanotubes, we have constructed single conical nanotubes with a built-in electromechanical mechanism that controls rectification of ionic currents based on the movement of charged DNA strands *(16)*. In these experiments, low-density tracked poly(carbonate) membranes were anisotropically etched to form conical nanopores. Membranes were masked in a manner that allowed the isolation and characterization of a single conical nanotube. Figure 17 shows SEM images of the large opening of a pore (Fig. 17A), and the small opening of a pore (Fig. 17B). In Fig. 17C, a scanning electron microscopy (SEM) image of a gold replica of a prepared pore is shown that demonstrates the conical geometry of the conical nanopore. Conical nanopores were plated with gold through electroless deposition, forming membranes possessing a single conical gold nanotube. After plating, conical gold nanotubes with small-diameter radii between 13 and 100 nm were obtained (Table 3). Thiolated DNA strands of varying base pair length and sequence were then chemisorbed to the surface of the gold nanotube. The DNA nanotubes prepared show an off state (low currents at positive potentials) and an on state (high currents at negative potentials) (Fig. 18A). We propose the rectification observed is due to electrophoretic movement of the DNA chains into (off state, Fig. 18C) and out of (on state, Fig. 18B) the nanotube mouth. The movement of the DNA chains into the nanotube mouth results in occlusion of the nanotube orifice, resulting in a higher ionic resistance. In Fig. 18, the effect of chain length on rectification can be clearly observed. That is to say, as DNA chain length increases, the

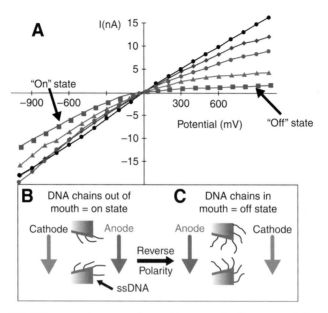

Fig. 18. (A) *I-V* curves for nanotubes with a mouth diameter of 40 nm containing no DNA (linear, circles) and attached 12-mer (diamonds), 15-mer (circles), 30-mer (triangles), and 45-mer (squares) DNAs. **(B,C)** Schematics showing electrode polarity and DNA chain positions for on (B) and off (C) states.

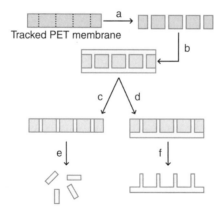

Fig. 19. Schematic of a method for template synthesis with nanoporous membranes. **A,** chemical etch of damage tracks; **B,** deposition of material to be templated (i.e., gold, silica); **C,** removal of material templated at the membrane faces through a tape stripping or mechanical abrasion; **D,** removal of material templated at one membrane face through tape stripping or mechanical abrasion; **E,** dissolution of the membrane and filtration of nanotubes; **F,** membrane removal through dissolution or oxygen plasma etch.

extent of rectification increases. It was found that an optimal length of DNA induces rectification based on the diameter of the small end of the nanotube. This work demonstrated the first example of a simple chemical (DNA chain length) or physical (nanotube pore size) method to control the extent of rectification of an artificial ion channel.

Studies of nanotubes and conical nanotubes that function as artificial ion channels are a relatively new endeavor in bioanalytical chemistry. We expect future applications of nanotube membranes to include highly sensitive and selective chemical sensors based on the design principles of Mother Nature.

5. TEMPLATE SYNTHESIS

Template synthesis is a powerful and elegant method capable of producing nanometer scale materials in a controlled fashion. Template synthesis involves the use of a template, or master, with nanometer scale features. Membranes, such as those described in the experimental section, have been our template of choice because they are convenient, versatile, and robust. A general scheme for template synthesis is shown in Fig. 19. Synthesis in the template involves the growth or deposition of materials inside the pores. The surrounding membrane material is then selectively removed, leaving nanomaterials that are negatives of the original membrane template. Depending on the conditions of membrane removal, the templated material can form a surface-bound array, or can be "freed" from the template to form individual nanoparticles. By controlling the parameters involved in the synthesis of a given material, a variety of geometries can be obtained. For instance, wires, tubes, and cones can be prepared with ease. Additionally, multi-component structures, such as segmented wires or coaxial tubes, can be prepared by modulating the materials templated. A diverse range of materials are amenable to template synthesis, including metals, semiconductors, and polymers. Further, we have demonstrated template-synthesized nanomaterials can be modified with or constructed using biochemical species.

5.1. Enzymatic Nanoreactors

Materials prepared through template synthesis can be used as nanometer-scale test tubes. One example of our use of these nano test tubes is the immobilization of enzymes *(17)*. In Fig. 20, a schematic of the process used to create nanometer scale enzymatic bioreactors is shown. This method makes use of a combination of electrochemical, chemical, and physical deposition methods. A polycarbonate membrane is first sputtered

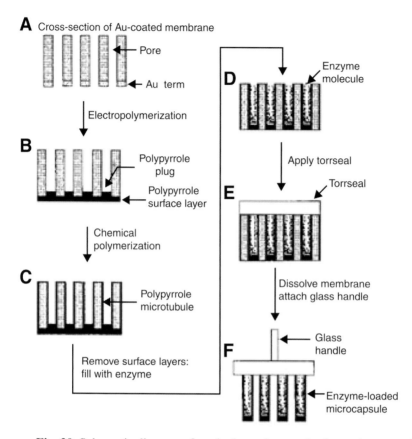

Fig. 20. Schematic diagram of methods used to synthesize and enzyme-load the capsule arrays. **A,** Au-coated template membrane; **B,** electropolymerization of polypyrrole film; **C,** Chemical polymerization of polypyrrole tubules; **D,** loading with enzyme; **E,** capping with epoxy; **F,** dissolution of the template membrane.

with a thin layer of gold (~50 nm) (Fig. 20A). This gold film serves as an electrode to electropolymerize a thin polypyrrole film across the membrane. A short polypyrrole plug is deposited in the pores as well (Fig. 20B). Additional polypyrrole is then chemically polymerized at the nanopore walls, forming a closed nanotubule, in effect a nano test tube (Fig. 20C). The thickness of the polypyrrole deposited can be controlled through the reaction conditions. Thickness is an important parameter, as the entrapment ability and permeability of the film depends greatly on the films' thickness.

These features are used to encapsulate an enzyme using the electropolymerized film as a filter. An enzyme is then loaded into the nano test tubes

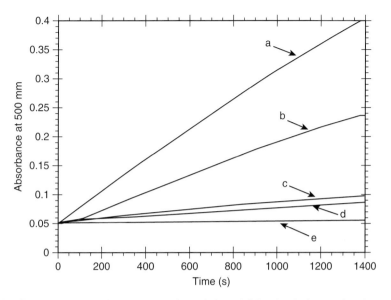

Fig. 21. Evaluation of the enzymatic activity of GOx-loaded capsules (curves A and B) and empty capsules (curve E). The standard *o*-dianisidine/peroxidase assay was used. A larger amount of GOx was loaded into the capsules used for curve A than in the capsules used for curve B. Curves C and D are for a competing GOx-immobilization methods, entrapment within a polypyrrole film.

by filtering a solution of the enzyme through the polypyrrole-modified membrane (Fig. 20D). The solvent can pass through the polymer coating, but the enzyme is too large to pass and is retained. After the nano test tube is loaded with the enzyme, a layer of Torrseal epoxy is applied to the membrane (Fig. 20E). The membrane is then dissolved in dichloromethane (Fig. 20F), leaving the enyzme loaded nano test tubes affixed to the epoxy backing in a random array.

Using this method, glucose oxidase, catalase, subtilisin, trypsin, and alcohol dehydrogenase have been successfully encapsulated. An example of the activity of glucose oxidase-filled nano test tubes is shown in Fig. 21. The enzymatic activity was evaluated using a standard *o*-dianisidine/peroxidase assay. In Fig. 21A,B, the catalytic activities of two different capsule arrays with different enzyme loadings are shown. In Fig. 21C,D, a competing encapsulation method—incorporation into a thin film of polypyrrole—is shown. In Fig. 21E, nano test tubes with no enzyme are shown. This work demonstrated that biochemical activity of templated materials could be retained, and that certain advantages, such as high surface area/volume ratio, can be obtained using template synthesis.

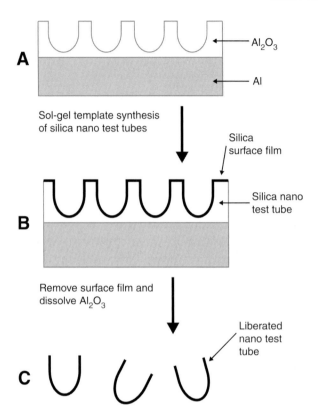

Fig. 22. Schematic of the template-synthesis method used to prepare the nano test tubes.

5.2. Nano Test Tubes

Another use of template synthesis with implications for biotechnology is the synthesis of nano test tubes free of a solid support *(18)*. Such materials have potential applications in drug delivery. For instance, if the void region of a nano test tube could be loaded with a specific payload, the open end could then be "capped," forming a nanometer-scale delivery vehicle. Molecular recognition chemistry could then be incorporated on the exterior of the capped nano test tube that would direct the nano test tube and payload to a specific portion of a cell. The cap could then be selectively released, or the nano test tube could degrade, releasing the payload at the targeted site.

Nano test tubes comprised of silica have been synthesized using an anodically oxidized alumina as the template. A schematic of the synthetic process is shown in Fig. 22. In the first step, Fig. 22A, an aluminum/alumina

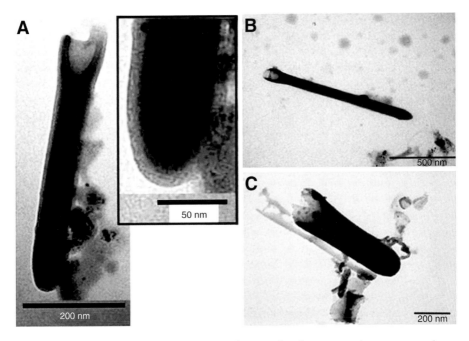

Fig. 23. (A) Transmission electron micrograph of a prepared nano test tube. The inset shows a close-up of the closed end of this nano test tube. **(B,C)** Transmission electron micrographs of nano test tubes prepared in membranes with different pore dimensions, demonstrating the variability in nano test tube size that can be templated.

template is produced by partial anodization of the aluminum substrate. This creates a template with one end open and one end closed, an appropriate configuration for the formation of nano test tubes. Silica is then deposited on the pore walls and surface of the template using a sol-gel method (Fig. 22B). The surface silica film is removed through a mechanical/chemical step with ethanol and polishing. The template membrane is then dissolved in a 25% (wt/wt) solution of H_3PO_4, liberating the templated nano test tubes (Fig. 22C). Transmission electron micrographs of silica nano test tubes prepared using this method are shown in Fig. 23. By varying the pore size and depth, the diameter and length of the prepared test tubes can be controlled. In Fig. 23A–C, nano test tubes prepared from membranes of differing geometries are shown. In the inset of Fig. 23A, it is clear that one end of the nano test tube is closed, as expected.

We have also shown that open-ended silica nanotubes can be prepared with a strategy similar to that used to prepare to prepare nano test tubes *(19)*.

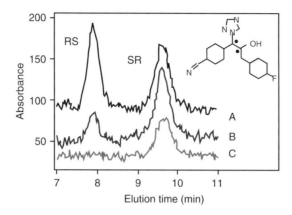

Fig. 24. Chiral high-performance liquid chromatograms for racemic mixtures of FTB before (**A**) and after (**B, C**) extraction with 18 mg/mL of 200-nm Fab-containing nanotubes. Solutions were 5% dimethyl sulfoxide in sodium phosphate buffer, pH 8.5.

These open-ended silica nanotubes can be selectively functionalized with different chemistries on the interior or exterior of the tubes. Template-synthesized, open-ended silica nanotubes are prepared by the sol-gel method previously described using either 60- or 200-nm-diameter alumina membranes. While still in the membrane, the tubes were exposed to a solution of a silane to selectively modify the interior of the tubes. Silica at the membrane faces is then removed by mechanically polishing the membrane on each side. The alumina membrane is then dissolved, liberating the silica nanotubes with the interiors selectively silylated. The liberated nanotubes with interior chemical modification are then exposed to a second solution of silane with a different chemical functionality, which only attaches to the previously unexposed nanotube exterior. In this manner, silanes with hydrophobic/hydrophilic character or silanes with molecular recognition capabilities can be selectively placed on the interior or exterior of a nanotube. This creates a functionalized nanotube with a specific chemistry on the tube interior and a potentially different chemistry on the tube exterior. These differentially functionalized nanotubes could be used for bioseparations and biocatalysis.

We have demonstrated the use of these materials as a smart nanophase extractor to remove molecules from solution was demonstrated. In these experiments, 5 mg of nanotubes having hydrophobic octadecyl silane coatings on the interior of the tubes and hydrophilic bare silica exteriors are suspended in a 1.0×10^{-5} M solution of aqueous 7,8-benzoquinoline (BQ). BQ is hydrophobic and has an octanol/water partition coefficient of 10. The suspension is stirred for 5 min and then filtered to recover the nanotubes. UV/vis spectroscopy of the filtrate solution showed as much as 82% of the BQ could be removed from solution. Control nanotubes with no coating showed less than 10% extraction of BQ.

The use of hydrophobic/hydrophilic interactions for extraction/separation is a general but nonspecific example of the use of functionalized nanotubes. The ability to use functionalized nanotubes for bioseparations in a highly specific manner has also been demonstrated. In these experiments, enantiomers of the drug 4-[3-(4-fluorophenyl)-2-hydroxy-1-[1,2,4]triazol-1-yl-propyl]-benzonitrile (FTB; Fig. 24) could be separated from a racemic mixture using RS enantiomer specific Fab antibody fragments. The Fab fragments are attached to the interior and exterior of the nanotubes using an aldhyde-terminated silane. The nanotubes are then suspended in a racemic mixture of FTB and stirred. Nanotubes are collected by filtration and the filtrate is assyed for the presence of the two enantiomers using chiral HPLC. Chromatograms of the filtrate are shown in Fig. 24. The top chromatogram (Fig. 24A) is a solution 20 μM in SR and RS enantiomers of FTB. The middle chromatogram (Fig. 24B) is the same solution as that present in Fig. 24A, but after exposure to the Fab-functionalized nanotubes. From integration of the peak ratios, 75% of the RS enantiomer, but none of the SR enantiomer, is removed. When the concentration of the initial racemic solution is lowered from 20 μM to 10 μM, all of the RS enatiomer could be removed (Fig. 24C). Unfunctionalized nanotubes do not remove any appreciable quantity of either enantiomer. Differential modification of the nanotube interiors with the RS specific Fab fragments also removed only RS FTB from solution, but at lower concentrations.

We have further demonstrated the ability to effect biocatalytic transformations with these modified nanotubes. The enzyme glucose oxidase (GOD) is immobilized on the interior and exterior of silica nanotubes using the same aldehyde silane coupling procedure used for the Fab fragments. The GOD nanotubes are suspended in a solution of glucose (90 mM) and the activity is assayed using a standard dianisidine-based assay. A GOD activity of 0.5 ± 0.2 units/mg is determined. When the nanotubes are filtered from solution, oxidation stopped, indicating that the enzyme does not leach from the nanotubes and that the enzyme retains biochemical activity when immobilized on the nanotubes.

5.3. Self-Assembly with Nano and Micro Tubes

We have also reported a method for preparing materials using template synthesis that can be self-assembled through modification with biomolecular recognition elements, namely streptavidin and biotin *(20)*. Using a modified

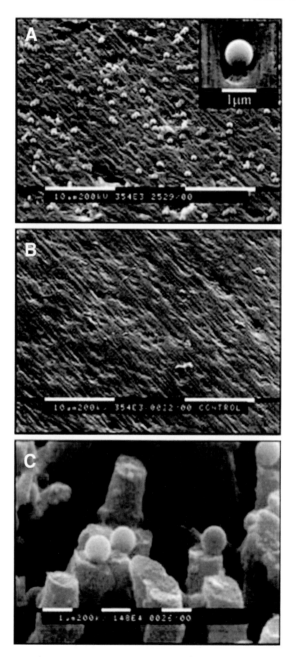

Fig. 25. Scanning electron micrographs. **(A)** The surface of a poly(AEPy) microwire-containing membrane after self-assembly of the latex particles to the ends of the microwires; the inset shows a higher magnification image of a single microwire/latex assembly, **(B)** the surface of an analogous membrane treated in the same way as in panel a but omitting the biotinylation step, and **(C)** Au/poly(AEPy) concentric tubular microwires after dissolution of the template membrane and self-assembly.

deposition procedure, nanowires comprised of poly[*N*-(2-aminoethyl)-2,5-di(2-thienyl)pyrrole] (poly(AEPy) or of poly(AEPy) coated with a thin film of gold could be produced. The membrane was then soaked in a solution of biotinyl-*N*-hydroxysuccinimide, resulting in coupling of biotin to the amine-containing polymer through amide bonds only at the tips of the nanowires. The array was then soaked in a solution of polystyrene beads coated with streptavidin (Spherotech). Scanning electron micrographs of such experiments are shown in Fig. 25. In Fig. 25A,B, membranes with nanowires of poly(AEPy) prior to membrane dissolution are shown. In Fig. 25A, the membrane has been biotinylated, whereas in Fig. 25B, the membrane has not. In both instances, the membranes are exposed to streptavidin-coated spheres, but only in the case of the biotinylated membrane, is specific irreversible adsorption observed. In Fig. 25C, free-standing (membrane dissolved) gold-coated biotinylated poly(AEPy) nanowires with streptavidin particles assembled specifically at the tip of the tubes are shown.

The examples given above demonstrate several important aspects of template-synthesis in the context of biotechnology. First, the activity of templated biochemical species, such as enzymes, can be retained. Second, biochemical recognition can be used for diverse functions, such as the assembly of templated materials or the separation of stereoisomers. Finally, templated materials of sizes appropriate for drug delivery applications can be synthesized. We believe that the template synthesis method affords the ability to prepare materials with unique properties, such as biodegradability, biocompatibility, ruggedness, size, and functionality. The ability to control these and other biomaterial design parameters will allow the investigation of new nanometer scale materials for biological applications.

6. CONCLUSIONS

In this chapter, we have reviewed our work related to nanotube membrane systems in biologically oriented or inspired settings. The ability to tune the material, size, and surface chemistries of the nanotubes affords a flexible venue to address a host of questions and problems at the forefront of bio-nanotechnology. "Smart" nanodelivery systems, artificial ion channels and separations platforms with unique selectivity are some of the immediate questions that we and others seek to address. An ultimate goal of nanotube membranes is to match or exceed the performance of transmembrane proteins found in living systems.

Although these technologies are largely tools available to the experimental nanotechnologist, mass production of templated materials is certainly possible. There is much to be done to optimize and expand the techniques of membrane-based nanotubes. Eventually, we hope to transition these important nanoscale systems to the biotechnology community in general.

Perhaps the most exciting venue for nanotube membranes lies in bio-mimetic strategies. The design and construction of membranes with nanometer-scale features that draw their fundamental inspirations from biological systems promises exciting possibilities for both fundamental and applied research in the nanotechnology community.

ACKNOWLEDGMENTS

C.R.M. and L.A.B. wish to acknowledge past and present group members whose work has contributed to this chapter. Aspects of this work have been funded by the National Science Foundation, Office of Naval Research and DARPA.

REFERENCES

1. Bayley H, Martin CR. Resistive-pulse sensing—from microbes to molecules. Chem Rev (Washington DC) 2000;100(7):2575–2594.
2. Martin CR, Kohli P. The emerging field of nanotube biotechnology. Nat Rev Drug Disc 2003;2(1):29–37.
3. Martin CR. Nanomaterials: a membrane-based synthetic approach. Science (Washington, DC) 1994;266(5193):1961–1966.
4. Sides CR, Martin CR. Unpublished results.
5. Kang M, Yu S, Li N, Martin CR. Nanowell-array surfaces. Small 2005;1(1):69–72.
6. Steinle ED, Mitchell DT, Wirtz M, Lee SB, Young VY, Martin CR. Ion channel mimetic micropore and nanotube membrane sensors. Anal Chem 2002; 74(10):2416–2422.
7. Martin CR, Nishizawa M, Jirage K, Kang M. Investigations of the transport properties of gold nanotubule membranes. J Phys Chem B 2001;105(10): 1925–1934.
8. Hulteen JC, Martin CR. A general template-based method for the preparation of nanomaterials. J Materials Chem 1997;7(7):1075–1087.
9. Yu S, Lee SB, Kang M, Martin CR. Size-based protein separations in poly(ethylene glycol)-derivatized gold nanotubule membranes. Nano Lett 2001;1(9):495–498.
10. Hulteen JC, Jirage K, Martin CR. Introducing chemical transport selectivity into gold nanotubule membranes. J Am Chem Soc 1998;120:6603–6604.
11. Lee SB, Martin CR. pH-switchable, ion-permselective gold nanotubule membrane based on chemisorbed cysteine. Anal Chem 2001;73(4):768–775.
12. Lakashmi BB, Martin CR. Enantioseparation using apoenzymes immobilized in a porous polymeric membrane. Nature 1997;338:758–760.
13. Lee SB, Mitchell DT, Trofin L, Nevanen TK, Soederlund H, Martin CR. Antibody-based bio-nanotube membranes for enantiomeric drug separations. Science (Washington, DC) 2002;296(5576):2198–2200.
14. Kohli P, Harrell CC, Cao Z, Gasparac R, Tan W, Martin CR. DNA-functionalized nanotube membranes with single-base mismatch selectivity. Science (Washington DC) 2004;305(5686):984–986.

15. Siwy Z, Heins E, Harrell CC, Kohli P, Martin CR. Conical-nanotube ion-current rectifiers: the role of surface charge. J Am Chem Soc 2004;126(35): 10,850–10,851.
16. Harrell CC, Kohli P, Siwy Z, Martin CR. DNA-nanotube artificial ion channels. J Am Chem Soc 2004;126(48):15,646–15,647.
17. Parthasarathy R, Martin CR. Synthesis of polymeric microcapsule arrays and their use for enzyme immobilization. Nature (London) 1994;369(6478): 298–301.
18. Gasparac R, Kohli P, Mota MO, Trofin L, Martin CR. Template synthesis of nano test tubes. Nano Lett 2004;4(3):513–516.
19. Mitchell DT, Lee SB, Trofin L, et al. Smart nanotubes for bioseparations and biocatalysis. J Am Chem Soc 2002;124(40):11,864–11,865.
20. Sapp SA, Mitchell DT, Martin CR. Using template-synthesized micro and nanowires as building blocks for self-assembly of supramolecular architectures. Chem Materials 1999;11:1183–1185.

18

Engineering a Molecular Railroad

Russell J. Stewart and Loren Limberis

Summary

It has been almost 20 yr since Spudich and his colleagues demonstrated purified fluorescent actin filaments writhing on a surface coated with purified myosin in an
inverted in vitro motility assay *(1)*. These assays, and the many techniques that have
been developed since to manipulate and modify filament-associated motors in vitro,
have led many to believe that biological motors will become important nanoengineering components. Still, only the first small steps have been taken toward packaging these
remarkable motors into practical and useful mechanisms. Motor proteins have several
engineering limitations, one of the most important being the limited range of chemical
and physical conditions under which they are stable and operative. With these limitations in mind, we believe that the first practical implementation of motor proteins may
be in microdevices implanted into physiological systems. The motors in such a device
may transport system components between compartments as on a conveyor belt, may
actuate gates or valves, or may power a microgenerator by scavenging energy from
biological hosts in the form of ATP. Critical hurdles to nanoengineering with motor proteins include the development of compatible interfaces with the cold hard materials of
conventional micro- and nanoengineering, the development of techniques to lay the fil-
amentous tracks in precisely defined arrays, the development of mechanisms to couple
cargos to the motors, and the development of control systems to throttle motor action
and to load and unload cargo. In this chapter, we review progress towards addressing
these challenges, focusing on the kinesin family of microtubule stepping motors.

Key Words: Active transport; kinesin; microdevices; microsystem engineering;
microtubule; protein immobilization; silicocompatibility.

1. INTRODUCTION

The fact that we can raise a glass in a toast with friends is everyday proof
of the existence of powerful, precisely controlled, long-lived, multilength
contraptions built with nanoscale biological motor proteins. These motors
power processes ranging from the powerful contraction of muscles to the

From: *NanoBioTechonology: BioInspired Devices and Materials of the Future*
Edited by: Oded Shoseyov and Ilan Levy © Humana Press Inc., Totowa, NJ

exquisitely precise sorting of chromosomes during cell division and the secretion of chemicals from highly polarized cells. During the last 20 yr or so, the remarkable biomachines that can read DNA and spin out a perfectly complementary copolymer, as well as much of the other machinery responsible for taking care of DNA, have been taken out of the cell and packaged into nearly foolproof kits for cloning and recombining DNA in vitro. The ability to manipulate DNA in the test tube so conveniently has had a profound effect on many industries. From this perspective, it seems entirely reasonable to imagine that the biomachines responsible for subcellular transport will be taken out of the cell and eventually packaged into easy to use molecular engine kits or molecular train sets. The widespread availability of efficient nanoscale transport machinery could lead to profound advances in micro- and nanosystem engineering by replacing mechanisms that depend on diffusion or bulk transport and conventional power sources.

2. MOLECULAR BIOLOGY OF MICROTUBULES AND KINESIN

Eukaryotic cells operate subcellular active transport systems that are highly analogous to railroad systems. The tracks are the filaments of the cytoskeleton and the engines are motor proteins of the myosin, dynein, and kinesin families. Myosin motors track on actin filaments, dyneins, and kinesins on microtubules. The microtubule transport network radiates from a small focused region (the microtubule organizing center) near the cell center out into every cell sector. The rolling cargo of this transport system depends on the cell type but includes vesicles destined for or returning from the cell surface, subcellular membranous organelles like mitochondria, the golgi, and the endoplasmic reticulum, nuclei, mRNAs, and during cell division, mitotic chromosomes *(2)*. The elaborate organization and complex subcellular commerce of eukaryotic cells depends on these actively powered, directional, and precisely regulated transport systems.

2.1. Microtubule Structure

Microtubules would be more accurately called nanotubes. They are hollow tubes, 24 nm in diameter, formed by the self-association of tubulin subunits (Fig. 1). The tubulin subunits, with dimensions of roughly $4 \times 4 \times 8$ nm *(3)*, are heterodimers of α- and β-tubulin proteins. The tubulin subunits associate in a head-to-tail fashion into protofilaments, which gives microtubules an 8-nm periodicity running along the protofilaments. Lateral association of parallel protofilaments with a 5-nm stagger forms hollow, unbranching, relatively rigid tubes with a helical subunit lattice. The persistence length of

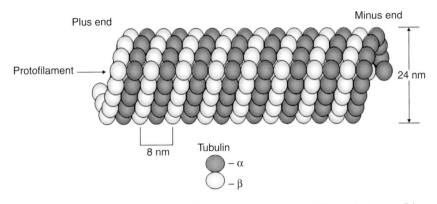

Fig. 1. Schematic representation of microtubule structure. Microtubules are 24-nm hollow tubes formed by the self-assembly of tubulin heterodimers. The asymmetry of the tubulin subunits gives the microtubule polymer an inherent structural polarity. One end of the microtubule terminates with α-tubulin subunits (minus-end), whereas the opposite end terminates with β-subunits (plus-end).

microtubules reassembled in vitro has been estimated to be between 2 and 10 mm *(4–6)*. In the cell, microtubules always contain 13 protofilaments, but microtubules reassembled in the test tube can contain 12–16 protofilaments, depending on the assembly conditions *(7)*. The head-to-tail arrangement of tubulin subunits and parallel arrangement of protofilaments results in an overall structural polarity of microtubules; one end has the β-tubulin subunit of each heterodimer exposed, whereas the α-tubulin subunit is exposed at the other end *(8)*.

2.2. Microtubule Assembly Dynamics

Microtubules are dynamic, self-assembling, and, importantly for their cellular functions, self-disassembling structures. Assembly requires that the exchangeable nucleotide binding site of β-tubulin be occupied by GTP (the nucleotide binding site of α-tubulin is nonexchangeable). After addition of a subunit to the end of a microtubule, the β-tubulin GTP is hydrolyzed, which destabilizes the assembled state of the microtubule. A few terminal subunits containing GTP may stabilize the entire microtubule. When these few subunits are lost stochastically, either through dissociation or by hydrolyzing their GTP, the microtubule disassembles rapidly by peeling apart at the ends *(9)*. The dynamic nature of microtubule assembly is intimately related to their cellular functions, particularly the dramatic cytoskeletal rearrangements that occur during cell division and accompany the assembly of the mitotic spindle. In the cell, assembled microtubules are stabilized against intrinsic disassembly by microtubule-associated proteins (MAPs) and perhaps other

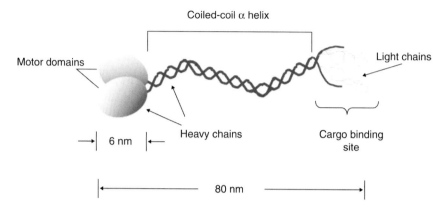

Fig. 2. Structural representation of the kinesin motor protein. Kinesin-1 is a heterote-tramer consisting of two heavy chains and two light chains. The heavy chain motor domain, located on the N-terminus of the polypeptide, is the site of mechanochemical energy transduction.

mechanisms. In the test tube, assembled microtubules can be stabilized with the plant alkaloid, taxol, purified from the Pacific Yew tree *(10)*.

The kinetics of tubulin subunit addition during microtubule assembly displays a polarity that is related to the structural polarity. The subunits add to the β-tubulin-terminated end faster than to the α-tubulin-terminated end of the microtubules. The consequence is that at steady-state, when the number of subunits adding to the microtubules is balanced by the number of subunits falling off the microtubules, there is a net addition of subunits at the β-end and a net loss of subunits at the α-end. Hence, the β-end is referred to as the plus-end and the α-end as the minus-end in the microtubule literature. In the cell, microtubule assembly is nucleated from the minus-end, resulting in a polarized organization with the plus-end distal to the cell center.

2.3. Kinesin Structure

The kinesins are a superfamily of proteins that are related by a conserved approx 340-amino-acid core domain that binds ATP and microtubules. The kinesin superfamily is subdivided into 14 families *(11)*. The first kinesin, now known as kinesin-1, was discovered in 1985 *(12)*. The kinesin-1 heavy chain is a multidomain protein containing, in addition to the N-terminal core domain, a short neck region, a coiled-coil stalk, and a C-terminal tail domain (Fig. 2). Together, the core domain and neck region constitute the force-generating motor domain. The coiled-coil stalk dimerizes two heavy chains and provides an extended spacer (~80 nm) between the motor and cargo. The tail domain binds light chains and hitches onto the kinesin-1 cellular cargo, which is a subset of neuronal vesicles *(13)*. The motor domain, which has a cross-sectional area on the order of 10 nm^2 *(14)*, is sufficient to generate force and motility in vitro *(15)*. Other kinesin families have diverse structural adaptations outside of the core domain that are related to their unique activities and cellular roles. The kinesin-5 family members, for example, are anti-parallel tetramers that bundle microtubules. The kinesin-14 family has C-terminal motor domains and move in the opposite direction as N-terminal kinesins, i.e., toward the microtubule's minus-end (although it is their unique neck regions that determine direction rather than the position of the motor domain) *(16–20)*. As one more example, the kinesin-13 family members have internal motor domains and function as microtubule depolymerizing ATPases rather than moving along intact microtubules like conventional motors *(21)*. The tail domains of the kinesin families specify their cellular cargoes and function.

2.4. Kinesin Motility

Many of the molecular details of kinesin-1 motility have been revealed through biochemical *(22,23)*, structural *(14,24,25)*, and, especially, functional studies using single-molecule motility assays. The development of in vitro motility assays for kinesin paralleled the development by Spudich and co-workers of these assays for myosin *(1,26)*. Kinesin motility can be investigated in either of two assay configurations: the "inverted" configuration, where kinesin-1 is adsorbed onto a glass surface and free microtubules are transported over the field of fixed motors *(12)*, or a configuration in which the microtubules are fixed to a glass surface and kinesin-1 is attached to free microspheres *(27)* (Fig. 3). Microtubules can be readily reassembled in vitro from purified tubulin, stabilized with taxol to inhibit their dynamic nature, and visualized by either enhanced differential interference contrast microscopy or fluorescence microscopy. To determine the direction a motor moves on microtubules, the polarity of reassembled microtubules can be marked by polymerizing dim fluorescently labeled tubulin onto brightly labeled microtubule seeds *(28)*. The polarity of the fluorescent microtubules is revealed by the asymmetric position of the bright seed since tubulin assembles onto the plus-end of the nucleating seed about three times faster than onto the minus-end. In either of the assay configurations, the motility of single kinesin-1 motors can be studied under a light microscope *(27,29)*. The bead configuration was particularly conducive to experiments using optical traps, which are created in the sample chamber of a light microscope by focusing a laser beam in the field of view. Optical traps can be used to position "motorized" microbeads onto microtubules and to apply calibrated loads to moving beads. Combined with split photodiode detectors, optical tweezers have been used to study the

A

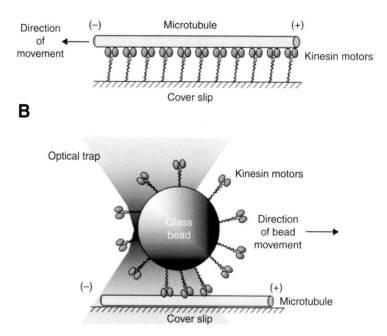

Fig. 3. Configurations for kinesin motility assays. (**A**) Schematic representation of a surface motility assay. Kinesin molecules are coated on a coverslip and interact with free microtubules in solution. The resultant motility is observed as microtubules "gliding" on the coverslip surface with the minus-end leading. (**B**) Schematic representation of a bead motility assay. Kinesin molecules are coated onto glass microspheres. The motors interact with immobilized microtubules on a coverslip surface and transport the bead towards the plus-end of the microtubule. Laser-based optical traps are used to "grab" the bead and place it on a microtubule. Optical traps are also used to measure motility characteristics of kinesin, such as step size and force production.

motility of single kinesin-1 proteins with nanometer spatial and kilohertz temporal resolution *(30–32)*.

The experimental approaches described above have culminated in the following understanding of kinesin motility. At top speed, kinesin-1 typically moves along its microtubule track with a velocity of about 0.8 μm/s. The velocity as a function of ATP concentration fits the Michaelis-Menten relationship, $v = V_{max} [ATP]/([ATP] + K_m)$, under both unloaded and loaded conditions *(30,33–35)*. The K_m for ATP is about 30 μ*M* *(36)*. Each time kinesin associates with a microtubule, it can go through hundreds of ATP hydrolysis and stepping cycles before detaching from the microtubule, which is a measure of its processivity *(27,29)*. Other kinesins, in contrast, like the ncd motor protein, are nonprocessive and let go of the microtubule after each force-generating event *(37)*. Mechanistically, kinesin is a stepping motor fueled by ATP. For each ATP molecule hydrolyzed, kinesin takes an 8-nm step from one tubulin dimer to the next along a single protofilament *(38,39)*. This stepping distance corresponds to the dimensions of the tubulin subunits. The highly processive stepping of kinesin is thought to occur as the side-by-side kinesin-1 motor domains alternately bind to the microtubule, such that one head is always attached as the other moves forward *(40,41)*. The coupling of ATP hydrolysis to stepping and force generation is diagrammed in Fig. 4. In load-velocity experiments using optical traps, single kinesin-1 molecules generated peak forces of 6 to 7 pN *(42,43)*. The maximum efficiency of kinesin-1 can be estimated from the work done at peak load (~7 pN × 8 nm = 56 pN·nm) compared to the free energy released by ATP hydrolysis under physiological conditions, which is about 20kT, where k is Boltzmann's constant and T is temperature in Kelvin, which corresponds to about 80 pN·nm *(44)*. Kinesin-1 efficiency is therefore about 70%. Others have estimated kinesin-1 work efficiency at about 50 to 60% *(45)*.

In the following sections, we summarize published work done with the goal of applying kinesin motors to microsystem engineering. Almost all of this enabling work has been done with kinesin-1. As this field progresses, it is likely that members of the many other kinesin motor families, each with unique specifications, will be particularly suitable for some engineering tasks. The kinesin superfamily and proteins of the microtubule cytoskeleton may become a rich parts catalog for building dynamic microsystems. For example, the minus-end-directed kinesin-14s, geared to go in reverse, will allow for force generation in both directions on arrayed microtubules. At 2.5 μm/s, some fungal kinesins have apparently been "geared-up" to move almost twice as fast as the original kinesin-1 *(46)*. Nonprocessive motors with short duty ratios (the fraction of the mechanochemical cycle in which they are tightly bound to the microtubule) would allow many motors to work together without gumming up the works by interfering with each other and may therefore be more appropriate in some applications. Kinesins from extremophilic organisms may be more stable and may widen somewhat the limited range of physical and chemical conditions under which microtubule motors can be employed.

3. ENGINEERING WITH KINESINS

3.1. Specifications

As components of a microsystem, kinesin motors can conceivably be used in the same way that motors are used in any system—to change one

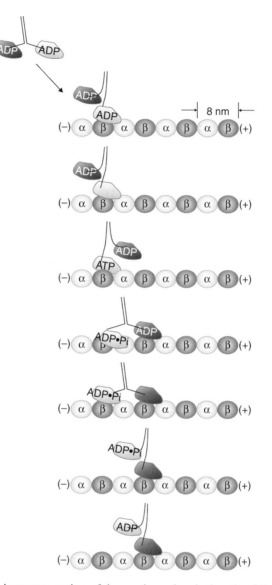

Fig. 4. Schematic representation of the mechanochemical cycle of the microtubule kinesin ATPase. The diagram illustrates one step in the cycle along a microtubule protofilament. In the absence of microtubules, kinesin has a low steady-state rate of ATP hydrolysis and predominantly has adenosine 5′-diphosphate (ADP) tightly bound in both heads. In the presence of microtubules, the ATPase rate of kinesin increases over 1000-fold *(22)*. The cycle starts with two ADP molecules bound to the kinesin heads prior to interacting with a microtubule. One head binds to a tubulin subunit, releases the ADP, and subsequently binds ATP causing the lagging head to move towards the plus-end of the microtubule. The ATP is hydrolyzed into ADP and Pi

form of energy into mechanical force so as to sort and transport components of the system, to actuate gates, valves, switches, shutters, and mirrors, or to power pumps or electrical generators. The difference is that the kinesin motor domain consumes a single molecule of "gas" per cycle and 100,000 of them would fit in a square micrometer. For perspective, about 8000 kinesin motor domains would fit on a single field effect transistor, the basic functional element of a computer chip. Producing 6 to 7 pN of force over an 8-nm displacement, the peak power produced by a single kinesin molecule per ATP hydrolysis cycle is on the order of 1×10^{-18} W. The power-to-weight ratio of a single kinesin-1 motor domain (17 W/g) is therefore around twice as much as a jet turbine (8 W/g). At a maximum packing density of 10^5 motor domains/μm^2, with each motor generating peak forces of up to7 pN, cumulative forces on the order of 650 nN/μm^2 could be generated at the theoretical upper limit with little waste heat. For comparison, the force density generated by high-force, densely arrayed electrostatic actuators is on the order of 1 nN/μm^2 *(47)*.

3.2. Stability and Operational Lifetime

The other difference between kinesins and conventional machines is that kinesins are designed for optimal operation in salty water that is buffered near neutral pH, with the correct metal cofactors, near the temperature of the organism it came from, at the intracellular redox potential of a eukaryotic cell, in the absence of photo-oxidizing radiation, firmly attached to cargo adapter proteins, in a thick soup of co-evolved macromolecules. In other words, kinesin-based devices are not likely to be the first choice for mission critical components of deep-space probes. Bohm and co-workers *(48,49)* have reported parametric studies of the effect of assay conditions including metal cofactors, pH, ionic strength, and temperature on in vitro kinesin-1 motility. Outside of a narrow range, all of these factors had profound effects on motility. In addition to direct effects on the motor, significant deviations from physiological conditions will affect other biological components, like the microtubule tracks, and biological interactions, like cargo coupling. Brunner et al. (50) investigated the operational lifetime of kinesin-1 using taxol-stabilized, fluorescently labeled microtubules in flowchambers fabricated from glass, polyurethane, polymethylmethacrylate, polydimethylsiloxane, and ethylene-vinyl alcohol copolymer. Under intense light exposure, even in the presence of an enzyme-based oxygen-scavenging system, the motility

---

Fig. 4. *(Continued)* inducing the free head to bind to the next tubulin subunit. The lagging head detaches from the tubulin subunit and releases the Pi. The cycle then repeats at this point.

lifetime was on the order of hours, limited by the oxidation and depolymerization of the fluorescently labeled microtubules. In the absence of strong illumination, kinesin-1 retained activity for 1 to 2 d. Verma et al. *(51)* have reported the effects of micro- and nanofabrication processing chemicals, such as lithography resists, developers, and removers, on the functionality of kinesin proteins and microtubules. Aside from tetramethylammonium hydroxide, kinesin can be used in conjunction with acetone and isopropyl alcohol solvents, methyl isobutyl ketone developer, and polymethyl-methacrylate and UV5 photoresists. On the other hand, microtubules are quite susceptible to these chemistries. Dilution of some developers in aqueous buffers was required to prevent microtubule depolymerization.

In a neuron it would take kinesin-1 traveling at 1 μm/s at least a million seconds (more than 11 days) to transport a vesicle from the cell body to the end of a meter-long axon. Because axons a meter long and longer are not unusual in nature (think of a giraffe), it seems likely that kinesin-1 has an intrinsic operational lifetime of at least several weeks and probably much longer in vivo. By designing kinesin-based devices to be put back into the environment for which they were optimized during evolution, at least some of the stability limitations may be solved. As nanocomponents of semi-permeable medical implants, for example, the motors would be operating at their optimal temperature, pH, ionic strength, and redox potential, protected from radiation. Perhaps most importantly, the implanted kinesin-powered mechanisms would have access to essentially unlimited fuel by scavenging ATP from its biological host. Storage and delivery of ATP fuel to the kinesin device would be unnecessary. Similarly, kinesin-based actuators or transporters in a biosensor could be "powered-up" by ATP in the biological sample itself. This could have important applications in remote medical diagnostics and remote sensing of biological signals. In addition to scale, potential power density, and efficiency, one of the most significant advantages of kinesin motors, compared to scaled-down conventional power sources, may be the ability to scavenge energy directly from a biological host, or biological sample.

3.3. Device Configurations

Like in-vitro motility assays, two configurations have been employed in demonstrating proof-of-principle kinesin devices. The "engines down" configuration, commonly referred to as molecular shuttles, is similar to the original inverted motility assay (Fig. 3A) in which the kinesin-1 is immobilized on the surface of a flowchamber and taxol-stabilized, fluorescently labeled microtubules float freely above the fixed motors. Microtubules land with random orientations and move in whatever direction that their minus-end is pointing. This is akin to building a railroad by planting the engines upside-down and hauling freight by moving the railroad tracks. A clever application of this configuration to imaging surface topography was described by Hess and co-workers *(52)*. By summing accumulated images of a few hundred fluorescent microtubules crawling in random directions on a kinesin-coated surface, a fluorescent image of the surface is gradually filled in (the opposite effect to the inchworm computer screen saver). The topography of the surface is revealed because the stiff microtubules cannot climb steep walls, leaving elevated structures dark in the bright surface image. In the "tracks down" device configuration, the microtubules are fixed to the surface of the flowchamber and the cargo is attached directly to the kinesin engines, as in the bead-motility assays (Fig. 3B). This configuration is more analogous to the natural transport system in the cell. Both device configurations are likely to have advantages and be better suited to some applications than to others. In the following, we address design issues that are common to both configurations.

3.4. Silicocompatibility

In much the same way that the biocompatibility of synthetic materials placed in a biological system is a major issue for the medical device industry *(53)*, the compatibility of protein machines with synthetic microdevice materials will be a major challenge. In their natural environment, proteins encounter conditions quite distinct from the hard surfaces of devices fabricated out of silicon, glass, ceramics, metal, or plastic. Controlling protein interactions at interfaces will be essential for maximizing device efficiency and operating lifetime both by preventing protein denaturation and by presenting proteins in a biologically meaningful orientation in the device. In the case of kinesin, this means with the motor domains away from the surface and accessible to microtubules.

As yet, developing well defined interfacial chemistries for immobilizing kinesins has not been a major emphasis in kinesin-1 microsystem prototype engineering. In most cases, kinesin-1 has been surface immobilized by precoating all device surfaces with the milk protein, casein, followed by nonspecific absorption of kinesin-1. Although effective and expedient for kinesin-1, this casein passivation method has not worked as well for other kinesin families. In the long run, a more generalized and defined method of motor immobilization will be needed. Several somewhat more specific immobilization schemes have been reported for in vitro motility studies but have not yet been applied in microdevices. These include genetic fusions of kinesin-1 with the biotinylated domain of the biotin carboxyl carrier protein, which are then immobilized on streptavidin-coated surfaces *(54,55)*; fusions with glutathione S-transferase (GST), which are immobilized on GST antibody-coated surfaces *(56)*; and fusion with green fluorescent protein (GFP), followed by immobilization on GFP-antibody coated surfaces *(16)*.

Another immobilization approach was based on Pluronic™ surfactants, which are tri-block copolymers of two hydrophilic poly(ethylene oxide) (PEO) chains connected by a hydrophobic poly(propylene oxide) (PPO) chain. The hydrophobic PPO block organizes the surfactant at hydrophobic interfaces into a protein-repelling layer of PEO blocks extending away from the surface. The anti-fouling properties of Pluronics and PEO have been well documented *(57–61)*. Pluronics have been end-group activated and fit with functional groups for conjugating proteins to the activity preserving interface provided by the PEO groups *(62–64)*. By adding metal-chelating nitrilotri-acetic acid (NTA) groups to Pluronic F108, in another embodiment of this approach, His_6-tagged proteins, like surface-sensitive firefly luciferase, could be immobilized with high specific activity *(65)*. When used in assays with a His_6-tagged ncd motor *(66)* and truncated *Drosophila* kinesin-1, the F108-NTA allowed quantitative motility experiments without blocking proteins. In the presence of Ni^{2+}, the rate that microtubules landed on the surface of the flow chamber and moved was proportional to the number of added motor proteins. In the absence of Ni^{2+}, no microtubules landed on the surface, even when high numbers of motors were added to the motility chamber. This and similar approaches may eventually lead to standardized methods for controlling motor interactions at device interfaces.

3.5. Finding Direction

Like any railroad system, the engines of the subcellular active transport system can only go where there are appropriate gauge tracks. The direction of kinesin transport in the cell is determined by the orientation of the micro-tubule tracks. As described above, the microtubule plus-ends grow out from nucleating sites (the microtubule organizing center or spindle poles). Therefore, plus-end motors haul their loads toward the cell periphery, minus-end motors transport cargo or generate forces directed toward the microtubule nucleating sites. In the same way that cellular function depends on appropriately directed kinesin forces, most kinesin devices will be useless without being able to efficiently direct the kinesin generated forces.

3.5.1. Laying the Tracks

In the "tracks down" configuration, as in the cell, the direction of kinesin transport will be guided by the immobilized microtubules. There will be no net transport on a randomly oriented array of fixed microtubules, whereas transport will be unidirectional on a polarized array. The ultimate functionality and efficiency of these kinesin-based devices will be determined by the ability to lay tracks of parallel microtubules with defined polarities. Immobilizing par-allel arrays of microtubules is straightforward: because microtubules have a net negative charge at neutral pH and bind strongly to positively charged surfaces, such as amino-silane-treated glass *(67,68)*, and because the shear forces on a solution of microtubules flowing into a chamber cause the microtubules to align in the direction of fluid flow, simply flowing a microtubule solution into an amino-silanized flowchamber creates a parallel array of surface-immobilized microtubules. The microtubules will be oriented with plus- and minus-ends pointing in both directions, however. To immobilize polarized parallel microtubule arrays is more challenging.

One method of polarizing an array is to subject microtubules gliding on a low-density field of immobilized kinesin-1 to fluid flow in one direction *(69)*. Microtubules gliding upstream into the flow eventually turned and glided downstream. After polarizing microtubules in this way, the array was fixed to the chamber surface by treatment with glutaraldehyde. The chemically fixed microtubule arrays still support kinesin motility *(70)* and can be stored in the refrigerator for several days *(71)*. Recently, Yokokawa and colleagues reported cryopreservation of immobilized microtubules on polydimethylsiloxane (PDMS)-glass surfaces for up to 30 d *(72)*. In a similar approach, Stracke *(73)* used an electric field to align microtubules gliding on a field of immobilized kinesin-1. Microtubules in suspension subjected to a static electric field moved toward the positive electrode but did not orient with one end preferentially pointing toward the electrode. However, microtubules gliding on the kinesin-1 surface re-oriented with the minus-end pointing toward the positive electrode depending on the kinesin-1 surface density. At high kinesin-1 surface densities, the rigid microtubules are attached to the surface at too many points to be influenced by the electric field. At low densities, on the other hand, the leading end (the minus-end) of the gliding microtubules were detached from the surface for sufficient time and for a sufficient length to be turned by the electric field.

The potential for using electric fields for controlling the direction of microtubule gliding was also investigated by Jia and colleagues *(74)*. As discussed by the authors, because each tubulin has a net negative charge of 48 e$^-$, corresponding to 84,000 e$^-$/μm of microtubule, microtubules should experience strong electrophorectic forces in a DC electric field. The charges, however, are shielded by counterions in the buffered physiological solutions used for kinesin motility. By measuring the velocity of microtubule movement in a static electric field, the effective charge was calculated to be 3×10^{-4} e$^-$ per tubulin dimer. In AC electric fields, microtubules were observed to accumulate by dielectrophoresis in regions of high electric field and possibly electrooriented between the electrodes.

Microtubules can also be polarized by holding onto one end of the microtubules while flowing fluid past them. Limberis and Stewart *(75)* used a single-chain antibody specific for the N-terminus of α-tubulin, which is exposed only at the minus-end of the microtubules, to specifically attach only the minus-end of microtubules to a flowchamber surface. The single-chain antibody was surface-immobilized through metal coordination bonds using F108-NTA. Subsequent fluid flow through the chamber aligned the microtubules with greater than 90% of the minus-ends pointing upstream. The oriented arrays were stabilized by adding a low concentration of methyl-cellulose to suppress rotational diffusion of the microtubules. With further development, the polarized arrays could also be fixed by photocrosslinking to the flowchamber surface.

In the cell, microtubule polarity is achieved by controlled nucleation from focused nucleating sites. This is the approach that Brown and Hancock *(76)* took to create polarized arrays of microtubules. Short segments of microtubules were immobilized in surface patterns and used as nucleating sites for the assembly of long microtubules. Because the critical concentration for sponta-neous microtubule nucleation is much higher than the critical concentration for tubulin assembly onto existing microtubules, by using low tubulin concentra-tions, microtubules were assembled only onto the existing seeds. Growth was limited to the plus-end by treating the tubulin with *N*-ethyl maleimide, which prevents tubulin from adding to the microtubule minus-end. This method has several advantages: (1) microtubules can be positioned precisely in desired locations or patterns; (2) an array can be grown within confined geometries; and (3) high densities of immobilized microtubules can be achieved.

Another promising approach for immobilizing and possibly polarizing microtubule arrays using reversible DNA hybridization was reported by Muthukrishnan and co-workers *(77)*. Microtubules were polymerized with biotinylated tubulin, and conjugated with single-stranded biotinylated oligonucleotides through neutravidin. Complementary biotinylated oligonu-cleotides were patterned on surfaces through neutravidin by microcontact printing or parylene dry lift-off *(78)*. The oligo-labeled microtubules were specifically and reversibly immobilized through hybridization between the complementary oligonucleotides. The oligonucleotide modifications did not interfere with kinesin motility on the immobilized microtubules. With fur-ther development using dip-pen lithography or microarray spotting tech-niques to create higher resolution patterns, and by making segmented microtubules with different oligonucleotides on each segment, it may be possible to pattern surfaces with complementary oligonucleotides that define the positions and orientation of microtubule arrays.

Another approach to manipulating microtubules, for use in a separation device, is reported by Yokokawa and colleagues *(79)*. Individual microtubules are polar aligned and immobilized within arrays of parallel submicrometer-wide channels fabricated on a PDMS membrane. Kinesin-coated surfaces transport microtubules from a loading chamber into the channels toward a separate collection chamber. Because microtubules enter from only one side of the microchannel array they are polar aligned within the array. The microtubules are immobilized within the channels by exposing the gliding assay to a mercury ultraviolet (UV) lamp. Surface kinesin is inactivated by the exposure but microtubules remain bound and active. By processing single microtubules in a channel, the transport direction is defined.

3.5.2. Working the Chain Gang

Although the "tracks down" configuration entails the extra problems of immobilizing polarized microtubule arrays for highest transport efficiencies, it has several important design advantages over the "engines down" configuration. Perhaps the most important advantage is that hundreds of thousands (any number really) of kinesin motors can be "ganged" on the surface of a large load to work together on a polarized microtubule array. The polarized array directs the cumulative force of the ganged motors. This approach does not limit us to the nanometer scale of the individual components when designing kinesin devices. Instead, as in biology, kinesin-powered devices useful at the millimeter and even larger length scales can be designed. Muscles, for example, are bundles of billions of myosin motors operating together to effect powerful movements on the meter-length scale. Engineering with kinesin motors is not limited to the nano- and microscale. Millimeter-scale devices could exploit kinesin's power density, efficiency, and especially its ability to use biological energy stores. These attributes are not easily achieved using scaled-down conventional engineering technologies.

The movement of relatively large loads for relatively long distances on fixed microtubule arrays has been reported. Kinesin-1 attached to 10 µm × 10 µm × 5 µm microfabricated silicon chips transported the chips over three or four parallel microtubules at approximately 0.8 µm/s, the speed of unloaded kinesin *(80)*. The microtubule array in this experiment was aligned but not polarized. Other microchips, interacting with two anti-parallel microtubules, rotated around a stationary point on the surface. These observations demonstrated, in principle, that kinesin-1 can actuate micromachine parts fabricated out of silicon and the direction of actuation was controlled by the orientation of fixed microtubules. Similarly, Bohm et al. *(71)* reported unidirectional kinesin-1 transport of micrometer-sized glass beads and glass shards (~24 µm × ~12 µm × 2–5 µm) on polar aligned microtubule arrays. Glass beads (3–10 µm in diameter) traveled distances of up to 2.2 mm by interacting simultaneously with several parallel microtubules. Individual microtubules in the array were 10 to 40 µm long, which is typical of

microtubules reassembled in vitro. In similar experiments, Yokokawa et al.
(81) reported immobilizing polar-aligned microtubules (10–30 μm) with
densities on the order of 5×10^4 filaments/mm^2, which supported unidirectional transport of kinesin-coated 320-nm diameter polystyrene beads
at densities of 2–3 $\times$ 10^5 beads/mm^2. These experiments demonstrated
that kinesin-based active transport of cargo across distances several times
longer than reassembled microtubules is feasible on polarized microtubule arrays.

3.5.3. Riding the Halfpipe

In the "engines down" configuration, the kinesin-1 motors are randomly
distributed on a flowchamber surface and transport microtubules in whatever
direction they happen to be pointing when they land on the surface. Because
individual surface-immobilized kinesin-1 are able to swivel through large
angles to interact with microtubules pointing in almost any direction *(82)*,
transport cannot be directed by orienting the engines. Instead, most efforts have
focused on physically constraining the microtubule shuttles within channels.
The problem of directing transport in the "engines down" configuration was
illustrated by Hess and co-workers *(83)*, who replica-molded polyurethane
transport channels onto coverslip surfaces. They reported that microtubules
moved predominantly within the channels if the angle of approach to the sidewalls was less than 20°. At greater angles, the microtubules climbed up and out
of the channel, driven by kinesin-1 absorbed to the channel walls. Efforts to
chemically guide microtubule movement by patterning kinesin-1 in strips were
about as effective *(84)*; about 80% of microtubules detached from the surface
at the boundary and only microtubules approaching the boundary at an angle of
10° or less were efficiently guided along kinesin-1 strips.

Better microtubule channeling has been achieved by limiting the absorption of kinesin-1 to the bottom of microchannels. Hiratsuka *(85)* photolithographically patterned channels in photoresist on glass. In an anionic detergent
and high ionic strength buffer, kinesin was selectively absorbed on the glass
bottoms of the channels, which effectively restrained the microtubules to
moving within the microchannels. Similarly, deep channels patterned in the
epoxy-based SU-8 photoresist efficiently restrained microtubule gliding to
the glass channel *(86)*. The advantages of the SU-8 photoresist are the high
aspect ratio surface channel features (1–200 mm) that are possible in a single
spin-coat process and the photoresist does not absorb functional kinesin-1,
preventing active motors from absorbing to the channel walls.

Another effective guiding approach has been to build "halfpipe" tracks that
redirect shuttling microtubules off the walls and back onto the channel floor
(87). The photolithographically patterned photoresist channels are 200-nm

high and have 1-μm deep sidewall undercuts that prevent microtubules from climbing the sidewalls. The halfpipe tracks effectively guided microtubules within channels without the need for treating the sidewalls to prevent kinesin absorption.

All of the techniques reported for localizing mictrotubule movements in microfabricated channels have used open-top systems. Although this configuration is optimal for studying microtubule motility in channels using fluorescence microscopy, microtubules that detach from the kinesin-coated surfaces simply diffuse away. Open channels also require the handling of bulk fluid above the channels, which complicates packaging for functional devices and presents problems with evaporation of the sample fluid. Huang et al. report the design and construction of capped channels with microfluidic connections for sample introduction *(88)*. SU-8 photoresist was used to fabricate open channels which were then encapsulated with a layer of dry-film photoresist, bisbenzocyclobutene (BCB). Motility assays in the capped channels showed an overall improvement in microtubule confinement and transport by preventing stalling at walls and diffusion of transiently detached filaments. Microtubules that detached from the channel floor rebounded to the surface, increasing the number of functional microtubules from a device standpoint.

3.5.4. Finding Directions

Although the methods described above effectively corral gliding microtubules within microchannels, they do not control the direction of shuttling; microtubules still go in both directions within the channel depending on which way they are pointing when they land on the channel surface. An elegant solution to this problem representing a big step toward practical kinesin-1 shuttle devices was reported by the Uyeda group *(85)*. Arrowhead shapes patterned into microchannels acted as physical rectifiers of bidirectional microtubule transport (Fig. 5). Microtubules traveling in the direction defined by the arrowheads continued around the channels in the same direction. Microtubules moving against the arrowhead direction, on the other hand, were redirected by the dead-end barbs of the arrowheads to travel in the opposite direction. Microtubules were turned around with up to 70% efficiency, depending on the arrowhead pattern. The technique was also used to create pinwheel ratchet patterns that effectively rectified the direction of shuttle movement *(89)*. The undercut walls and directional rectifiers culminated in a recent investigation of the efficiency of several types of undercut channel structures that might be useful for sorting in shuttle-type microdevices *(90)*. The structures included blind alleys meant to reverse the direction of gliding microtubules, "figure eights" with orthogonally crossing junctions or tangentially arranged curving channels, and one-way culs-de-sac that microtubules

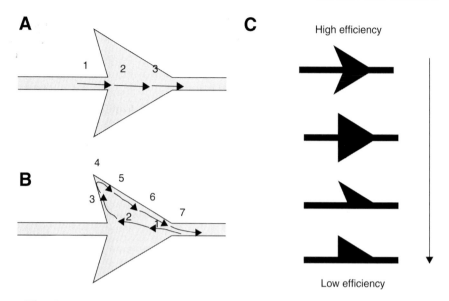

Fig. 5. Arrowhead rectifiers for bidirectional microtubule movement. **(A)** A microtubule entering the arrowhead in the direction of the arrow passes through the pattern. **(B)** Schematic demonstration of a microtubule entering the arrowhead in the opposite direction and turning around. **(C)** Relative efficiencies of arrowhead patterns as rectifiers. For additional information, *see* ref. *85*.

could enter but never leave. Slightly different arrowhead-type rectifier geometries, best described as Valentine's Day hearts, were reported by van den Heuvel et al. *(91)*. By longitudinally offsetting the entrance and exit channels through the rectifier and curving the exit channel, microtubule rectification efficiency up to 92% was achieved with 97% overall unidirectional movement. Combinations of these structures can be used to guide microtubules around elaborate channel networks. It will be interesting to see what the first real cargo and application of these microtubule shuttle devices might be.

3.6. Unnatural Freight

As subcellular transporters, kinesins have been adapted to haul dozens of distinct cellular cargos through natural fusions of the motor domain with unique protein domains that specifically couple the motor to a cargo. Mimicking this natural approach for addressing cellular freight, it is possible to genetically fuse freight-coupling domains directly to recombinant kinesin-1. For example, if kinesin-1 were fused to protein A, a staphylococcal protein that binds the Fc region of mammalian IgG antibodies *(92,93)*, the kinesin cargo

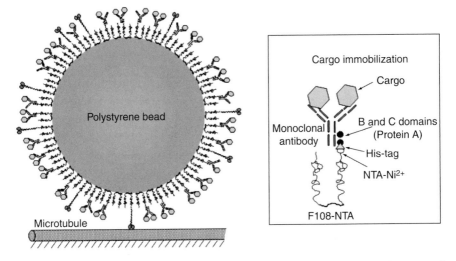

Fig. 6. Schematic representation of motorized antibodies. His_6-tagged truncated kinesin and His_6-tagged B and C IgG-binding domains of staphylococcal protein A are specifically co-immobilized on polystyrene beads through NTA-derivatized Pluronic™ F108 surfactant. The histidine-tagged proteins bind to metal ions chelated by NTA at the poly(ethylene oxide) (PEO) termini of the F108 triblock copolymer. The hydrophobic poly(propylene oxide) (PPO) segment of the triblock copolymer interacts with the hydrophobic polystyrene surface, leaving the hydrophilic PEO chains to extend into solution. Large quantities of antibodies, as well as their antigens, are coupled indirectly to a few kinesin molecules through protein A and can be transported along microtubule tracks.

could be the antigen of IgG monoclonal antibodies *(94)*. This is a very large category of potential cargos relevant to biosensor applications and medical diagnostics. To generalize this concept further, recombinant His_6-tagged kinesin and a recombinant His_6-tagged segment of protein A were indirectly coupled by co-immobilization on polystyrene microbeads *(95)* (Fig. 6). The advantage of this configuration is that only a handful of motors could transport hundreds or thousands of antibodies and their antigens (analytes) as part of an active separation device or biosensor.

The molecular shuttle configuration requires that the freight be attached to the microtubule tracks. In several reported examples, this has been accomplished using biotin and streptavidin linkages. Microtubules are reassembled spiked with biotinylated tubulin, which can be uniformly distributed along the microtubule or limited to smaller regions by seeded assembly. Cargo coated with the biotin-binding streptavidin protein bind to and are transported by the biotinylated microtubules gliding on a kinesin-1 surface. Shuttled cargos have included superparamagnetic polystyrene beads *(83)*, DNA *(96)*, and quantum dots *(97)*. DNA linked by one end to a flowchamber surface and the other end to a microtubule was stretched as the microtubule translocated on a kinesin-1 surface. It was envisioned that the kinesin-1-stretched DNA molecules could be used as precisely positioned templates for metallization into nanowires *(98)*. With much more development, the system might be useful for assembling components, like nanowires, into multidimensional structures. Another intriguing observation was that in some cases, the DNA restrained the microtubule to gliding in circles, like a goat tethered to a stake in a pasture. DNA tethering may be another useful means of controlling microtubule transport direction. As another example of a shuttled cargo, quantum dots have also been coupled to microtubules using biotin and streptavidin. If streptavidin-coated quantum dots were coupled along the entire length of biotinylated microtubules, microtubule surface binding and motility was significantly inhibited, likely because of the 10- to 15-nm dimensions of the dots. On the other hand, quantum dots coupled to short biotinylated microtubule segments did not interfere significantly with microtubule transport. Active transport of quantum dots was interesting because of the potential for assembling dynamic photonic materials. Eventually, these approaches and combinations of these enabling techniques may become practical approaches for moving and positioning the components of microscale structures using biological motors.

3.7. Throttles and Brakes

Perhaps the biggest challenge to practical kinesin devices will be developing systems to control the starting, stopping, and speed of kinesin-based transport mechanisms. One potential solution is to throttle the flow of fuel, $Mg^{2+}ATP$, to the kinesin engines. At concentrations above approx 200 µM $Mg^{2+}ATP$, kinesin-1 steps at its maximum rate (V_{max}, around 1 µm/s). Below the saturating $Mg^{2+}ATP$ concentration, kinesin moves at a rate proportional to the $Mg^{2+}ATP$ concentration, limited by the diffusion of $Mg^{2+}ATP$ into the active site. As a throttle mechanism, Hess and his colleagues *(83,99)* used UV irradiation to release pulses of caged ATP. The concentration of ATP created by photocleavage was proportional to the intensity of UV radiation. To brake the motors, hexokinase was included in the system to quickly consume ATP released by the UV irradiation to stop transport. The authors reported that the rate of microtubule motility would rise suddenly after exposure to UV light and would decline exponentially as the hexokinase competed with kinesin-1 for ATP. Repeated flashes of UV light could be used to move the microtubules a few micrometers at a time.

Dielectrophoresis seems to be a promising second force, in addition to kinesin motor forces, for manipulating microtubules in microdevices *(74)*.

Perhaps dielectrophoretic counterforces could be used in some circumstances as a control mechanism to brake kinesin-generated forces on microtubules. Kinesin velocity decreases linearly with increasing load up to a stall force *(42,43)*. Dielectric force fields could temporarily stop microtubules at loading or unloading stations if the counterforce and the kinesin surface density were balanced. The microtubules would be underway again when the field was switched off. Or, as suggested by the authors, dielectric fields could function as switching stations to selectively shunt some microtubules and their cargo onto sidetracks. It may also be possible to selectively control the speed of certain cargos as an on-the-run sorting mechanism.

Another significant control problem for the shuttling transport devices proposed so far in the literature will be the spatial and/or temporal loading and unloading of cargos. For biotin-streptavidin-coupled cargos, Hess et al. *(83)* have proposed using "smart" streptavidin, which binds and unbinds in response to changes in pH or other environmental parameters *(100)*, or photocleavable biotin, allowing cargo to be dumped by exposure to UV light. The known sensitivity of kinesin-1 and microtubules to UV irradiation suggests that photochemistry will not be the best mechanism for controlling either ATP concentration or cargo loading for the long haul. The natural control systems have not been fully worked out, but at least in part involve kinases and phosphatases working as cellular loadmasters. An analogous approach for some applications may be to station kinases in loading areas to ticket freight for specific destinations by marking it up with phosphate groups. Phosphatases stationed at transfer and destination points could unload the marked freight at the correct station.

4. CONCLUSION

We are optimistic that the microtubule transport system of eukaryotic cells will find practical, but probably limited applications in microsystem engineering. Perhaps the biggest step forward so far has been the invention of directional rectifiers *(85)*, which have given microtubule shuttle-type devices a means for controlling direction. What will the next big step forward be? As more research groups enter the field from different backgrounds, with different perspectives and new tools, kinesin devices will likely look less and less like elaborate in vitro motility assays. There remains plenty of room for invention in control system engineering, materials compatibility, enhancing operational lifetimes, and in packaging and plumbing. Expectations about potential applications should be realistic. It will eventually take a significant market force to drive continued development of kinesin technology. In the end, perhaps the practical applications of kinesins will be narrow, limited to short-lived, disposable devices. Of course, someone has probably said similar things about other nascent technologies.

REFERENCES

1. Kron SJ, Spudich JA. Fluorescent actin filaments move on myosin fixed to a glass surface. Proc Natl Acad Sci 1986;83:6272–6276.
2. Alberts B, Johnson A, Lewis J, Raff M, Roberts K, Walter P. Molecular Biology of the Cell, 4th ed. New York: Garland Science, 2001.
3. Nogales E, Wolf SG, Downing KH. Structures of the αβ tubulin dimer by electron crystallography. Nature 1998;391:199–203.
4. Kurz JC, Williams RC. Microtubule-associated proteins and the flexibility of microtubules. Biochemistry 1995;34:13,374–13,380.
5. Venier P, Maggs AC, Carlier MF, Pantaloni DJ. Analysis of microtubule rigidity using hydrodynamic flow and thermal fluctuations. Biol Chem 1994;269:13,353–13,360.
6. Gittes F, Mickey B, Nettleton J, Howard JJ. Flexural rigidity of microtubules and actin filaments measured from thermal fluctuations in shape. Cell Biol 1993;120:923–934.
7. Chretien D, Metoz F, Verde F, Karsenti E, Wade RH. Lattice defects in microtubules: protofilament numbers vary with individual microtubules. J Cell Biol 1992;117:1031–1040.
8. Fan J, Griffiths AD, Lockhart A, Cross RA, Amos LA. Microtubule minus ends can be labeled with a phage display antibody specific to alpha-tubulin. J Mol Biol 1996;259:325–330.
9. Mandelkow EV, Mandelkow E, Milligan RA. Microtubules dynamics and microtubules caps: a time-resolved cryo-electron microscopy study. J Cell Biol 1991;114:977–991.
10. Guenard D, Guerite-Voegelein F, Poteir P. Taxol and taxotere: discovery, chemistry, and structure-activity relationships. Acc Chem Res 1993;26:160–167.
11. http://www.proweb.org/kinesin/
12. Vale RD, Reese TS, Sheetz MP. Identification of a novel force-generating protein, kinesin, involved in microtubule-based motility. Cell 1985;42:39–50.
13. Saxton WM, Hicks J, Goldstein LSB, Raff EC. Kinesin heavy chain is essential for viability and neuromuscular functions in drosophila, but mutants show no defects in mitosis. Cell 1991;64:1093–1102.
14. Kull FJ, Sablin EP, Lau R, Fletterick RJ, Vale RD. Crystal structure of the kinesin motor domain reveals a structural similarity to myosin. Nature 1996;380:550–559.
15. Yang JT, Saxton WM, Stewart RJ, Raff EC, Goldstein LSB. Evidence that the head of kinesin is sufficient for force generation and motility in vitro. Science 1990;249:42–47.
16. Case RB, Pierce DW, Hom-Booher N, Hart CL, Vale RD. The directional preference of kinesin motors is specified by an element outside of the motor catalytic domain. Cell 1997;90:959–966.
17. Endow SA, Waligora KW. Determinants of kinesin motor polarity. Science 1998;281:1200–1202.
18. Henningsen U, Schliwa M. Reversal in the direction of movement of a molecular motor. Nature 1997;389:93–96.

19. Sablin EP, Case RB, Dai SC, et al. Direction determination in the minus-end-directed kinesin motor ncd. Nature 1998;395:813–816.

20. Endow SA, Higuchi H. A mutant of the motor protein kinesin that moves in both directions on microtubules. Nature 2000;406:913–916.

21. Desai A, Veram S, Mitchison TJ, Walczak CE. Kin I kinesins are microtubule-destabilizing enzymes. Cell 1999;96:69–78.

22. Hackney DD. Evidence for alternating head catalysis by kinesin during microtubule-stimulated ATP hydrolysis. Proc Natl Acad Sci 1994;91: 6865–6869.

23. Ma YZ, Taylor EW. Mechanism of microtubule kinesin ATPase. Biochemistry 1995;34:13,242–13,251.

24. Hoenger A, Milligan RA. Motor domains of kinesin and ncd interact with microtubule protofilaments with the same binding geometry. J Mol Biol 1997;265:553–564.

25. Hirose K, Lowe J, Alonso M, Cross RA, Amos LA. Congruent docking of dimeric kinesin and ncd into three dimensional electron cyromicroscopy of microtubule-motor ADP complexes. Mol Biol Cell 1999;10:2063–2074.

26. Spudich JA, Kron SJ, Sheetz MP. Movement of myosin-coated beads on oriented filaments reconstituted from purified actin. Nature 1985; 315:584–586.

27. Block SM, Goldstein LS, Schnapp BJ. Bead movement by single kinesin molecules studied with optical tweezers. Nature 1990;348:348–352.

28. Hyman AA. Preparation of marked microtubules for the assay of the polarity of microtubule-based motors by fluorescence. J Cell Sci Suppl 1991;14: 125–127.

29. Howard J, Hudspeth AJ, Vale RD. Movement of microtubules by single kinesin molecules. Nature 1989;342:154–158.

30. Visscher K, Schnitzer MJ, Block SM. Single kinesin molecules studied with a molecular force clamp. Nature 1999;400:184–189.

31. Kojima H, Muto E, Higuchi H, and Yanagida T. Mechanics of single kinesin molecules measured by optical trapping nanometry. Biophys J 1997;73: 2012–2022.

32. Coppin CM, Peirce DW, Hsu L, Vale RD. The load dependence of kinesin's mechanical cycle. Proc Natl Acad Sci USA 1997;94:8539–8544.

33. Hua W, Young EC, Fleming ML, Gelles J. Coupling of kinesin steps to ATP hydrolysis. Nature 1997;388:390–393.

34. Schnitzer MJ, Block SM. Kinesin hydrolyses one ATP per 8 nm step. Nature 1997;388:386–390.

35. Block SM, Asbury CL, Shaevitz JW, Lang MJ. Probing the kinesin reaction cycle with a 2D optical force clamp, Proc Natl Acad Sci USA 2003;100: 2351–2356.

36. Gilbert SP, Johnson KA. Expression, purification, and characterization of the Drosophila kinesin motor domain produced in *Escherichia coli*. Biochemistry 1993;32:4677–4684.

37. deCastro MJ, Fondecave RM, Clarke LA, Schmidt CF, Stewart RJ. Working strokes by single molecules of the kinesin-related microtubule motor ncd. Nat Cell Biol 2000;10:724–729.

38. Svoboda K, Schmidt CF, Schnapp BJ, Block SM. Direct observation of kinesin stepping by optical trapping interferometry. Nature 1993;365: 721–727.
39. Ray S, Meyhofer E, Milligan RA, Howard J. Kinesin follows the microtubule's protofilament axis. J Cell Biol 1993;121:1083–1093.
40. Schief WR, Clark RH, Crevenna AH, Howard J. Inhibition of kinesin motility by ADP and phosphate supports a hand-over-hand mechanism. Proc Natl Acad Sci USA 2004;101:1183–1188.
41. Yildiz A, Tomishige M, Vale RD, Selvin PR. Kinesin walks hand-over-hand. Science 2004;303:676–678.
42. Svoboda K, Block SM. Force and velocity measured for single kinesin molecules. Cell 1994;77:773–784.
43. Hunt AJ, Gittes F, Howard J. The force exerted by a single kinesin molecule against a viscous load. Biophysical J 1994;67:766–781.
44. Cross RA. Molecular motors: the natural economy of kinesin. Curr Biol 1997;7:R631–R633.
45. Vale RD, Milligan RA. The way things move: looking under the hood of molecular motor proteins. Science 2000;288:88–95.
46. Kirchner J, Woehlke G, Schliwa M. Universal and unique features of kinesin motors: insights from a comparison of fungal and animal conventional kinesins. Biol Chem 1999;380:915–921.
47. Rodgers MS, Kota S, Hetrick J, Li A, et al. A New Class of High Force, Low-Voltage, Compliant Actuation Systems Presented at Solid-State Sensor and Actuator Workshop. Hilton Head Island, South Carolina, June 4–8, 2000.
48. Bohm KJ, Stracke R, Unger E. Speeding up kinesin-driven microtubule gliding in vitro by variation of cofactor composition and physiochemical parameters. Cell Biol Intl 2000;24:335–341.
49. Bohm KJ, Stracke R, Baum M, Zieren M, Unger E. Effect of temperature on kinesin-driven microtubule gliding and kinesin ATPase activity. FEBS Lett 2000;466:59–62.
50. Brunner C, Ernst K-H, Hess H, Vogel V. Lifetime of biomolecules in polymer-based hybrid nanodevices. Nanotechnology 2004;15:S540–S548.
51. Verma V, Hancock WO, Catchmark JM. Micro- and nanofabrication processes for hybrid synthetic and biological system fabrication. IEEE Trans Adv Packaging 2005;28:584–593.
52. Hess H, Clemmens J, Howard J, Vogel V. Surface imaging by self-propelled nanoscale probes. Nano Lett 2002;2:113–116.
53. Park JB, Lakes RS. Biomaterials, an Introduction, 2nd ed. New York: Plenum Press, 1992.
54. Berliner E, Mahtani HK, Karki S, Chu LF, Cronan JE, Gelles J. Microtubule movement by a biotinated kinesin bound to streptavidin-coated surface. J Biol Chem 1992;269:8610–8615.
55. Romet-Lemonne G, VanDuijn M, Dogterom, M. Three-dimensional control of protein patterning in microfabricated devices. Nano Lett 2005;5:2350–2354.
56. Stewart RJ, Thaler JP, Goldstein LSB. Direction of microtubule movement is an intrinsic property of the motor domains of kinesin heavy chain and Drosophila ncd protein. Proc Natl Acad Sci USA 1993;90:5209–5213.

57. Andrade JD, Hlady V, Jeon SI. In: Glass JE, ed. ACS Advances in Chemistry Series No. 248 Washington, DC: American Chemical Society, 1996:51.
58. Li J-T, Carlsson J, Huang S-C, Caldwell K.D. Adsorption of poly(ethylene oxide)-containing block copolymers. In: J.E. Glass, ed. ACS Advances in Chemistry Series No. 248 Washington, DC: American Chemical Society, 1996:61–78.
59. Bridgett MJ, Davies MC, Denyer SP. Control of staphylococcal adhesion to polystyrene surfaces by polymer surface modification with surfactants. Biomaterials 1992;13:411–416.
60. Gombotz WR, Guanhgui W, Horbett TA, Hoffman AS. Protein adsorption to poly(ethylene oxide) surfaces. J Biomed Mater Res 1991;25:1547–1562.
61. Lee J, Martic PA, Tan JS. Protein adsorption on pluronic copolymer-coated polystyrene particles. J Colloid Int Sci 1989;131:252–266.
62. Li JT, Carlsson J, Lin JN, Caldwell KD. Chemical modification of surface active poly(ethylene oxide)-poly (propylene oxide) triblock copolymers. Bioconjug Chem 1996;7:592–599.
63. Neff JA, Caldwell KD, Tresco PA. A novel method for surface modification to promote cell attachment to hydrophobic substrates. J Biomed Mater Res 1998;40:511–519.
64. Neff JA, Tresco PA, Caldwell KD. Surface modification for controlled studies of cell-ligand interactions. Biomaterials 1999;20:2377–2393.
65. Ho C–H, Limberis L, Caldwell KD, Stewart RJ. A metal-chelating Pluronic for immobilization of histidine-tagged proteins at interfaces: immobilization of firefly luciferase on polystyrene beads. Langmuir 1998;14:3889–3894.
66. deCastro MJ, Ho CH, Stewart RJ. Motility of dimeric ncd on a metal-chelating surfactant: evidence that ncd is not processive. Biochemistry 1999;38:5076–5081.
67. Turner DC, Chang C, Fang K, Brandow SL, Murphy DB. Selective adhesion of functional microtubules to patterned silane surfaces. Biophys J 1995;69:2782–2789.
68. Kacher CM, Weiss IM, Stewart RJ, et al. Imaging microtubules and kinesin decorated microtubules using tapping mode atomic force microscopy in fluids. Eur Biophys J 2000;28:611–620.
69. Stracke R, Bohm KJ, Burgold J, Schacht H-J, Unger E. Physical and technical parameters determining the functioning of a kinesin-based cell-free motor system. Nanotechnology 2000;11:52–56.
70. Turner D, Chang CY, Fang K, Cuomo P, Murphy D. Kinesin movement on glutaraldehyde-fixed microtubules. Anal Biochem 1996;242:20–25.
71. Bohm KJ, Stracke R, Muhlig P, Unger E. Motor protein-driven unidirectional transport of micrometer-sized cargoes across isopolar microtubule arrays. Nanotechnology 2001;12:238–244.
72. Yokokawa R, Yoshida Y, Takeuchi S, Kon T, Sutoh K, Fujita H. Evaluation of cryopreserved microtubules immobilized in microfluidic channels for a bead-assay-based transportation system. IEEE Trans Adv Packaging 2005;28:577–583.
73. Stracke R, Bohm KJ, Wollweber L, Tuszynski JA, Unger E. Analysis of the migration behaviour of single microtubules in electric fields. Biochem Biophys Res Comm 2002;293:602–609.

19

Water-Based Nanotechnology
What if We Could Dope Water?

Andreas Kage and Eran Gabbai

Summary

For many years, scientists around the world have studied water, the basis of all life on our planet. Expecting to reach disruptive benefits similar to the huge value created by doping silicon and glass fibers, some scientists have been relentlessly searching for ways to dope water and bring about a quantum leap in life sciences similar to those seen in microelectronics and communications. Although early failures such as the now infamous "polywater" discovery discouraged and disappointed some in the scientific community, these temporary setbacks have not halted the effort. The quest to dope water and to create a revolutionary material upon which to base a new generation of aqueous material useful for therapeutics has continued unabated. Recent developments in nanotechnology, which focuses on materials at their nanometer scale, are enabling scientists to experiment with and understand a new class of water-based nanotechnology materials that are poised to become the first generation of "doped-water" materials.

In the first section of this revealing chapter, the reader is invited to take a closer look at some of the unique packaging and other properties of the material that we call "water." Following a brief primer on nanotechnology, we describe a novel water-based nanotechnology approach that links the unique properties of nanoparticles (i.e., large surface-to-volume ratio) with those of water (i.e., density anomaly point) as a viable path to reach the goal of doping water. We close the chapter by summarizing some preliminary experimental results obtained by applying a first-generation class of doped-water material, branded as "Neowater™," to biotechnology, and specifically to molecular biology. We postulate that the benefits from these early demonstrated successes integrating doped-water materials into biotechnology applications are just the "tip of the iceberg," and the best is yet to come from water-based nanotechnology.

Key Words: Nanobubbles; nanotechnology; structured water; wetting; hydrophobic effect; PCR; transfection; nanostructures.

From: *NanoBioTechonology: BioInspired Devices and Materials of the Future*
Edited by: Oded Shoseyov and Ilan Levy © Humana Press Inc., Totowa, NJ

1. INTRODUCTION: THE UNIQUE PROPERTIES OF WATER

Water is the third most common molecule in the universe, the most abundant substance on earth, and the only naturally occurring inorganic liquid. Liquid water is unique, and as much as 37 anomalous properties are reported *(1)*. Because of the central role that water plays in all living things, it has been very well studied. In this section, we present some of the essential water properties as well as an overview of water packaging.

Water has the molecular formula H_2O, but the hydrogen atoms are constantly exchanging as a result of protonation/deprotonation processes at a millisecond rate for pH 7.0 and at higher rates for different pH values. Note that this has to do with the fundamental principles of symmetry, which do not enable a stationary state with electric dipole. There are two mirror planes of symmetry and a twofold rotation axis where the hydrogen atoms can be parallel (ortho-water) or antiparallel (para-water) (*see* Fig. 1) *(2)*.

Most of the properties of water are attributed to the above-mentioned hydrogen bonding between singular water molecules occurring when a hydrogen atom is attracted by rather strong forces to two oxygen atoms (as opposed to one), so that it can be considered to be acting as a binding between the two atoms. These hydrogen bondings cause special properties such as high surface tension, high viscosity, and the capability of forming ordered hexagonal, pentagonal, or dodecahedral water clusters by themselves or around other substances *(3)*. The strength of hydrogen bonding raises the melting point of water to 100 K or higher if compared to other molecules with similar molecular weight. In the hexagonal ice phase of the water (the normal form of ice and snow), all water molecules participate in four hydrogen bonds (two as donor and two as acceptor) and are held relatively static *(4)*. The free energy change must be zero at the melting point *(5)*. During the melting process, hydrogen bonds are broken to allow the molecules to move around. The amount of energy required for breaking these bonds is much larger than the amount of energy required for the change in volume. Melting will only occur when there is a sufficient entropy change to provide the energy required for the bond breaking *(6)*. The low entropy (high organization) of liquid water causes this melting point to be high *(6)*.

Water has its highest density in the liquid phase at a temperature of 3.984°C *(7)*. This phenomenon is known as the density anomaly of water, and can be explained by the cohesive nature of the hydrogen-bonded network resulting in dodecahedral molecule clusters. This reduces the free volume and ensures a relatively high density, compensating for the partial open nature of the hydrogen-bonded network *(8–10)*.

The hydrogen-bonded network can also form other structures described as hexagonal or pentagonal clusters of water molecules *(2,11)*. A rise in

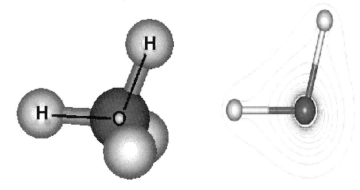

Fig. 1. The water molecule. Graphic depiction of (ortho-water) or antiparallel (para-water). (Courtesy of M. F. Chaplin [http://www.lsbu.ac.uk/water].)

temperature is accompanied by positive changes in entropy and enthalpy and a collapsing of those clusters *(1)*; the ordering effects of hydrogen bonding are neutralized by disordering kinetic effects *(13,14)*.

Most of the ordered structures of liquefied water are on a short-range scale, typically about 1 nm *(25)*. Principally, hydrogen-bonding enables the formation of large, ordered icosahedral water clusters consisting of hundreds of water molecule clusters when the water is in its liquid phase *(23,26–30)*. However, such long-range order has an extremely low probability of occuring spontaneously, because molecules in a liquid state are in constant thermal motion *(26)*.

As known, water molecules can form ordered structures and superstructures with other molecules. For example, shells of ordered water form around various bio-molecules such as proteins and carbohydrates *(15,16)*. The ordered water environment around these bio-molecules plays a significant role in biological function with regard to intracellular function including, for example, signal transduction from receptors to cell nuclei *(17,18,19)*. These water structures are stable and can protect the surface of the molecule *(20–24)*.

Other properties of water include a high boiling point, a high critical point, reduction of melting point with pressure (the pressure anomaly), compressibility that decreases with increasing temperature up to a minimum at about 46°C, and the like *(31,32)*.

2. A NANOTECHNOLOGY PRIMER

Nanoscience (or nanotechnology) is the science of small particles of materials and is one of the most important research frontiers in modern science. These small particles are of interest from a fundamental viewpoint because many properties of a material, such as its melting point and its electronic and optical properties, change when the size of the particles that make up the material becomes nanoscopic. One of the key characteristics of small particles in general, and of nanoparticles in particular, is that the surface area-to-volume ratio is orders of magnitude larger than that of millimetric-sized particles.

With new properties come new opportunities for technological and commercial development, and applications of nanoparticles have been shown or proposed in areas as diverse as micro- and nanoelectronics, nanofluidics, coatings and paints, and life sciences.

Much industrial and academic effort is presently directed toward the development of integrated micro devices or systems combining electrical, mechanical, and/or optical/electro-optical components, commonly known as micro electromechanical systems (MEMS). MEMS are fabricated using integrated circuit batch processing techniques and can range in size from micrometers to millimeters. These systems can sense, control, and actuate on the micro scale, and are able to function individually or in arrays to generate effects on the macro scale.

In the biotechnology area, nanoparticles are frequently used in nanometer-scale equipment for probing the real-space structure and function of biological molecules. Auxiliary nanoparticles, such as calcium alginate nanospheres, have also been used to help improve gene transfection protocols.

In metal nanoparticles, resonant collective oscillations of conduction electrons, also known as particle plasmons, are excited by optical fields. The resonance frequency of a particle plasmon is determined mainly by the dielectric function of the metal and the surrounding medium and by the shape of the particle. Resonance leads to a narrow, spectrally selective absorption and an enhancement of the local field confined on and close to the surface of the metal particle. When the laser wavelength is tuned to the plasmon resonance frequency of the particle, the local electric field in proximity to the nanoparticles can be enhanced by several orders of magnitude. Hence, nanoparticles are used for absorbing or refocusing electromagnetic radiation in proximity to a cell or a molecule, e.g., for the purpose of identification of individual molecules in biological tissue samples, in a similar fashion to the traditional fluorescent labeling.

An additional area in which nanoscience can play a role is heat transfer. It is well known that materials in solid form have thermal conductivities that are orders of magnitude larger than those of fluids. Low thermal conductivity is a primary limitation in the development of energy-efficient heat transfer fluids required in many industrial applications. Numerous theoretical and experimental studies of the effective thermal conductivity of dispersions containing particles have been conducted since Maxwell's theoretical work was published more than 100 yr ago. Maxwell's model shows, for millimeter and micrometer particles, that the effective thermal conductivity of suspensions containing spherical particles increases with the volume fraction of the solid particles. However, thermal conductivity of suspensions also increases with the ratio of the surface area to volume of the particle. Because the surface area to volume ratio is 1000 times larger for particles with a 10-nm diameter than for particles with a 10-mm diameter, a much more dramatic improvement in effective thermal conductivity is expected as a result of decreasing the particle size in a solution than can obtained by altering the particle shapes of large particles. Consequently, a new class of heat transfer fluids called nanofluids has been developed. These nanofluids are typically liquid compositions in which a considerable amount of nanoparticles are suspended in liquids such as water, oil, or ethylene glycol. The resulting nanofluids possess extremely high thermal conductivities compared to the liquids without dispersed nanoparticles.

Traditionally, nanoparticles are synthesized from a molecular level up, by the application of arc discharge, laser evaporation, pyrolysis process, use of plasma or sol gel, and the like. Widely used nanoparticles are the fullerene carbon nanotubes, which are broadly defined as objects having a diameter below about 1 μm. In a narrower sense, a material having the carbon hexagonal mesh sheet of carbon in parallel with the axis is called a carbon nanotube, and a material with amorphous carbon surrounding a carbon nanotube is also included within the category of carbon nanotube.

Nanoshells are nanoparticles that have a dielectric core and a conducting shell layer. Similarly to carbon nanotubes, nanoshells are manufactured from a molecular level up, for example, by bonding atoms of metal on a dielectric substrate. Nanoshells are particularly useful in applications in which exploitation of the above-mentioned optical field enhancement phenomenon is desired. Nanoshells, however, are known to be useful only in cases of near-infrared wavelengths applications.

It is recognized that nanoparticles produced from a molecular level up tend to lose the physical properties characterizing the bulk, unless further treatment is involved in the production process. As can be understood from the above nonexhaustive list of potential applications in which nanoparticles are already in demand, there is a large diversity of physical properties to be considered when producing nanoparticles. In particular, nanoparticles retaining physical properties of larger, micro-sized particles are of utmost importance.

3. WATER-BASED NANOTECHNOLOGY: COMBINING NANOPARTICLES AND WATER

Several implementations of combining nanoparticles and water have been reported. As an example of a straightforward combination, we have already mentioned nanofluids, and other examples may include sonification

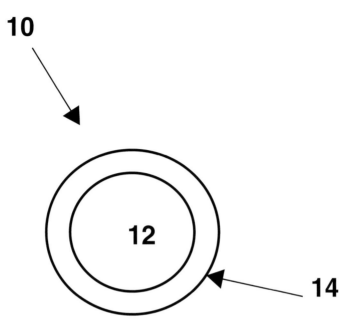

Fig. 2. Nanostructure of a first-generation class of doped-water material. (http://-www.wipo.int/ipdl/IPDL-CIMAGES/images3.jsp?WEEK=35/2005&-DOC=05/079153&TYPE=A2&TIME=1139352476).

of nanoparticles and water to create a dispersion of the former in the latter.

Recently, a new, first-generation class of doped-water material has been manufactured using a novel "top-down" process that creates nanoparticles in water from an insoluble, micron-sized crystal aggregate. The process involves heating the crystals in an oven, and quenching them in reverse osmosis (RO) water maintained just below the density anomaly temperature of water. The quenching process takes place while a radiofrequency (RF) signal (cold rf) is applied to create gas nanobubbles inside the liquid water. The nanoparticles, which are in very low concentration inside the water (10^{15} particles/L), modify the physical properties of the water around them by organizing the nanobubbles and packaging the surrounding water. Thus, the average distance between the nanostructures in the composition is rather large, in the order of microns.

Figure 2 illustrates such a nanostructure 10 comprising a core material 12 of a nanometric size, surrounded by an envelope 14 of ordered water molecules. The core material 12 and envelope 14 are in a steady physical state, meaning that objects or molecules are bound by any potential having at least a local minimum. Representative examples of such a potential include, but are not limited to, van der Waals potential, Yukawa potential, Lenard-Jones potential, and the like. Other potentials are also possible.

As demonstrated in Fig. 3, because envelope 14 comprises a gaseous material, the nanostructure is capable of floating when subjected to sufficient g-forces. Core material 12 is not limited to a certain type or family of crystal materials, and can be selected in accordance with the application for which the nanostructure is designed. Representative examples include ferroelectric material, a ferromagnetic material, and a piezoelectric material. The nanostructure 10 has a particular feature in which macro scale physical properties are brought into a nanoscale environment.

The nanostructure 10 is capable of clustering with at least one additional nanostructure. More specifically, when a certain concentration of nanostructure 10 is mixed in a liquid (e.g., water), attractive electrostatic forces between several nanostructures may cause adherence among each other so as to form a cluster of nanostructures. Even when the distance between the nanostructures prevents cluster formation, nanostructure 10 is capable of maintaining long range interaction (about 0.5–10 μm) with the other nanostructures. Long-range interactions between nanostructures present in a liquid induce unique characteristics in the liquid, which can be exploited in many applications, such as, but not limited to, biological and chemical assays.

The formation of the nanostructures in the water may be explained as follows. The combination of cold water and RF radiation (i.e., highly oscillating electromagnetic field) influences the interface between the solid phase and the water, thereby breaking the solid material and water molecule clusters into singular water molecules, which envelope the (nano-sized) debris of the solid material. Being at a low temperature, the singular water molecules and the debris enter a steady physical state. The attraction of the singular water molecules to the nanostructures can be understood from the relatively small size of the nanostructures compared to the correlation length of the water molecules. It has been argued that a small-sized perturbation may contribute to a pure Casimir effect, which is manifested by long-range interactions *(33)*. The long-range interaction and thereby the long-range order of the liquid composition allows the liquid composition self-organize, so as to adjust to different environmental conditions such as, but not limited to, different temperatures, electrical currents, radiation, and the like.

4. EXPERIMENTAL OBSERVATIONS OF APPLICATIONS WITH A DOPED-WATER MATERIAL

One of the characteristics of the nanoparticle-doped water material is the small contact angle if attached to a solid surface. The contact angle between the doped-water material and the surface is significantly smaller than a contact

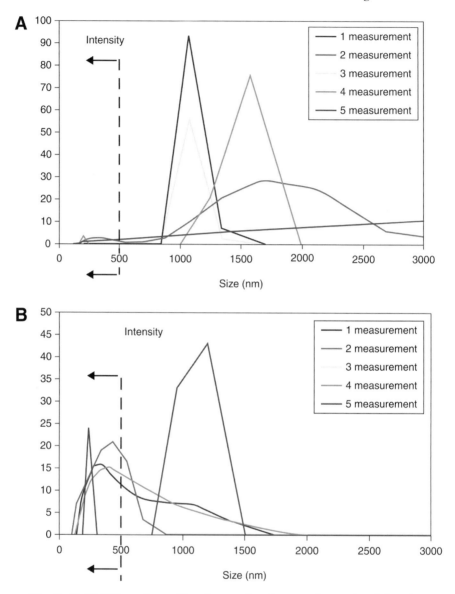

Fig. 3. (A,B) Effect of centrifugation on doped water. A sample of doped water was centrifuged (about 30 Kg). Figure 3A–B shows results of five integrated light scattering (ILS) measurements of the sample (LC1) after centrifugation. Figure 3A shows signals recorded at the lower portion of the tube. As shown, no signal from structures less that 0.5 mm was recorded from the lower portion. Figure 3B shows signals recorded at the upper portion of the tube. A clear presence of structures less

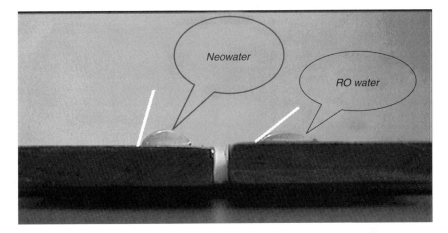

Fig. 4. Contact angle comparison. Surface wetting after 5 min shows that reverse-osmosis water spreads more over a hydrophilic surface than doped-water. Drops placed on aluminum surface. *(White lines drawn to show contact angle.)* (Courtesy of Do-Coop Technologies Ltd.)

angle between water (without the nanoparticles) and the surface. It can thus be appreciated that a smaller contact angle allows the liquid composition to "wet" the surface in a larger extent (Fig. 4).

As an example of the benefits of doped-water materials in molecular biology applications, we provide below experimental results based on a novel type of nanoparticle-doped water material branded as Neowater™. This particular doped-water material is capable of improving the efficiency of a nucleic acid amplification process, namely, enhancing the catalytic activity of a DNA polymerase in PCR procedures, increasing the sensitivity and/or reliability of the amplification process, and/or reducing the reaction volume of the amplification reaction (Fig. 5).

As a second example of the benefits of doped-water materials in molecular biology applications, below are experimental results of the same doped-water material applied to DNA transfection experiments (Fig. 6).

Fig. 3. *(Continued)* than 0.5 mm is shown. In all the measurements, the location of the peaks are consistent with nanostructures of about 200–300 nm. This experiment demonstrates that the specific gravity of the nanostructures is lower than the specific gravity of water. (Courtesy of Do-Coop Technologies Ltd.)

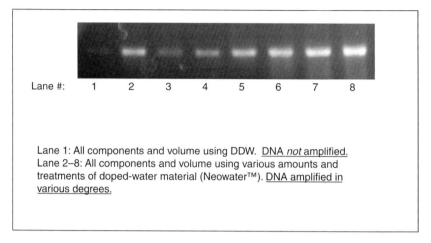

Fig. 5. 1.5% Agarose gel result of a 69% GC-rich PCR with double-distilled water/undoped water (Lane 1) and with doped-water material (lanes 2–8) under identical PCR program conditions. (Courtesy of Do-Coop Technologies Ltd.)

5. CONCLUSIONS

We have shown in this chapter that atypical properties of water are related to the formation of molecule clusters, which are stabilized by hydrogen bonding. The spontaneous occurrence of larger clusters is a very rare event in nature. By doping water with nanoparticles, we can induce the formation of very large clusters of water molecules that display beneficial effects in molecular biology experiments.

Although a final proof is difficult, we propose that these effects and many others not shown in this chapter are caused by a perturbation effect of the liquid water by nanostructures, which consists of singular nanoparticle attached air bubbles (nanobubbles). Because the air–water interface is hydrophobic, water molecules form clathrate shells with an "ice-like" structure around the nanobubbles, which can induce nano-scale ordering in the water that can have singular effect on the convection patterns and the formation of the double layer at the solid–liquid interface.

We have all enjoyed the benefits in our daily lives from advances in the information technology (IT) and telecommunications industries with their evolving applications based on the value of doped materials such as silicon and glass fiber. We invite the reader to closely follow the unfolding of the quest to dope water in general, and the new developments in the field of

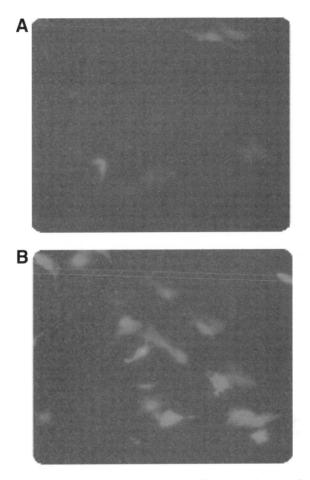

Fig. 6. Example of enhancement of transfection of human primary culture cells. **(A)** Human primary culture cells transfected with standard Lipofectamine 2000 transfection reagent (Invitrogen's TM). **(B)** Transfection of the same human primary culture cells using a mix of Neowater™ and just **12.5%** of the vendor's recommended Lipofectamine 2000 reagent amount. (Photo taken 48 h posttransfection.) (Courtesy of O. Karnieli.)

water-based nanotechnology in particular. The benefits of breakthrough advances in this field are as large, wide, and deep as the reader's imagination.

ACKNOWLEDGMENTS

The authors thank M.F. Chaplin, O. Karnieli, Tel Aviv University and Do-Coop Technologies Ltd. for useful material, experiments, suggestions, and discussion.

REFERENCES

1a. Eisenberg D, Kauzmann W. The structure and properties of water. London, Oxford University Press, 1969.

1b. Pauling L. The structure of water. In: Hadzi D, Thompson HW, eds. Hydrogen Bonding. London, Pergamon Press Ltd: 1959; pp 1–6.

2. Kropman MF, Bakker HJ. Dynamics of water molecules in aqueous solvation shells. Science 2001;291:2118–2120.

3. Stanley HE, Teixeira J. Interpretation of the unusual behavior of H_2O and D_2O at low temperature: tests of a percolation model. J Chem Phys 1980;73: 3404–3422.

4. Dore JC. Hydrogen-bond networks in supercooled liquid water and amorphous vitreous ices. J Mol Struct 1990;237:221–232.

5. Luck WAP. The importance of cooperativity for the properties of liquid water. J Mol Struct 1998;448:131–142.

6. Khan A. A liquid water model: Density variation from supercooling to superheating states, prediction pf H-bonds, and temperature limits. J Phys Chem 2000;104:11,268–11,274.

7. Cho CH, Singh S, Robinson GW. Understanding all of water's anomalies with a nonlocal potential. J Chem Phys 1997;107:7979–7988.

8. Tanaka H, Simple physical explanation of the unusual thermodynamic behavior of liquid water. Phys Rev Lett. 1998;80:5750–5753.

9. Sobott F, Wattenberg A, Barth HD, Brutschy B. Ionic clathrates from aqueous solutions detected with laser induced liquid beam ionization/desorption mass spectrometry. Int J Mass Spectr 1999;185–7:271–279.

10. Graziano G. On the size dependence of hydrophobic hydration. J Chem Soc Faraday Trans 1998;94:3345–3352.

11. Chaplin MF. A proposal for the structuring of water. Biophys Chem 2000;- 83:211–221.

12. Müller A, Bögge H, Diemann E. Structure of a cavity-encapsulated nanodrop of water. Inorg Chem Commun 2003;6:52–53.

13. Hodge IM. Strong and fragile liquids—a brief critique. J Non-Cryst Solids 1996;202:164–172.

14. Tanaka H. A new scenario of the apparent fragile-to-strong transition in tetrahedral liquids: water as an example. J Phys: Condens Matter 2003;15: L703–L711.

15. Franks F. Protein stability: the value of 'old literature.' Biophys Chem 2002;- 96:117–127.

16. Parsegian VA. Protein-water interactions. Int Rev Cytology 2002;215:1–31.

17. Berenden HJC. Discussion. Phil Trans R Soc Lond B 2004;359:1266–1267.

18a. Ling GN. Life at the cell and below-cell level. The hidden history of a functional revolution in Biology. New York, Pacific Press: 2001.

18b. Ling GN. A convergence of experimental and theoretical breakthroughs affirms the PM theory of dynamically structured cell water on the theory's 40th birthday. In: Pollack GH, Cameron IL, Wheatley DN, eds. Water and the Cell. Dordrecht, Springer: 2006; pp 1– 52.

19. Pollack GH. Cells, gels and the engines of life; a new unifying approach to cell function. Washington, Ebner and Sons Publishers: 2001.

20. Bandyopadhyay S, Chakraborty S, Balasubramanian S, Bagchi B. Sensitivity of polar solvation dynamics to the secondary structures of aqueous proteins and the role of surface exposure of the probe. J Am Chem Soc 2005;127: 4071–4075.

21. Bandyopadhyay S, Chakraborty S, Bagchi B. Secondary structure sensitivity of hydrogen bond lifetime dynamics in the protein hydration layer. J Am Chem Soc 2005;127:16,660–16,667.

22. Kurkal V, Daniel RM, Finney J, Tehei M, Dunn RV, Smith JC. Low frequency enzyme dynamics as a function of temperature and hydration: A neutron scattering study. Chem Phys 2005;317:267–273.

23. Barbour LJ, Orr GW, and Atwood JL. An intermolecular $(H_2O)_{10}$ cluster in a solid-state supramolecular complex. Nature 1998;393:671–673.

24. Atwood L, Barbour LJ, Ness TJ, Raston CL, Raston PL. A well-resolved ice-like $(H_2O)_8$ cluster in an organic supramolecular complex. J Am Chem Soc 2001;123:7192–7193.

25. Murrell JN, Jenkins AD. Properties of Liquids and solutions, 2nd Ed. Chichester, John Wiley & Sons: 1994.

26. Cho CH, Urquidi J, Singh S, Wilse Robinson G. Thermal offset viscosities of liquid H_2O, D_2O, and T_2O. J Phys Chem B 1999;103:1991–1994.

27 Doye JPK, Wales DJ. Polytetrahedral clusters. Phys Rev Lett 2001;86: 5719–5722.

28. Atwood L, Barbour LJ, Ness TJ, Raston CL, Raston PL. A well-resolved ice-like $(H_2O)_8$ cluster in an organic supramolecular complex. J Am Chem Soc 2001;123:7192–7193.

29. Hajdu F. A model of liquid water Tetragonal clusters: description and determination of parameters. Acta Chim 1977;93:371–394.

30. Tanaka H. Simple physical model of liquid water. J Chem Phys 2000;112: 799–809.

31. Kell GS. Density, thermal expansivity, and compressibility of liquid water from $0°$ to $150°C$: correlations and tables for atmospheric pressure and saturation reviewed and expressed on 1968 temperature scale. J Chem Eng Data 1975;20: 97–105.

32. Angell CA, Bressel RD, Hemmati M, Sare EJ, Tucker JC. Water and its anomalies in perspective: tetrahedral liquids with and without liquid-liquid phase transitions. Phys Chem Chem Phys 2000;2:1559–1566.

33. Bartolo D, Long D, Fournier JB. Long-range Casimir interactions between impurities in nematic liquid crystals and the collapse of polymer chains in such solvents. Europhys Lett 2000;49:729–734.

Index

A

ADH, *see* Alcohol dehydrogenase
AFM, *see* Atomic force microscopy
Alcohol dehydrogenase (ADH),
    enzymatic molecular recognition
    with nanotube membranes, 410, 412
Antisense-layered double hydroxide
    hybrids, delivery and toxicity test-
    ing, 358–361
Artificial enzyme
  overview, 370, 371
  polymer–metal–polymer complexes
    and biosensing, 371–373
Atomic force microscopy (AFM)
  DNA immobilized on substrates,
    148–150, 154, 156–158
  DNA on metal surfaces, 194
  quadruple helical DNA and
    conductivity enhancement,
    167, 169
ATP
  biosensing with artificial enzyme,
    371, 372
  layered double hydroxide hybrid
    delivery and properties, 356–358
  molecular motors, *see* Kinesin;
    Microtubule
Azurin, nanoelectronic devices, 146

B

Bacterial cell surface layer proteins,
    see S-layer proteins
Bacteriorhodopsin
  function, 123
  optimization
    directed evolution, 122, 123,
      129, 130

photophysical optimization,
    131–135
  thermal optimization, 135, 136
  photocycle, 123, 124
  rationale for optoelectronic and
    photovoltaic devices, 122, 123
  structure, 123
Bionanotechnology, *see*
    Nanobiotechnology
Biosensing, *see* Artificial enzyme;
    Cellular biosensing; Nanotube
    membranes; Tissue biosensing
Bone, scaffolds, 318, 319
Bottom-up assembly, 6, 141,
    385–387
BrdU, *see* Bromodeoxyuridine
Bromodeoxyuridine (BrdU),
    photochemical assay for
    DNA reductive electron
    transport, 97, 98

C

Carbon nanotubule (CNT)
  advantages and limitations in
    nanobiotechnology, 387, 388
  diameter range, 387
Catenanes, electrochemical switches,
    144, 145
Cellular biosensing
  applications, 373
  bioanalytical technology, 378–381
  glutamate receptor engineering,
    373–375
  microarray electrode, 379, 380
  nitric oxide, 379–381
Ceramic nanoparticle, drug
    delivery, 335

475

R

Reductive electron transport, *see* DNA electron transport

Rhodopsin, limitations for optoelectronic and photovoltaic devices, 122, 123

Ribosome, assembly, 143

Rotaxanes, electrochemical switches, 144, 145

S

SbpA, *see* S-layer proteins

SbsB, *see* S-layer proteins

SbsC, *see* S-layer proteins

Scaffolds

DNA scaffolds

immobile DNA junction, 46

order, 46, 47

single-stranded DNA tethers, 43–46

organic chains and polymers, 42, 43

protein scaffolds

closed protein scaffolds, 47–49

open protein scaffolds, 47

quantum dots, 41, 42

Scanning electrochemical microscopy (SECM), biological activity imaging, 379

Scanning tunneling microscopy (STM)

DNA immobilized on substrates, 150, 151, 58, 159

DNA on metal surfaces, 194

SECM, *see* Scanning electrochemical microscopy

Self-assembly, *see* DNA self-assembly; Peptide self-assembly

Silicocompatibility, kinesins, 443, 444

Simulation of exclusive disjunction (XOR), DNA self-assembly computing, 221, 222, 268

S-layer proteins

biophysical properties, 60, 61

functions, 56, 57, 63

genes and engineering, 61–63

molecular construction kits with fusion proteins and heteropolysaccharides

fusion protein design, 68–73

heteropolysaccharide functionalization of solid supports, 66–68

overview, 63, 64

SbpA, 64, 65

SbsB, 65, 66

SbsC, 66

nanoparticle array templates

in situ synthesis of nanoparticles on S-layers, 75, 76

preformed nanoparticle binding on S-layers in ordered arrays, 76–78

overview, 6, 7, 55, 56

prospects, 78, 79

recrystallization

on liposomes, 73–75

properties, 59, 60

self-assembly, 59

structural analysis of lattices, 57–59

Sleep apnea, cardiac monitoring, 316

SLN, *see* Solid lipid nanoparticle

Sol-gel deposition, nanotube membranes, 402, 403

Solid lipid nanoparticle (SLN), drug delivery, 330

SPR, *see* Surface plasmon resonance

Spurious nucleation error, DNA self-assembly computing, 225

Sticky ends, DNA, 218

STM, *see* Scanning tunneling microscopy

Surface plasmon resonance (SPR), DNA on metal surfaces, 195, 197

T

Thermal conductivity, nanofluids, 464, 465